Physical Activity and Type 2 Diabetes

Therapeutic Effects and Mechanisms of Action

John A. Hawley, PhD, FACSM

RMIT University

Juleen R. Zierath, PhD

Karolinska Institutet

Editors

Human Kinetics

Library of Congress Cataloging-in-Publication Data

Physical activity and type 2 diabetes : therapeutic effects and mechanisms of action / John A. Hawley, Juleen R. Zierath, editors.

    p. ; cm.

  Includes bibliographical references and index.

  ISBN-13: 978-0-7360-6479-8 (hard cover : alk. paper)

  ISBN-10: 0-7360-6479-6 (hard cover : alk. paper)

  1. Diabetes--Exercise therapy. 2. Insulin resistance--Exercise therapy. 3. Diabetes--Prevention. I. Hawley, John A. II. Zierath, Juleen R.

  [DNLM: 1. Diabetes Mellitus, Type 2--prevention & control. 2. Exercise. 3. Insulin Resistance--physiology. WK 810 P5785 2008]

  RC661.E94P49 2008

  616.4'62062--dc22

                        2007043954

ISBN-10: 0-7360-6479-6

ISBN-13: 978-0-7360-6479-8

The Web addresses cited in this text were current as of January 10, 2007, unless otherwise noted.

Acquisitions Editor: Michael S. Bahrke, PhD; **Managing Editors:** Heather M. Tanner and Lee Alexander; **Copyeditor:** Jocelyn Engman; **Proofreader:** Pamela Johnson; **Indexer:** Sharon Duffy; **Permission Manager:** Carly Breeding; **Graphic Designer:** Joe Buck; **Graphic Artist:** Denise Lowry; **Cover Designer:** Bob Reuther; **Art Manager:** Kelly Hendren; **Associate Art Manager:** Alan L. Wilborn; **Illustrator:** Accurate Art; **Printer:** Sheridan Books

Printed in the United States of America   10 9 8 7 6 5 4 3 2 1

Human Kinetics

Web site: www.HumanKinetics.com

United States: Human Kinetics
P.O. Box 5076, Champaign, IL 61825-5076
800-747-4457
e-mail: humank@hkusa.com

Canada: Human Kinetics
475 Devonshire Road Unit 100, Windsor, ON N8Y 2L5
800-465-7301 (in Canada only)
e-mail: info@hkcanada.com

Europe: Human Kinetics
107 Bradford Road, Stanningley, Leeds LS28 6AT, United Kingdom
+44 (0) 113 255 5665
e-mail: hk@hkeurope.com

Australia: Human Kinetics
57A Price Avenue, Lower Mitcham, South Australia 5062
08 8372 0999
e-mail: info@hkaustralia.com

New Zealand: Human Kinetics
Division of Sports Distributors NZ Ltd.
P.O. Box 300 226 Albany, North Shore City, Auckland
0064 9 448 1207
e-mail: info@humankinetics.co.nz

Contents

Part III   Prevention of Type 2 Diabetes Through Exercise Training

Chapter 8   Transcription Factors Regulating Exercise Adaptation   107

David Kitz Krämer, PhD; and Anna Krook, PhD

Chapter 9   Exercise and Calorie Restriction Use Different Mechanisms to Improve Insulin Sensitivity   119

Gregory D. Cartee, PhD

Chapter 10   Mitochondrial Oxidative Capacity and Insulin Resistance   135

Kevin R. Short, PhD

Contributors

Jason R. Berggren, PhD
East Carolina University

Arend Bonen, PhD
University of Guelph, Ontario

Frank W. Booth, PhD
University of Missouri at Columbia

Gregory D. Cartee, PhD
University of Michigan

Adrian Chabowski, MD, PhD
Medical University of Bialystok, Poland

Manu V. Chakravarthy, MD, PhD
Washington University School of Medicine

Leslie A. Consitt, PhD
East Carolina University

Sarah Crunkhorn, PhD
Joslin Diabetes Center and Harvard Medical School at Boston

Flemming Dela, MD
University of Copenhagen

Jan F.C. Glatz, PhD
Maastricht University, Netherlands

Bret H. Goodpaster, PhD
University of Pittsburgh

Edward W. Gregg, PhD
Centers for Disease Control and Prevention

John A. Hawley, PhD, FACSM
RMIT University

Joseph A. Houmard, PhD
East Carolina University

David E. Kelley, MD
University of Pittsburgh

David Kitz Krämer, PhD
Karolinska Institute, Stockholm

Andrea K. Kriska, PhD
University of Pittsburgh

Anna Krook, PhD
Karolinska Institute, Stockholm

Matthew J. Laye, BS
University of Missouri at Columbia

Sarah J. Lessard, PhD
RMIT University

Joost J.F.P. Luiken, PhD
Maastricht University, Netherlands

André Marette, PhD
Department of Anatomy-Physiology and Lipid Research Unit, Laval University Hospital Research Centre, Quebec

Mary Elizabeth Patti, MD
Joslin Diabetes Center and Harvard Medical School at Boston

Henriette Pilegaard, PhD
University of Copenhagen

Erik A. Richter, MD, DMSci
University of Copenhagen

Carsten Schmitz-Peiffer, PhD
Insulin Signalling Group, Diabetes and Obesity Program, Garvan Institute of Medical Research

Kevin R. Short, PhD
Department of Pediatrics, Section of Diabetes and Endocrinology, University of Oklahoma Health Science Center

D.M. Thomson, PhD
Department of Physiology and Developmental Biology, Brigham Young University

Phillip James White, MS
Lipid Research Unit, Laval University Hospital Research Centre, Quebec

W.W. Winder, PhD
Department of Physiology and Developmental Biology, Brigham Young University

Jørgen F.P. Wojtaszewski, PhD
University of Copenhagen

Preface

Over the past 50 years, the prevalence of a cluster of interrelated metabolic disease states, including obesity, insulin resistance, and type 2 diabetes, has increased dramatically, reaching epidemic proportions. In modern, westernized nations, the population-based prevalence of insulin resistance is approaching 20%, and the incidence of type 2 diabetes in adults ranges from 5% to 10%, making it the most common endocrine disorder in developed societies. In the United States there was a sixfold increase in the prevalence of type 2 diabetes between 1958 and 1993, a trend only partially explained by an aging population. Indeed, between 1982 and 1994 there was a 10-fold rise in type 2 diabetes in adolescents, with about 30% of all newly diagnosed cases occurring among persons aged between 10 and 19 years.

Physical inactivity elevates the risk of type 2 diabetes in individuals of normal weight, while insulin resistance directly relates to the degree of an individual's habitual physical activity. As such, physical inactivity is now regarded as an independent risk factor for insulin resistance and type 2 diabetes. While the primary defects in the development of whole-body insulin resistance remain unclear, during the past decade significant progress was made toward an understanding of the molecular basis underlying the beneficial effects of exercise training in stimulating the entry of glucose into insulin-sensitive tissues. Indeed, it is now well accepted that regular physical exercise offers an effective therapeutic intervention to improve insulin action in skeletal muscle of individuals who are insulin resistant. This book provides a series of independent but related reviews that present state-of-the-art knowledge in diabetes research on some of the mechanisms by which exercise training alleviates the development of insulin resistance in skeletal muscle. Such information is essential in order to define the precise variations in physical activity (i.e., intensity, duration, and frequency) that achieve the desired effects on targeted risk factors and to bridge the gap between science and practice. This book is intended as a valuable reference for graduate students, research fellows, basic academic scientists, and pharmacological scientists as well as clinical investigators.

This book is divided into four parts. Part I consists of two chapters that describe the scope and extent of the diabesity epidemic. In chapter 1, Dr. Gregg and Dr. Kriska outline the risk factors for diabetes and the underlying causes of the epidemic and look ahead at some of the potential consequences that might befall. Dr. Booth and colleagues have been major players in propagating North America's war against diabetes, and in chapter 2 they outline their battle plans for the prevention of diabetes through exercise biology. Part II of the book examines metabolism defects that occur in individuals with insulin resistance. Dysregulated lipid metabolism is a hallmark of metabolic syndrome, and in chapter 3, Dr. Bonen and colleagues discuss the regulation of a protein-mediated fatty acid transport system as well as the role of fatty acid transporters in insulin resistance. Altered lipid metabolism profoundly affects insulin action, and in chapter 4 Dr. Berggren, Dr. Consitt, and Dr. Houmard discuss the effects of lipid oversupply on insulin signaling pathways. In chapter 5, Dr. Goodpaster and Dr. Kelley focus on metabolic inflexibility of substrate oxidation in skeletal muscle, a feature implicated in obesity, insulin resistance, and type 2 diabetes. In chapter 6, Dr. Crunkhorn and Dr. Patti highlight nutrient-sensing pathways in normal physiology that, when disrupted by long-term overnutrition, contribute to insulin resistance, insulin secretory dysfunction, and type 2 diabetes. The final chapter in this section, chapter 7 by Dr. White and Dr. Marette, reviews the recent evidence linking obesity and inflammation and the underlying events in the progression of this interaction that eventually lead to type 2 diabetes and cardiovascular diseases.

Part III of the book examines the effect of exercise training in preventing type 2 diabetes. In chapter 8, Dr. Krämer and Dr. Krook review the transcriptional factors that mediate some of the exercise-induced adaptations to exercise training. The effects of exercise and caloric restriction and their influence on insulin sensitivity are examined by Dr. Cartee in chapter 9. In chapter 10, Dr. Short discusses the role of mitochondrial oxidative capacity in insulin resistance, while in chapter 11 Dr. Richter and Dr. Wojtaszewski examine the effects of a single bout of exercise on insulin signaling pathways and insulin action. In chapter 12, Dr. Wojtaszewski and colleagues examine the evidence for the benefits of resistance exercise training in the management and treatment of type 2 diabetes.

Part IV presents three chapters that focus on novel molecular targets and pathways that may provide therapies for the prevention of type 2 diabetes. In chapter 13, Dr. Winder and Dr. Thomson examine evidence for the role of AMPK in insulin resistance and explain how its chronic activation may help prevent diabetes. In chapter 14, Dr. Schmitz-Peiffer discusses the role of protein kinase C in defective glucose disposal, and in chapter 15, Dr. Lessard and Dr. Hawley examine the molecular evidence for prescribing exercise training in the treatment of insulin resistance and the potential for combined exercise and drug therapy to overcome this disorder.

All the authors who have contributed chapters to this text are world leaders in their respective fields and have extensive research experience in their areas. The challenge now is for the biomedical and research community along with the responsible government agencies to translate this information into prescriptive therapeutic tools to alleviate the global epidemic of insulin resistance.

Part I

Aetiology of Insulin Resistance and Type 2 Diabetes

Prevalence and Consequences of the "Diabesity" Epidemic

Chapter 1

The Increasing Burden of Type 2 Diabetes
Magnitude, Causes, and Implications of the Epidemic

Edward W. Gregg, PhD; and Andrea K. Kriska, PhD

Type 2 diabetes is one of the fastest growing chronic disease threats to the health of industrialized populations (Diamond 2003; Zimmet, Alberti, and Shaw 2001). Considerable media and political attention have followed the epidemic due to its wide-reaching ramifications for the public, for health care, and for private industry. The high lifetime risk of diabetes makes it a potential health threat to the majority of the industrialized population, while the extensive complications of diabetes make it a major source of costs for the health care systems that provide its treatment and prevention. Furthermore, diabetes affects private enterprise, including the food, beverage, and health industries, as politicians debate public health strategies to alter trends in diabetes and obesity. In this chapter, we describe the dramatic changes in the prevalence of type 2 diabetes that have occurred over recent decades, briefly summarize the factors underlying this increasing prevalence, and describe the implications of increased diabetes prevalence for the long-term health of the industrialized population.

Diabetes Prevalence and Incidence in Adults

Diabetes has emerged in many areas of the world as one of the few major chronic diseases that are becoming more common over time. U.S. national surveillance estimates show that the prevalence of diagnosed diabetes in the adult population has tripled during the past 40 y, growing from about 2% to about 6% (Cowie et al. 2006; Gregg et al. 2004;

figure 1.1). Because many people with diabetes do not know they have the disease, in 1976 to 1980 U.S. national surveillance programs began collecting information on fasting glucose levels to determine the prevalence of undiagnosed diabetes, currently defined as a fasting glucose level that is ≥ 126 mg/dl. These data revealed that the total diabetes prevalence (i.e., prevalence of diagnosed plus undiagnosed diabetes) has increased about 55% during the past 26 y (1976 to 2002), growing from 5.3% to 8.2% of the adult population aged 20 to 74 y (Gregg et al. 2004; figure 1.1). Recent estimates extending through the year 2004 for the entire U.S. adult population (20 y and older) indicate that the prevalence of diabetes is now 9.3%, affecting about 19 million Americans (13 million diagnosed and 6 million undiagnosed; Cowie et al. 2006). Statistical projection models suggest that this number will increase to almost 50 million by the year 2050 (Narayan et al. 2006).

Estimates of the diabetes prevalence in the general U.S. population obscure the considerable variation that exists among specific demographic groups (Centers for Disease Control and Prevention 2006; Cowie et al. 2006). Age is the most important demographic factor in the Untied States. Prevalence increases from 2.4% of young adults (aged 20-39 y) to about 9.8% of middle-aged adults (aged 40-59 y) to about 21.1% of older adults (aged 60 y and older). Men have a slightly higher prevalence than women have (10.6% versus 8.2%), and prevalence is almost twice as high among minority populations such as non-Hispanic Blacks (14.6%) and Mexican

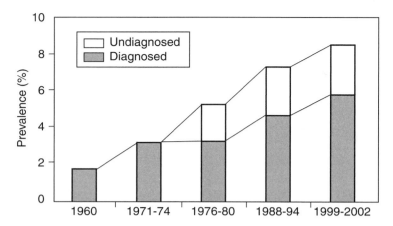

Figure 1.1   Prevalence of total diabetes from 1960 to 2002 measured among U.S. adults aged 20 to 74 y. Undiagnosed diabetes was not assessed in 1960 and in 1971 to 1974.

Created from data in E. Gregg et al., 2004, "Trends in the prevalence and ratio of diagnosed to undiagnosed diabetes according to obesity levels in the U.S.," *Diabetes Care* 27(12): 2806-2812.

Americans (13.5%) than it is among Whites (7.8%). Prevalence is even higher—about 33%—among older Blacks and Mexican Americans.

Recent evidence that diabetes may be prevented or delayed has raised interest in the prevalence of impaired fasting glucose (IFG), which is defined by a fasting plasma glucose level between 100 and 125 mg/dl. Of the U.S. adult population, 26% have IFG, and this number ranges from about 15% of the population aged 20 to 39 y to about 38% of those aged 65 y and older. Curiously, although IFG prevalence is notably higher among men (around 33%) than it is among women (around 20%), the racial and ethnic differences that exist for diabetes prevalence rates are less apparent for IFG.

International Comparisons of Type 2 Diabetes

Estimates of the international prevalence of diabetes evoke concerns similar to those raised by the U.S. experience. Recent national estimates from Australia and Canada were essentially equivalent to those observed in the United States, albeit with a larger proportion of undiagnosed cases (Dunstan et al. 2002). Like the United States, Australia has undergone a large increase in diabetes prevalence, which has approximately doubled over the past 20 y (Dunstan et al. 2002). Prevalence estimates from European populations have reported rates for women that are slightly lower than the U.S. estimates and have reported rates for men that are similar or slightly lower than the U.S. rates

(DECODE Study Group 2003). Several areas of Latin America, including Mexico and Central America, have reported diabetes prevalence rates similar to or greater than those reported in the United States (Barcelo and Rajpathak 2001; Aguilar-Salinas et al. 2003), whereas urban Chinese and Japanese populations appear to have a diabetes prevalence that is higher than that found in the United States (Qiao et al. 2003). Areas of the greatest recent concern include populations of the Middle East and India (King, Aubert, and Herman 1998; Qiao et al. 2003).

Some of the most dramatic examples of population-wide increases in diabetes prevalence come from specific populations, including the Nauru, the Pima Indians in the United States, the Australian Aborigines, and the urban Samoans. In these cases, which are described in further detail later in this chapter, the prevalence has increased by an astounding magnitude—severalfold within a mere 1 to 2 generations—and has accompanied fundamental and rapid changes in the typical lifestyle (figure 1.2). Even moderate increases in diabetes prevalence among the sizable populations of India, China, and the rest of Southeast Asia have ominous implications for the control of chronic disease in these regions (King, Aubert, and Herman 1998).

Incidence of Diabetes

Although prevalence studies are important in gauging the magnitude of the diabetes epidemic, prevalence is influenced by several factors, includ-

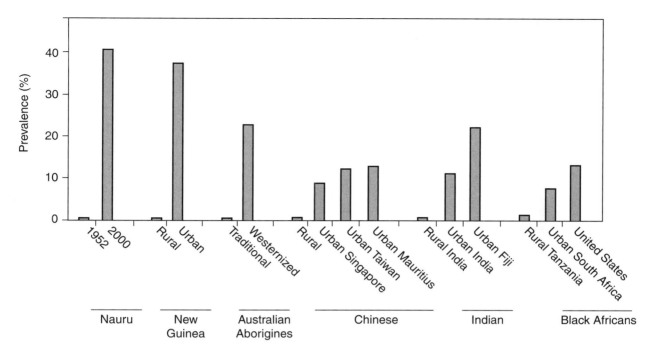

Figure 1.2   Ecological examinations of the association between westernization and diabetes prevalence in susceptible populations.
Created from data in J. Diamond, 2003, "The double puzzle of diabetes," *Nature* 423(6940): 599-602.

ing (1) the incidence, or the rate of development of new cases; (2) the detection rate of new cases; (3) the changes in definitions over time, which together with easier diagnosis can influence detection; (4) the mortality rate among people who already have diabetes; and (5) the changes in the age and ethnic composition of the population, which affect the incidence rate (Leibson et al. 1997). For example, increasing the life span among the population with diabetes increases the prevalence of diabetes. This observation has led some to argue that the diabetes incidence rates are stable and that the observed increases in prevalence are explained largely by decreasing mortality (Green et al. 2005). A recent report from a provincial registry of Ontario, Canada (see figure 1.3), demonstrated how the opposing trends of increasing incidence in the overall population and decreasing mortality in the population with diabetes can result in consistent increases in the diabetes prevalence (Lipscomb and Hux 2007). Whether these trends produce the same result in the United States is not clear. Although regional studies (Fox et al. 2004; Thomas et al. 2003; Tierney et al. 2004) have suggested that mortality has declined, this finding has not been observed in national studies (Gu, Cowie, and Harris 1999).

Trends in diabetes detection are also unclear. In Rochester, Minnesota, there was essentially no change in the percentage of the population receiving blood glucose testing between 1987 and 1994 (Burke et al. 2002). In the United States, the proportion of total people who have diabetes who know they have the disease increased from 62% to 70% during the past 25 y (Gregg et al. 2004). Diagnosed diabetes has shown a greater increase in prevalence (76% relative increase) than total diabetes (diagnosed and undiagnosed combined) has shown (55% relative increase). This suggests that increasing detection is a contributor, albeit not a full explanation, for the increasing diabetes prevalence. Internationally, the proportion of cases diagnosed varies tremendously (DECODE Study Group 2003; Qiao et al. 2003), making it very difficult to interpret studies of changes in prevalence.

Because of the interrelated factors that influence prevalence, examining incidence trends over time is a preferred way to investigate the status of an epidemic. In a study of medical records of adults in Rochester, Minnesota, the age-adjusted diabetes incidence increased 40% between 1970 and 1990 (Thomas et al. 2003). A more recent analysis of data from the National Health Interview Survey (NHIS),

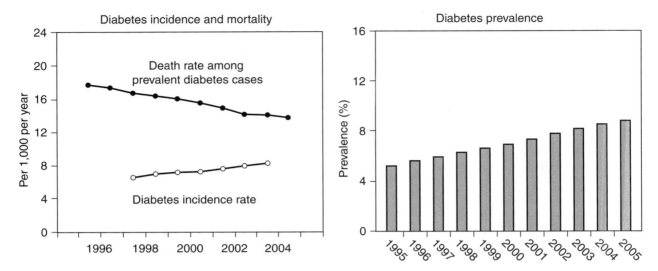

Figure 1.3 Trends in the incidence of diabetes diagnosis, the mortality among prevalent diabetes cases, and the diabetes prevalence in Ontario, Canada, from 1995 to 2005.

Created from data in L.L. Lipscombe and J.E. Hux, 2007, "Trends in diabetes prevalence, incidence, and mortality in Ontario, Canada 1995-2005: a population-based study," *Lancet* 369(9563): 750-756.

a nationally representative sample in the United States, indicated a similar magnitude of increase (37%, from 5.2 to 7.1 cases per 1,000) during the 7 y from 1997 to 2003 (Geiss et al. 2006). This trend was observed across all age groups, as incidence increased by 19% (from 2.1 to 2.5 cases per 1,000) among adults aged 18 to 44 y, by 30% (from 8.6 to 11.2 cases per 1,000) among persons aged 45 to 64 y, and by about 60% (from 10.2 to 16.8 cases per 1,000) among those aged 65 to 79 y. Unfortunately, as the NHIS is based on self-reported survey data and the Rochester, Minnesota, study is based on medical record review, neither assessed the prevalence of undiagnosed diabetes.

A way to avoid these limitations and to truly gauge the diabetes incidence is by regular, repeated interviews and blood assessments among a population-based cohort (Burke et al. 2002; Burke, Hazuda, and Stern 2000; Fox et al. 2006; Leibson et al. 1997). The two most recent U.S. studies to have done this yielded similar results. The Framingham study (Fox et al. 2006) observed that from the 1970s to the 1990s the diabetes incidence increased by 85% among women (from 2% to 3.7% over 8 y) and by 114% among men (from 2.7% to 5.8% over 8 y). Adjusting for body mass index (BMI) attenuated the increasing trend by about 30%. The San Antonio Heart Study found a twofold increase in incidence during an early 8 y period, but limited by a smaller sample, found statistically significant results only

among Mexican Americans (Burke et al. 1999). These two regional studies suggest that at least part of the increase in the diabetes prevalence observed nationally indeed reflects an increasing incidence of diabetes in the United States.

Recently published results based on statistical models of lifetime diabetes incidence provide a particularly ominous view of the problem. The average likelihood of developing diabetes over the lifetime is estimated at 33% for men and 39% for women (Gregg et al. 2004; Narayan et al. 2003). For people who make it to age 50 free of diabetes, the lifetime risk is about 25%, and from age 60, the remaining lifetime risk is about 20%.

Gestational Diabetes

Gestational diabetes (GDM), defined as glucose intolerance whose onset or first recognition occurs during pregnancy, provides another view of the diabetes epidemic. First, GDM is an important health threat in its own right because it increases the risk for birth defects, childhood growth problems, and childhood obesity and diabetes. The effect of maternal diabetes on childhood diabetes risk has been identified as an ominous cycle that is perpetuating the diabetes epidemic among Native Americans (Dabelea et al. 1998). In other words, a high prevalence of diabetes in pregnancy increases the risk of diabetes in youths, and this risk persists as youths move into childbearing age (Pettitt and Knowler 1998). Second, GDM serves as a sensitive indicator of diabetes in the population in that the hormonal changes during pregnancy serve as a natural glucose tolerance test for women. A recent population-based study in Colorado found a 10% increase in GDM prevalence per year that was evident among several ethnic groups (Dabelea et al. 2005). For example, over a 5 y period the GDM prevalence increased from 1.9% to 3.4% among non-Hispanic Whites and grew by an even greater magnitude among Hispanics (from 2.8% to 5.1%), African Americans (from 2.5% to 4.6%), and Asians (from 6.3% to 8.6%). Similar increases have been reported from northern California and Australia (Beischer et al. 1991; Ferrara et al. 2004).

Type 2 Diabetes in Youths and Adolescents

Diabetes that occurs during childhood and adolescence is traditionally thought to be type 1, defined as an absolute insulin deficiency due to autoimmune-mediated destruction of pancreatic beta cells. However, there is increasing concern about type 2 diabetes occurring in youths and adolescents (Dabelea et al. 1999; Fagot-Campagna et al. 2000). This concern has been fueled largely by case series conducted in specific populations, including Native Americans, First Nations Canadians (i.e., indigenous Canadians), and selected urban U.S. populations. A multicenter population-based study found the prevalence of diabetes in U.S. youths aged 10 to 19 y to be about 0.3% (or 3 cases per 1,000; SEARCH for Diabetes in Youth Study Group 2006). About 20% of the affected children had type 2 diabetes, but among Native American children, about 80% of the cases were type 2. Diabetes of either type was extremely rare among children less than 10 y of age. Although the SEARCH study found that the prevalence of type 2 diabetes in U.S. youths remains relatively low, the prevalence of IFG (fasting glucose >100 mg/dl) among U.S. adolescents was recently estimated at 7% (4% of girls and 10% of boys) and the prevalence of overweight among adolescents has increased dramatically (Williams et al. 2005; Ogden et al. 1997; Ogden et al. 2002). These observations of increasing IFG and obesity raise concerns about future increases in diabetes in youths.

Risk Factors for Diabetes and Causes of the Epidemic

Type 2 diabetes is thought to result from a gradual development of insulin resistance and deterioration of the ability to transport glucose from circulating blood across muscle, liver, or other tissue cells. Typically these are followed by a decline in pancreatic beta-cell function, which plays out as an inability to secrete insulin in response to changes in blood glucose levels and ultimately as a failure to produce insulin. However, type 2 diabetes is increasingly regarded as a heterogeneous disorder that varies substantially in pathogenesis as well as time course. Persons with type 2 diabetes are believed to pass through prediabetes, a stage currently defined as impaired glucose tolerance (i.e., glucose concentration >140 mg/dl and <200 mg/dl 2 h after a glucose challenge) or IFG (fasting glucose >100 and <126 mg/dl; Meigs et al. 2003). Persons with prediabetes have a yearly diabetes risk that is roughly 10 times that of the general population (Edelstein et al. 1997).

Aside from impaired glucose tolerance or prediabetes, a history of GDM is probably the strongest single risk factor for diabetes. A meta-analysis of diabetes incidence among women with a history of GDM showed considerable variation across studies, with 5% to 25% of women with GDM developing diabetes over the subsequent 5 y and 20% to 50% developing diabetes over the next 10 y (Kim, Newton, and Knopp 2002; Ogden et al. 1997; Ogden et al. 2002). Age and family history are major nonmodifiable risk factors for diabetes. Obesity and weight gain throughout life, particularly in the form of intra-abdominal fat, are regarded as the most prominent modifiable risk factors due to their combined high prevalence and magnitude of risk (Ford, Williamson, and Liu 1997). Among U.S. adults, each unit of BMI is associated with a 12% increase in the incidence of type 2 diabetes. Compared to persons with a BMI that is less than 22 kg/m^2, persons who are overweight (BMI of 25-30 kg/m^2) have 3 to 5 times the risk of diabetes, persons with a BMI between 30 and 35 kg/m^2 have 5 to 10 times the risk, and persons with class II or class III obesity have more than 10 times the risk. Furthermore, a 10 kg increase in body weight over a 10 y period during adulthood is associated with a 45% increase in diabetes risk (Ford, Williamson, and Liu 1997). Findings from a lifetime diabetes risk model suggest that 30% of men and 35% of women who are overweight will develop diabetes during their lifetime. In addition, about 55% of persons who are moderately obese (BMI of 30-35 kg/m^2) and more than 70% of persons who are very obese (BMI > 35 kg/m^2) will develop diabetes during their lifetime (Narayan et al. 2007).

Other lifestyle factors acting through obesity, independently of obesity, or both may play an important role in diabetes risk. For example, numerous observational studies have associated higher levels of physical activity with a lower incidence of diabetes. A recent meta-analysis of 10 prospective cohort studies showed that participating regularly in moderate-intensity physical activity or in a brisk walking regimen reduced diabetes incidence by 30% (Jeon et al. 2007). These studies have tended to find strong dose–response relationships between the amount of physical activity and the reduction in risk. Results from most observational studies suggest that this relationship is independent of body weight and weight gain; the review by Jeon and colleagues (2007) found that controlling for BMI attenuates the association between physical inactivity and diabetes incidence by about one-fourth. However, it remains unclear whether the measurements of physical activity and body composition in large cohort studies are sufficiently precise to determine if physical activity affects diabetes incidence independently of change in body weight. Another area of debate is if the intensity of physical activity, whether quantified by cardiorespiratory vigor or resistance load, has any effect on diabetes incidence above and beyond the total energy that is expended in physical activity. Most observational studies suggest that energy expenditure is the most important attribute of physical activity in terms of lowering diabetes risk and suggest that public health recommendations should encourage regular, moderate-intensity physical activity as opposed to high-intensity exercise.

Several dietary elements have been identified as potential modifiable risk factors for diabetes. Greater intakes of saturated fat, sugar-sweetened soft drinks, and fast food have been associated with increased risk for diabetes (Hu, van Dam, and Liu 2001). A high proportion of polyunsaturated fats in the diet and higher intake of fish oils, whole-grain and cereal fibers, and dairy have each been associated with a reduced risk of diabetes (Hu, van Dam, and Liu 2001). Similarly, higher levels of caffeine and coffee intake and moderate alcohol consumption decreased the diabetes risk in prospective population-based studies (Howard, Arnsten, and Gourevitch 2004; van Dam and Hu 2005). However, as in the case of physical activity, randomized controlled trials have not been designed to examine how specific aspects of the diet may prevent or delay diabetes.

The combined benefits of weight maintenance, physical activity, and healthy diet, however, have been demonstrated further by a recent series of diabetes prevention trials conducted in China, Finland, and the United States (Knowler et al. 2002; Pan et al. 1997; Tuomilehto et al. 2001). Each of these studies recruited persons with impaired glucose tolerance and randomized them to long-term lifestyle intervention programs. The study in Daqing, China, randomized people to clinics that carried out a dietary intervention condition, a physical activity condition, or both (Pan et al. 1997). The Finnish and American studies combined all lifestyle changes (reduced caloric intake, reduced fat intake, increased intake of fiber and whole grains, increased frequency and duration of moderate-intensity physical activity) into a single multidisciplinary program (Tuomilehto et al. 2001). These programs were implemented through small groups or one-on-one counseling by dietitians, psychologists, and exercise specialists. The Chinese, Finnish, and American studies found a 30%, 58%, and 58% reduced incidence of diabetes, respectively, among persons enrolled in the lifestyle interventions. In the U.S. Diabetes Prevention Program (DPP), there was also a 31% reduced incidence of diabetes among a group that was randomized to receive metformin therapy (Knowler et al. 2002). In follow-up analyses, the number of goals achieved (i.e., smaller caloric intake, increased physical activity, reduced weight) significantly affected the subsequent diabetes risk. In the Finnish Diabetes Prevention Study, weight and level of physical activity were independently associated with diabetes risk (Laaksonen et al. 2005). In the DPP, the risk in diabetes incidence dropped by 46% among participants who at 1 y did not meet the weight loss goal but did meet the physical activity goal (150 min of moderately intense physical activity, which is equivalent to a brisk walk). However, weight loss appeared to be the more dominant factor influencing diabetes risk (Hamman et al. 2006).

Interest has gathered around the possibility that environmental contaminants may play a role in the diabetes epidemic. A recent analysis of National Health and Nutrition Examination Survey (NHANES) data found that the prevalence of diabetes was substantially elevated among persons with higher levels of organic pollutants such as biphenyls and dioxins (Lee et al. 2006). Interestingly, the association between pollutants and diabetes risk was driven primarily by persons who were obese. The authors of the analysis noted that concentrations of pollutants have generally declined over recent decades, and they speculated that fat may serve as a depot for environmental contaminants. This study was cross-sectional, however, and could not rule out the possibility that diabetes somehow affects the clearance or concentration of pollutants. Thus, prospective studies are needed to clarify whether environmental pollutants are truly a risk factor for diabetes.

Determinants of Recent Diabetes Trends

The risk factors underlying the trends in diabetes prevalence and incidence are subtly different from the diabetes risk factors affecting individuals. For example, although family history, age, GDM, smoking, and even environmental pollutants may be important risk factors to individuals, it is unlikely that any of these factors has contributed to the observed increase in diabetes prevalence. When exploring the risk factors affecting *trends* in the diabetes epidemic, increasing obesity and lifestyle changes emerge as the overwhelming factors. Several classic ecological studies have shown that populations moving to westernized environments, populations moving to urban areas, or populations undergoing rapid westernization have an increased prevalence of diabetes (figure 1.2). The most striking migration studies have been observed among Asian Indians who emigrated to Fiji, Mauritius, and Singapore as well as among Chinese who emigrated to Hong Kong, Mauritius, Singapore, and Taiwan. These migrant populations have diabetes prevalence rates that are 2 to 3 times those of their homeland relatives (Diamond 2003; Zimmet, Alberti, and Shaw 2001). A recent study of type 2

diabetes among the Pima Indians living in Arizona, the United States, compared to type 2 diabetes among the genetically related Pima Indians living the traditional lifestyle in Sonora, Mexico, found that the diabetes prevalence was approximately 5 to 6 times higher on the U.S. side (34% of men and 41% of women) than it was on the Mexican side (6% of men and 8.5% of women; Diamond 2003; Schulz et al. 2006; Zimmet, Alberti, and Shaw 2001).

In other cases, diabetes prevalence greatly increased among populations who, due to environmental, political, or economic changes, underwent substantial alteration in typical employment, food availability, food type, and level of physical activity. These include the people of the Nauru Islands, where economic boon from phosphate mining led from agricultural and fishing-based employment to sedentary lifestyle and sugar intake (Diamond 2003). In these examples of the diabetes epidemic, although environmental factors are clearly profound, lifestyle change likely interacted with genetic predisposition. This idea has led to the thrifty gene hypothesis, which suggests that many populations evolved the ability to deposit fat and glucose more efficiently due to highly fluctuating food availability and periodic famine (Diamond 2003; Zimmet, Alberti, and Shaw 2001). An additional theory termed the *thrifty phenotype hypothesis* suggests that intrauterine malnutrition followed by low birth weight creates permanent anatomical or physiological changes that increase the risk of diabetes and other chronic conditions later in life.

Within the United States, obesity prevalence grew from 14% to 31% during the past 40 y, with the largest increases starting in the 1980s and continuing unabated through 2004 (Diamond 2003; Ogden et al. 2004; Ogden et al. 2006; Zimmet, Alberti, and Shaw 2001). The rise in obesity is most pronounced at the far right side of the obesity distribution, as the prevalence of class II or III obesity (BMI >35 kg/m^2) has tripled (from 4% to 13%) and the prevalence of class III obesity (BMI >40 kg/m^2) has quintupled (from 1% to 5%). Similar increases have been observed in waist circumference (Ford, Mokdad, and Giles 2003).

A recent examination of U.S. national data found that increases in diabetes prevalence during the past 25 y were dominated by the minority of the population in the most extreme BMI categories

(Gregg et al. 2007). Of the growing number of prevalent cases during the past 25 y, roughly one-fourth came from the overweight population (BMI of 25-30 kg/m²) and one-fourth came from persons with class I obesity (BMI of 30-35 kg/m²). Surprisingly, however, about one-half came from the segment of the population with a BMI greater than 35 kg/m² (class II and III obesity combined), a group that represents about 13% of the general population. Even more remarkably, about one-fourth of the increase in prevalent cases came from persons with class III obesity, or a BMI greater than 40 kg/m². Although the prevalence of total diabetes in the overall population (all BMI groups combined) has increased, there has been little change in diabetes prevalence within BMI strata (Gregg et al. 2004). In other words, today persons who are obese have a diabetes prevalence that is similar to what persons who are obese had 25 y ago. This finding suggests that the factors influencing the trends in diabetes are working primarily through obesity.

Several specific trends in dietary consumption could be contributing to the growing diabetes prevalence. Data from U.S. national surveys suggest that the average number of calories Americans consume daily has increased by 168 for men and 335 for women. Among men, saturated and total fat intake decreased when expressed as calories of intake, or as a proportion of the total caloric intake. Among women, the proportion of calories consumed from total and saturated fat modestly decreased, but the absolute number of calories consumed as fat actually increased. Carbohydrate intake increased for both men and women (Centers for Disease Control and Prevention 2004; Nielsen, Siega-Riz, and Popkin 2002; Popkin, Haines, and Reidy 1989; Popkin, Siega-Riz, and Haines 1996). More recent data from the Behavioral Risk Factor Surveillance System suggest that the number of servings of fruits and vegetables consumed in the American diet has also increased. Ostensibly, these changes follow dietary recommendations to reduce dietary fat and represent positive trends in the American diet. They may also be contributors to the falling cholesterol levels and decrease in mortality from cardiovascular disease seen in Americans. On the other hand, carbohydrates are heterogeneous, and much of the increased carbohydrate proportion in the diet has been driven by greater consumption of refined sugars and high

fructose corn syrup as opposed to vegetables, fruits, and whole grains (Gross et al. 2004; Popkin and Nielsen 2003). Portion sizes of fast foods and snacks have grown, and a greater number of meals and snacks are eaten outside of the home. Although the cooking oils used as well as the availability of healthy foods have improved in fast-food restaurants, and labeling and information for consumers have improved, the degree to which these positive trends have countered the larger portion sizes is unknown. On the whole, American diets appear to have become more obesogenic and diabetogenic but not necessarily more atherogenic.

Trends in level of physical activity, another major contributor to the increases in obesity and diabetes, are less clear. Recent estimates from the Behavior Risk Factor Surveillance System suggest that sedentary behavior, defined by a lack of participation in leisure-time physical activity, walking, and exercise, has been relatively stable during the most recent decade and may be declining slightly (Simpson et al. 2003; Brownson, Boehmer, and Luke 2005). Unfortunately, physical activity surveillance estimates are limited by the lack of reliable information about the amount of energy expended in nonleisure time, such as time spent in job, household, and daily activities. Numerous sources of evidence, ranging from the number of hours spent watching television each day to the number of errands made by car instead of by foot, provide indirect evidence for a lower energy expenditure (Brownson, Boehmer, and Luke 2005). Thus, it is possible that the population has slightly increased its leisure-time physical activity while decreasing its nonleisure activity, leading to a net reduction in physical activity. If true, this reduction would be detrimental to diabetes risk because total energy expenditure seems to be the attribute of physical activity most relevant for obesity and diabetes risk.

Anticipated Consequences of Diabetes and the Outlook for Prevention

Diabetes imposes a large burden on the population in part because of its acute complications and the extensive personal and medical attention needed to optimally manage the disease on a daily basis. For many people, however, diabetes creates its greatest burden through long-term vascular complications (Engelgau et al. 2004; Nathan 1993). Common microvascular complications include diabetic retinopathy, peripheral neuropathy, and diabetic nephropathy. Diabetic retinopathy, along with glaucoma, cataract, and age-related macular degeneration, which are all more common among people with diabetes, may lead to vision loss. Diabetic neuropathy increases the risk for foot ulcers and amputations, particularly in the presence of peripheral arterial disease, while diabetic nephropathy can result in end-stage renal disease and the need for either kidney dialysis or transplant.

Common macrovascular complications of diabetes include coronary artery disease, stroke, and peripheral arterial disease. Diabetes is associated with a two- to fourfold increased risk of coronary artery disease, myocardial infarction, congestive heart failure, and stroke, and ultimately about 40% of deaths among persons with diabetes are due to cardiovascular disease (Engelgau et al. 2004; Nathan 1993; Tierney et al. 2001). Among older adults, diabetes increases the risk of falls and fractures, cognitive decline, and dementia, and due to the cumulative effect of its complications, diabetes doubles the risk of physical disability (Gregg et al. 2002; Gregg, Engelgau, and Narayan 2002). Among people diagnosed at the age of 40, diabetes reduces life expectancy by an average of 11.6 y for men and 14.3 y for women (Narayan et al. 2003).

Fortunately, numerous advances have been made in the clinical management and prevention of diabetes. Randomized controlled trials, particularly those conducted during the 1990s, showed that tightly controlling glycemia, blood pressure, and lipid levels, as well as receiving regular screening for eye, kidney, and foot problems, can reduce the effects of diabetes complications (Bowman et al. 2003; Narayan et al. 2000). Clinical and public health programs have responded by emphasizing quality of diabetes care as well as by implementing disease management programs to improve the delivery of these interventions (Bowman et al. 2003; Wang et al. 2006). As a result, the United States has seen a modest improvement in screening for eye conditions, foot conditions, and glycosylated hemoglobin (Hb_{A1c}) levels and a substantial improvement in cardiovascular risk factors among the population with diabetes (Geiss et al. 2005; Saaddine et al. 2002; Saaddine et al. 2006). Rates

of death or hospitalization due to diabetic ketoacidosis have improved. After several decades of increase, the rates of lower-extremity amputation and end-stage renal disease have decreased among the population with diabetes (Centers for Disease Control and Prevention 2006; Wang et al. 2006). Unfortunately, when rates of kidney disease and amputation are expressed in terms of events per 1,000 of the overall population (i.e., diabetic and nondiabetic), these rates have continued to increase owing to the growing prevalence of diabetes (Burrows et al. 2005; Geiss et al. 2005; Centers for Disease Control and Prevention 2006). This characterizes the broader threat and challenge for public health efforts against diabetes: Despite the success of secondary prevention, the burden of diabetes-related disease will continue to grow until approaches to preventing or delaying the onset of diabetes are successful.

Concluding Remarks

During the past 25 y, the prevention and control of diabetes have emerged as a major public health priority because diabetes is arguably the most prominent chronic condition that is worsening over time. In virtually all other indicators of chronic disease, ranging from heart disease to cancer, the health of the United States and industrialized populations has improved due to education, healthier environments, and better medical care. Diabetes may threaten these improvements, however, as its prevalence and complications affect a progressively larger segment of the population. The primary hope to limit the disease lies in the continued success of secondary prevention and minimizing diabetes complications in order to improve life with the disease and in the prevention or delay of diabetes.

References

Aguilar-Salinas, C.A., M.O. Velazquez, F.J. Gomez-Perez, C.A. Gonzalez, A.L. Esqueda, C. Molina, J.A. Rull-Rodrigo, and C.R. Tapia. 2003. Characteristics of patients with type 2 diabetes in Mexico: Results from a large population-based nationwide survey. *Diabetes Care* 26:2021-6.

Barcelo, A., and S. Rajpathak. 2001. Incidence and prevalence of diabetes mellitus in the Americas. *Revista Panamericana de Salud Public* 10:300-8.

Beischer, N.A., J.N. Oats, O.A. Henry, M.T. Sheedy, and J.E. Walstab. 1991. Incidence and severity of gestational diabetes mellitus according to country of birth in women living in Australia. *Diabetes* 40:35-8.

Bowman, B.A., E.W. Gregg, D.E. Williams, M.M. Engelmau, and L. Jack Jr. 2003. Translating the science of primary, secondary, and tertiary prevention to inform the public health response to diabetes. *J Publ Health Manag Pract* November (suppl): S8-14.

Brownson, R.C., T.K. Boehmer, and D.A. Luke. 2005. Declining rates of physical activity in the United States: What are the contributors? *Annu Rev Publ Health* 26:421-43.

Burke, J.P., H.P. Hazuda, and M.P. Stern. 2000. Rising trend in obesity in Mexican Americans and non-Hispanic Whites: Is it due to cigarette smoking cessation? *Int J Obes Relat Metab Disord* 24:1689-1694.

Burke, J.P., P. O'Brien, J. Ransom, P.J. Palumbo, E. Lydick, B.P. Yawn, M.L. Joseph III, and C.L. Leibson. 2002. Impact of case ascertainment on recent trends in diabetes incidence in Rochester, Minnesota. *Am J Epidemiol* 155:859-65.

Burke, J.P., K. Williams, S.P. Gaskill, H.P. Hazuda, S.M. Haffner, and M.P. Stern. 1999. Rapid rise in the incidence of type 2 diabetes from 1987 to 1996: Results from the San Antonio Heart Study. *Arch Intern Med* 159:1450-6.

Burrows, N.R., A.S. Narva, L.S. Geiss, M.M. Engelgau, and K.J. Acton. 2005. End-stage renal disease due to diabetes among southwestern American Indians, 1990-2001. *Diabetes Care* 28:1041-4.

Centers for Disease Control and Prevention. 2004. Trends in intake of energy and macronutrients—United States, 1971-2000. *MMWR Morb Mortal Wkly Rep* 53:80-2.

Centers for Disease Control and Prevention. 2006. Diabetes surveillance system: Diabetes prevalence. 2004. Available: www.cdc.gov/diabetes/statistics/prev/national/index.htm.

Cowie, C.C., K.F. Rust, D.D. Byrd-Holt, M.S. Eberhardt, K.M. Flegal, M.M. Engelgau, S.H. Saydah et al. 2006. Prevalence of diabetes and impaired fasting glucose in adults in the U.S. population: National Health and Nutrition Examination Survey 1999-2002. *Diabetes Care* 29:1263-8.

Dabelea, D., R.L. Hanson, P.H. Bennett, J. Roumain, W.C. Knowler, and D.J. Pettitt. 1998. Increasing prevalence of type II diabetes in American Indian children. *Diabetologia* 41:904-10.

Dabelea, D., D.J. Pettitt, K.L. Jones, and S.A. Arslanian. 1999. Type 2 diabetes mellitus in minority children and adolescents. An emerging problem. *Endocrinol Metabol Clin N Am* 28:709-29.

Dabelea, D., J.K. Snell-Bergeon, C.L. Hartsfield, K.J. Bischoff, R.F. Hamman, and R.S. McDuffie. 2005. Increasing prevalence of gestational diabetes mellitus (GDM) over time and by birth cohort: Kaiser Permanente of Colorado GDM Screening Program. *Diabetes Care* 28:579-84.

DECODE Study Group. 2003. Age- and sex-specific prevalences of diabetes and impaired glucose regulation in 13 European cohorts. *Diabetes Care* 26:61-9.

Diamond, J. 2003. The double puzzle of diabetes. *Nature* 423:599-602.

Dunstan, D.W., P.Z. Zimmet, T.A. Welborn, M.P. de Courten, A.J. Cameron, R.A. Sicree, T. Dwyer et al. 2002. The rising prevalence of diabetes and impaired glucose tolerance: The Australian Diabetes, Obesity and Lifestyle Study. *Diabetes Care* 25:829-34.

Edelstein, S.L., W.C. Knowler, R.P. Bain, R. Andres, E.L. Barrett-Connor, G.K. Dowse, S.M. Haffner et al. 1997. Predictors of progression from impaired glucose tolerance to NIDDM: An analysis of six prospective studies. *Diabetes* 46:701-10.

Engelgau, M.M., L.S. Geiss, J.B. Saaddine, J.P. Boyle, S.M. Benjamin, E.W. Gregg, E.F. Tierney et al. 2004. The evolving diabetes burden in the United States. *Ann Intern Med* 140:945-50.

Fagot-Campagna, A., D.J. Pettitt, M.M. Engelgau, N.R. Burrows, L.S. Geiss, R. Valdez, G.L. Beckles et al. 2000. Type 2 diabetes among North American children and adolescents: An epidemiologic review and a public health perspective. *J Pediatr* 136:664-72.

Ferrara, A., H.S. Kahn, C.P. Quesenberry, C. Riley, and M.M. Hedderson. 2004. An increase in the incidence of gestational diabetes mellitus: Northern California, 1991-2000. *Obstet Gynecol* 103:526-33.

Ford, E.S., A.H. Mokdad, and W.H. Giles. 2003. Trends in waist circumference among U.S. adults. *Obes Res* 11:1223-31.

Ford, E.S., D.F. Williamson, and S. Liu. 1997. Weight change and diabetes incidence: Findings from a national cohort of US adults. *Am J Epidemiol* 146:214-22.

Fox, C.S., S. Coady, P.D. Sorlie, D. Levy, J.B. Meigs, R.B. D'Agostino Sr., P.W. Wilson, and P.J. Savage. 2004. Trends in cardiovascular complications of diabetes. *JAMA* 292:2495-9.

Fox, C.S., M.J. Pencina, J.B. Meigs, R.S. Vasan, Y.S. Levitzky, and R.B. D'Agostino Sr. 2006. Trends in the incidence of type 2 diabetes mellitus from the 1970s to the 1990s: The Framingham Heart Study. *Circulation* 113:2914-8.

Geiss, L., M. Engelgau, L. Pogach, K. Acton, B. Fleming, S. Roman, L. Han, J. Wang, and F. Vinicor. 2005. A national progress report on diabetes: Successes and challenges. *Diabetes Tech Therapeut* 7:198-203.

Geiss, L.S., L. Pan, B. Cadwell, E.W. Gregg, S.M. Benjamin, and M.M. Engelgau. 2006. Changes in incidence of diabetes in U.S. adults, 1997-2003. *Am J Prev Med* 30:371-7.

Green, A., H. Stovring, M. Andersen, and H. Beck-Nielsen. 2005. The epidemic of type 2 diabetes is a statistical artefact. *Diabetologia* 48:1456-8.

Gregg, E.W., B.L. Cadwell, Y.J. Cheng, C.C. Cowie, D.E. Williams, L. Geiss, M.M. Engelgau, and F. Vinicor. 2004. Trends in the prevalence and ratio of diagnosed to undiagnosed diabetes according to obesity levels in the U.S. *Diabetes Care* 27:2806-12.

Gregg, E.W., Y.J. Cheng, K.M. Narayan, T.J. Thompson, and D.F. Williamson. 2007. The contributions of increases in the prevalence of overweight, obesity, and severe obesity to the increased prevalence of diabetes in the United States: 1976-2004. (Abstract). *Diabetes*. In press.

Gregg, E.W., M.M. Engelgau, and V. Narayan. 2002. Complications of diabetes in elderly people. *Br Med J* 325:916-7.

Gregg, E.W., C.M. Mangione, J.A. Cauley, T.J. Thompson, A.V. Schwartz, K.E. Ensrud, and M.C. Nevitt. 2002. Diabetes and incidence of functional disability in older women. *Diabetes Care* 25:61-7.

Gross, L.S., L. Li, E.S. Ford, and S. Liu. 2004. Increased consumption of refined carbohydrates and the epidemic of type 2 diabetes in the United States: An ecologic assessment. *Am J Clin Nutr* 79:774-9.

Gu, K., C.C. Cowie, and M.I. Harris. 1999. Diabetes and decline in heart disease mortality in US adults. *JAMA* 281:1291-7.

Hamman, R.F., R.R. Wing, S.L. Edelstein, J.M. Lachin, G.A. Bray, L. Delahanty, M. Hoskin et al. 2006. Effect of weight loss with lifestyle intervention on risk of diabetes. *Diabetes Care* 29:2102-7.

Howard, A.A., J.H. Arnsten, and M.N. Gourevitch. 2004. Effect of alcohol consumption on diabetes mellitus: A systematic review. *Ann Intern Med* 140:211-9.

Hu, F.B., R.M. van Dam, and S. Liu. 2001. Diet and risk of type II diabetes: The role of types of fat and carbohydrate. *Diabetologia* 44:805-17.

Jeon, C.Y., R. Lokken, F.B. Hu, and R.M. van Dam. 2007. Physical activity of moderate intensity and risk of type 2 diabetes. A systematic review. *Diabetes Care* 30:744-52.

Kim, C., K.M. Newton, and R.H. Knopp. 2002. Gestational diabetes and the incidence of type 2 diabetes: A systematic review. *Diabetes Care* 25:1862-8.

King, H., R.E. Aubert, and W.H. Herman. 1998. Global burden of diabetes, 1995-2025: Prevalence, numerical estimates, and projections. *Diabetes Care* 21:1414-31.

Knowler, W.C., E. Barrett-Connor, S.E. Fowler, R.F. Hamman, J.M. Lachin, E.A. Walker, and D.M. Nathan. 2002. Reduction in the incidence of type 2 diabetes with lifestyle intervention or metformin. *New Engl J Med* 346:393-403.

Laaksonen, D.E., J. Lindstrom, T.A. Lakka, J.G. Eriksson, L. Niskanen, K. Wikstrom, S. Aunola et al. 2005. Physical activity in the prevention of type 2 diabetes: The Finnish diabetes prevention study. *Diabetes* 54:158-65.

Lee, D.H., I.K. Lee, K. Song, M. Steffes, W. Toscano, B.A. Baker, and D.R. Jacobs Jr. 2006. A strong dose-response relation between serum concentrations of persistent organic pollutants and diabetes: Results from the National Health and Examination Survey 1999-2002. *Diabetes Care* 29:1638-44.

Leibson, C.L., P.C. O'Brien, E. Atkinson, P.J. Palumbo, and L.J. Melton III. 1997. Relative contributions of incidence and survival to increasing prevalence of adult-onset diabetes mellitus: A population-based study. *Am J Epidemiol* 146:12-22.

Lipscombe, L.L., and J.E. Hux. 2007. Trends in diabetes prevalence, incidence, and mortality in Ontario, Canada 1995-2005: A population-based study. *Lancet* 369:750-6.

Meigs, J.B., D.C. Muller, D.M. Nathan, D.R. Blake, and R. Andres. 2003. The natural history of progression from normal glucose tolerance to type 2 diabetes in the Baltimore Longitudinal Study of Aging. *Diabetes* 52:1475-84.

Narayan, K.M., J.P. Boyle, L.S. Geiss, J.B. Saaddine, and T.J. Thompson. 2006. Impact of recent increase in incidence on future diabetes burden: U.S., 2005-2050. *Diabetes Care* 29:2114-6.

Narayan, K.M., J.P. Boyle, T.J. Thompson, E.W. Gregg, and D.F. Williamson. 2007. Effect of body mass index on lifetime risk for diabetes in the United States. *Diabetes Care* 30:1562-6.

Narayan, K.M., J.P. Boyle, T.J. Thompson, S.W. Sorensen, and D.F. Williamson. 2003. Lifetime risk for diabetes mellitus in the United States. *JAMA* 290:1884-90.

Narayan, K.M., E.W. Gregg, A. Fagot-Campagna, M.M. Engelgau, and F. Vinicor. 2000. Diabetes—a common, growing, serious, costly, and potentially preventable public health problem. *Diabetes Res Clin Pract* 50:S77-84.

Nathan, D.M. 1993. Long-term complications of diabetes mellitus. *New Engl J Med* 328:1676-85.

Nielsen, S.J., A.M. Siega-Riz, and B.M. Popkin. 2002. Trends in energy intake in U.S. between 1977 and 1996: Similar shifts seen across age groups. *Obes Res* 10:370-8.

Ogden, C.L., M.D. Carroll, L.R. Curtin, M.A. McDowell, C.J. Tabak, and K.M. Flegal. 2006. Prevalence of overweight and obesity in the United States, 1999-2004. *JAMA* 295:1549-55.

Ogden, C.L., K.M. Flegal, M.D. Carroll, and C.L. Johnson. 2002. Prevalence and trends in overweight among US children and adolescents, 1999-2000. *JAMA* 288:1728-32.

Ogden, C.L., C.D. Fryar, M.D. Carroll, and K.M. Flegal. 2004. Mean body weight, height, and body mass index, United States 1960-2002. *Adv Data* 347:1-17.

Ogden, C.L., R.P. Troiano, R.R. Briefel, R.J. Kuczmarski, K.M. Flegal, and C.L. Johnson. 1997. Prevalence of overweight among preschool children in the United States, 1971 through 1994. *Pediatrics* 99:E1.

Pan, X.R., G.W. Li, Y.H. Hu, J.X. Wang, W.Y. Yang, Z.X. An, Z.X. Hu et al. 1997. Effects of diet and exercise in preventing NIDDM in people with impaired glucose tolerance. The Da Qing IGT and Diabetes Study. *Diabetes Care* 20:537-44.

Pettitt, D.J., and W.C. Knowler. 1998. Long-term effects of the intrauterine environment, birth weight, and breast-feeding in Pima Indians. *Diabetes Care* 21:B138-41.

Popkin, B.M., P.S. Haines, and K.C. Reidy. 1989. Food consumption trends of US women: Patterns and determinants between 1977 and 1985. *Am J Clin Nutr* 49:1307-19.

Popkin, B.M., and S.J. Nielsen. 2003. The sweetening of the world's diet. *Obes Res* 11:1325-32.

Popkin, B.M., A.M. Siega-Riz, and P.S. Haines. 1996. A comparison of dietary trends among racial and socio-economic groups in the United States. *New Engl J Med* 335:716-20.

Qiao, Q., G. Hu, J. Tuomilehto, T. Nakagami, B. Balkau, K. Borch-Johnsen, A. Ramachandran et al. 2003. Age- and sex-specific prevalence of diabetes and impaired glucose regulation in 11 Asian cohorts. *Diabetes Care* 26:1770-80.

Saaddine, J.B., B. Cadwell, E.W. Gregg, M.M. Engelgau, F. Vinicor, G. Imperatore, and K.M. Narayan. 2006. Improvements in diabetes processes of care and intermediate outcomes: United States, 1988-2002. *Ann Intern Med* 144:465-74.

Saaddine, J.B., M.M. Engelgau, G.L. Beckles, E.W. Gregg, T.J. Thompson, and K.M. Narayan. 2002. A diabetes report card for the United States: Quality of care in the 1990s. *Ann Intern Med* 136:565-74.

Schulz, L.O., P.H. Bennett, E. Ravussin, J.R. Kidd, K.K. Kidd, J. Esparza, and M.E. Valencia. 2006. Effects of traditional and western environments on prevalence of type 2 diabetes in Pima Indians in Mexico and the U.S. *Diabetes Care* 29:1866-71.

SEARCH for Diabetes in Youth Study Group. 2006. The burden of diabetes among U.S. youth: Prevalence estimates from the SEARCH for Diabetes in Youth Study. *Pediatrics.* 118(4):1510-18.

Simpson, M.E., M. Serdula, D.A. Galuska, C. Gillespie, R. Donehoo, C. Macera, and K. Mack. 2003. Walking trends among U.S. adults: The Behavioral Risk Factor Surveillance System, 1987-2000. *Am J Prev Med* 25:95-100.

Thomas, R.J., P.J. Palumbo, L.J. Melton III, V.L. Roger, J. Ransom, P.C. O'Brien, and C.L. Leibson. 2003. Trends in the mortality burden associated with diabetes mellitus: A population-based study in Rochester, Minn, 1970-1994. *Arch Intern Med* 163:445-51.

Tierney, E.F., B.L. Cadwell, M.M. Engelgau, L. Shireley, S.L. Parsons, K. Moum, and L.S. Geiss. 2004. Declining mortality rate among people with diabetes in North Dakota, 1997-2002. *Diabetes Care* 27:2723-5.

Tierney, E.F., L.S. Geiss, M.M. Engelgau, T.J. Thompson, D. Schaubert, L.A. Shireley, P.J. Vukelic, and S.L. McDonough.

2001. Population-based estimates of mortality associated with diabetes: Use of a death certificate check box in North Dakota. *Am J Publ Health* 91:84-92.

Tuomilehto, J., J. Lindstrom, J.G. Eriksson, T.T. Valle, H. Hamalainen, P. Ilanne-Parikka, S. Keinanen-Kiukaanniemi et al. 2001. Prevention of type 2 diabetes mellitus by changes in lifestyle among subjects with impaired glucose tolerance. *New Engl J Med* 344:1343-50.

van Dam, R.M. and F.B. Hu. 2005. Coffee consumption and risk of type 2 diabetes: A systematic review. *JAMA* 294:97-104.

Wang, J., D.E. Williams, K.M. Narayan, and L.S. Geiss. 2006. Declining death rates from hyperglycemic crisis among adults with diabetes, U.S., 1985-2002. *Diabetes Care* 29:2018-22.

Williams, D.E., B.L. Cadwell, Y.J. Cheng, C.C. Cowie, E.W. Gregg, L.S. Geiss, M.M. Engelgau, K.M. Narayan, and G. Imperatore. 2005. Prevalence of impaired fasting glucose and its relationship with cardiovascular disease risk factors in US adolescents, 1999-2000. *Pediatrics* 116:1122-6.

Zimmet, P., K.G. Alberti, and J. Shaw. 2001. Global and societal implications of the diabetes epidemic. *Nature* 414:782-7.

Chapter 2

Waging War on Type 2 Diabetes
Primary Prevention Through Exercise Biology

Frank W. Booth, PhD; Manu V. Chakravarthy, MD, PhD; and Matthew J. Laye, BS

A word such as *pandemic* usually arouses a strong response. For example, since 2003 the outbreak of avian flu is projected to become a pandemic, and governments across the globe are mobilizing massive infrastructural, monetary, and human resources to fight it. Not surprisingly, the media give such events immense airtime on national news programs, educating the public of the dos and don'ts of the disease and providing other related information such that, at the end of the day, the public is highly aware of the problem and prepared to tackle the impending pandemic. All of this is exactly as it should be.

Therefore, we find it ironic that similar propaganda, mobilization of resources, support, and education are strikingly lacking to help fight a pandemic that we have been in the midst of since the last decade. We are, of course, referring to the pandemic of diabetes and obesity. Since the 2001 declaration by the director of the Centers for Disease Control and Prevention (CDC) that the United States is facing an epidemic of obesity and diabetes, several other countries, including developing nations, have reported similar alarming trends. We are in the thick of a public health war against several burgeoning chronic diseases, many of which are inextricably linked to lifestyle changes, especially a lack of routine physical activity. Hence, understanding the biological basis of these inactivity-mediated chronic conditions as well as of physical inactivity itself is critical if we are to win the war on the diabetes and obesity pandemic and help alleviate the burden of human disease. Success in the fight against inactivity-mediated disease

requires that we provide the same overwhelming mobilization of resources and support that we are providing for other pandemics.

Scope of the Problem

For the purposes of this chapter, the term *diabetes* will refer to type 2 diabetes, the form of the disease that was previously called *non-insulin-dependent diabetes mellitus* (NIDDM) or *adult-onset diabetes*. Type 2 diabetes differs from type 1 diabetes, previously called *insulin-dependent diabetes mellitus* (IDDM), which usually manifests much earlier in life with a distinct pathogenetic profile. Until just before 2000, type 2 diabetes was regarded as a disease of middle-aged and elderly individuals (hence the name *adult-onset diabetes*). However, once teenagers began displaying clinical cases of type 2 diabetes, largely due to concurrent increases in childhood obesity and sedentary lifestyle, the terminology *adult-onset* was discarded, for it was no longer a disease confined to adults. Noting the escalating increases in sedentary living (Brownson, Boehmer & Luke, 2005), the 2004 International Diabetes Federation Consensus Workshop (Alberti et al., 2004) indicated that within 10 y type 2 diabetes will be the predominant form of diabetes in many ethnic groups of children worldwide, surpassing type 1 diabetes in prevalence in children. The global figure of all people with diabetes, including adults, is skyrocketing. It is now 150 million and is predicted to rise to 300 million in 2025, with 75% of the cases occurring in developing countries (King, Aubet & Herman, 1998).

This chapter was written while supported by a Life Sciences Fellowship (MJL).

According to the U.S. National Institutes of Health (NIH), type 2 diabetes directly contributes to the following conditions:

- Heart disease and stroke. Adults with diabetes have death rates due to heart disease that are 2 to 4 times greater than rates for adults without diabetes. The risk for stroke is 2 to 4 times higher among individuals with diabetes.
- High blood pressure. About 73% of adults with diabetes have blood pressures greater than 130/80 mmHg.
- Blindness. Diabetes is the leading cause of new cases of blindness among adults aged 20 to 74.
- Kidney disease. Diabetes is the leading cause of kidney failure; 150,000 people with diabetes live on chronic dialysis or with a kidney transplant.
- Nervous system disease. Over 50% of individuals with diabetes have impaired sensation or pain in the feet or hands, slowed digestion in the stomach, carpal tunnel syndrome, and other nerve problems. A severe form of diabetic nerve disease is a major factor in lower-extremity amputations.
- Amputation. Of nontraumatic lower-limb amputations, 60% occur among individuals with diabetes.
- Dental disease. About one-third of individuals with diabetes have severe periodontal diseases with loss of gum attachment to the teeth measuring 5 mm or more.
- Pregnancy complication. Mothers with diabetes have a greater number of spontaneous abortions, and their babies have an increased risk of major birth defects and of developing diabetes later in life.
- Immune system disorder. People with diabetes have a reduced ability to reject bacterial and viral infections and are more likely to die from pneumonia or influenza than are people who do not have diabetes.

Besides resulting in associated health costs, diabetes creates tremendous economic costs. In 2002, the American Diabetes Association estimated that the indirect costs of diabetes in the United States were $132 billion U.S. (Hogan, Dall & Nikolov, 2003). Direct expenditures were $92 billion U.S., or about 5% of the nearly $1.9 trillion U.S. annual total health care costs. The medical expenditure for an individual with diabetes was $13,243 U.S., as compared to $2,560 U.S. for those without diabetes. Even when adjusted for differences in age, sex, and race or ethnicity, medical expenditures were about 2.5 times higher in an individual with diabetes than in a person without diabetes. This economic strain leads to other adverse consequences, such as diverting scarce monies from research in order to pay for health care. Biomedical research (and hence funding for the research) is needed to help alleviate the diabetes burden, but at the same time, funds are also needed to treat patients who already have the disease, setting up a vicious cycle. In the United States, the health care industry consumes about $1 in every $6 spent. By 2015, health care will use $1 of every $5 spent, which is a 20% jump. Increasing health care costs will redistribute monies from other areas (including research on diabetes), further lowering the quality of life for most individuals.

Eliminating or minimizing the health problems produced by diabetes could significantly improve the quality of life for patients with diabetes and their families while at the same time diminish health care costs and thus enhance economic productivity (Hogan, Dall & Nikolov, 2003, 10). It is in this arena that the old axiom that prevention is better than cure rings truer than ever. One powerful weapon for prevention is to reintroduce physical activity into daily living in order to curb the rise in health care costs.

Rationale for Action

Knowing that wars cost money as well as quantity and quality of life, what is the justification to have scientists and other health care professionals go to war against chronic diseases, type 2 diabetes, and, in particular, physical inactivity? Considering the information given in the preceding section, preventing type 2 diabetes is an endeavor worth accomplishing. In fact, when considered carefully in light of the current scientific data, it is a condition that is highly amenable to primary prevention. Incremental increases in physical activity are not linearly related to incremental decreases in diabetes: Sedentary individuals who moderately increase their level of physical activity drastically reduce the prevalence of diabetes. However, on average, a threshold of moderate physical activity (about 30 min of brisk walking daily) exists before the prevalence of type 2 diabetes declines.

The CDC has stated that physical inactivity is an actual cause of chronic diseases (Mokdad, Marks, Stroup & Gerberding, 2004). Thus, one major weapon against diabetes is physical activity. As the current thought seems to be that consuming excess calories is the sole environmental factor responsible for diabetes, it must be emphasized that physical inactivity itself is an independent diabetes risk factor. Among individuals who are overweight or obese, the disease is less prevalent in the physically active than it is in the inactive (Wei et al., 1999). Therefore, we must combat physical inactivity if we are to progress against the prediction that the incidence of type 2 diabetes will almost double to 300 million individuals in 2025 (Zimmet, Alberti & Shaw, 2001).

As mentioned, type 2 diabetes is preventable. From 1980 to 1996, 91% of diabetes cases among 84,941 female nurses could be attributed to habits that did not conform to a low-risk lifestyle. A low-risk lifestyle was defined according to a combination of five variables: a BMI of less than 25 kg/m^2, a diet high in cereal fiber and polyunsaturated fat and low in trans fat and glycemic load, engagement in moderate to vigorous physical activity for at least half an hour per day, no current smoking, and moderate alcohol consumption.

We describe the conversion from a high-risk to a low-risk lifestyle as a *moderate* change. Others have mislabeled the modification as an *intense* lifestyle change. Using the term *intense* with an individual who is completely sedentary could lead to even poorer adherence to a healthier lifestyle. We define *intense* physical activity as replacing all modern conveniences with the hard physical labor needed to grow and harvest your own food and build your own house. Intense physical activity could also be running marathons. From this perspective, we contend that adding 30 min of brisk walking to each day and eating better can hardly be classified as an intense lifestyle modification. Most importantly, the conclusions from three large randomized clinical trials amply support the notion that moderate-intensity activ-

ity, such as brisk walking, is sufficient to prevent at least 58% of cases of diabetes from occurring in people with prediabetes.

The Finnish Diabetes Prevention Study consisted of subjects who have abnormal glucose tolerance. In this study, the cumulative incidence of diabetes was reduced by 58% in the lifestyle intervention group (which consisted of individualized counseling aimed at reducing weight, total fat intake, and saturated fat intake while increasing fiber intake and participation in physical activity; Tuomilehto et al., 2001). Some of the subjects continued in a 4.1 y follow-up to the initial 3.2 y of the trial; during this follow-up participation in physical activity was recorded and the data were then adjusted for age, gender, smoking status, major risk factors for diabetes at baseline (BMI, fasting and 2 h plasma glucose and insulin levels, and family history of diabetes), and baseline total leisure-time physical activity (LTPA). Participants in the upper third of the change in total LTPA were 66% less likely to develop diabetes than were those in the lower third (Laaksonen et al., 2005). Furthermore, when the same subjects were analyzed for walking adherence, individuals who walked more than 2.5 h/wk (an average of 22 min/d) were 69% less likely to develop diabetes than were those who walked less than 1 h/wk.

During the 2.8 y of the U.S. Diabetes Prevention Program (DPP), the incidence of newly diagnosed diabetes in individuals who started the trial with abnormal glucose tolerance tests was 11.0% and 4.8% in the placebo and lifestyle groups, respectively (lifestyle intervention resulted in a 58% reduction in diabetes). The lifestyle group underwent individualized counseling aimed at lowering initial body weight through a healthy low-calorie, low-fat diet and by engaging in physical activity of moderate intensity, such as brisk walking, for at least 150 min/wk (Knowler et al., 2002). The proportion of participants who met this goal was 58% at the most recent visit to the DPP. However, the data were presented as intent to treat, so that both the 58% who adhered to the exercise goal and the 42% who did not were combined to obtain the 58% reduction in diabetes compared to the placebo group. What was not presented in the published paper was the percentage reduction in diabetes among the 58% who made the goal of exercising 150 min/wk. While health care professionals desire intent to treat so that they can decide on the efficacy of treatment, failing to present the results for the 58% who did walk 150 min/wk is falling short of a balanced data presentation that includes the success rate of those who complied with the treatment. However, even in the 495 participants not meeting the weight loss goal at year 1 of the follow-up from 2.8 y in the U.S. DPP, those who achieved the physical activity goal had a 44% lower diabetes incidence (Hamman et al. 2006).

Lastly, in subjects with impaired glucose tolerance who entered the Daqing study (Pan et al., 1997), the risk of developing diabetes was reduced by 31%, 46%, and 42% after adjusting for BMI difference in the diet-only group, the exercise-only group, and the combined diet-plus-exercise intervention group, respectively.

Taken together, the results from these large-scale clinical trials highlight two critical points: First, identifying people with prediabetes has prognostic significance, and prevention of overt diabetes is achievable. It is because of such trials that organizations like the American Diabetes Association dropped the cutoff for normal fasting blood glucose from 110 mg/dl to 100 mg/dl. This means that a value of 100 mg/dl or greater leads to a diagnosis of IFG, which is included in the term *prediabetes*. While the large-scale clinical trials have demonstrated that without appropriate intervention people with prediabetes have a strong propensity for developing overt diabetes, other studies have shown that prediabetes also carries an increased risk of cardiovascular disease as compared to having a fasting blood glucose of less than 100 mg/dl (Schnell & Standl, 2006). Second, modest physical activity, such as brisk walking for 2.5 h/wk (22 min/d), is a powerful weapon that prevents the onset of overt disease by at least 58% in sedentary individuals with prediabetes.

Despite these simple conclusions, the majority of adults do not engage in moderate physical activity each day. Several reasons, which are beyond the scope of this chapter, have been given for the inability to perform this seemingly trivial amount of activity. Readers interested in the many barriers to participating in daily physical activity and the means to overcome them are referred to another text (Chakravarthy & Booth, 2003). The bottom line here is that engaging in the war against diabetes is overwhelmingly justified, given the highly favorable benefit-to-risk ratio.

Role of Physical Inactivity in Diabetes

Sedentary behavior is a rapid initiator of insulin resistance. Insulin resistance in turn is a major initiator of diabetes. Therefore, this vicious cycle of inactivity needs to be broken, and this can be achieved by incorporating physical activity into daily living.

The following examples highlight the rapid induction of insulin insensitivity by physical inactivity. In highly trained athletes, the metabolic clearance rates of glucose as determined by euglycemic clamps were 15.6, 10.1, and 8.5 ml $\cdot$ kg^{-1} $\cdot$ min^{-1} at 12, 60, and 168 h of no exercise, respectively, following the last exercise bout (Burstein et al., 1985). Metabolic clearance rates reached the values of the sedentary group (not athletically active; 7.8 ml $\cdot$ kg^{-1} $\cdot$ min^{-1}) at 60 and 168 h of inactivity. Insulin receptor binding to insulin in young erythrocytes also fell during 48 h of no exercise. The rates of insulin-mediated glucose uptake (glucose disposal) in a euglycemic clamp were 9.40, 7.78, 6.82, and 7.11 ml $\cdot$ kg^{-1} $\cdot$ min^{-1} for athletes at 14, 38, 86, and 144 h following the last exercise bout; no differences existed for untrained subjects (6.80 ml $\cdot$ kg^{-1} $\cdot$ min^{-1}) at 38, 86, or 144 h (Oshida et al., 1991). These human data are remarkably similar to data from rats that stopped 21 d of voluntary running. Four measures of insulin signaling in the epitrochlearis muscle that had increased over sedentary levels at 5 h of inactivity following the 21 d of running returned to sedentary levels by 53 h of inactivity. These measures were (1) submaximal insulin-stimulated glucose uptake, (2) insulin binding to muscle homogenates, (3) insulin receptor beta subunit protein, and (4) insulin receptor beta subunit tyrosine phosphorylation (Kump & Booth, 2005a).

Additionally, in subjects who stopped training for 10 d after engaging in at least 45 min of exercise 5 to 7 d/wk for the preceding 6 mo (Heath et al., 1983), the maximal increase in plasma insulin concentration after an oral glucose challenge was 100% higher after 10 d without exercise than it was before the 10 d of no exercise, despite no changes in maximal oxygen uptake, percent body fat, or body weight. Even with the increased insulin levels following the oral glucose load, blood glucose concentrations remained higher in the group of subjects who stopped their exercise regimen for 10 d. Remarkably, just one bout of exercise after 11 d without physical activity returned the blood insulin and glucose responses to almost the trained value during an oral glucose test performed on day 11. Likewise, other studies have shown a rapid induction of insulin resistance with inactivity. Continuous bed rest for 3 d increased the level of blood insulin needed to remove an oral glucose load (Lipman et al., 1972; Smorawinski et al., 2000). The soleus muscle lost all insulin-stimulated glucose uptake after 24 h of hind-limb immobilization as compared to pair-fed mice (Seider, Nicholson & Booth, 1982).

Abrupt cessation of voluntary physical activity rapidly increases abdominal obesity, which consequently results in insulin resistance. Upon ending 21 d of voluntary running, rats showed 25% and 48% increases in epididymal and omental fat (Kump & Booth, 2005b). Kohrt and Holloszy (1995) contend that most of the decreased glucose tolerance seen in older humans is due to accumulation of abdominal fat with development of insulin resistance. We suggest that decreased glucose tolerance in older humans is the consequence of physical inactivity directly and indirectly (via abdominal fat) inducing insulin resistance in skeletal muscle.

Skeletal muscle increases its insulin sensitivity for glucose uptake from the blood only during contraction and during postexercise just long enough to replenish the stores of intramyocellular glycogen used in the exercise, thereby conserving blood glucose when the muscle is inactive. Consequently, rises in insulin sensitivity with exercise are fleeting. While insulin sensitivity does not change after a meal, the increased uptake of glucose into skeletal muscle is also transitory after the meal. Thus it appears that chronically inactive skeletal muscle has no immediate need for blood glucose. Due to the lower insulin sensitivity of the muscle, glucose is partitioned to the liver (for storage as glycogen) and adipocytes (for conversion and storage as triacylglycerol) when muscle is inactive.

The concept of metabolic cycling of fuels is evoked to explain why skeletal muscle does not adapt well to physical inactivity (Chakravarthy & Booth, 2004). *Metabolic networks evolved to support a cycling of insulin sensitivity at skeletal muscle in response to cycles of feast and famine and exercise and rest.* The key initiator for the

cycling of insulin sensitivity in skeletal muscle is substrate oxidation by muscle. It is the cycling of substrates through the mitochondria that elicits the signals for mitochondrial biogenesis and increases the capacity of skeletal muscle to oxidize fatty acids and produce adenosine triphosphate (ATP) in the electron transport chain (Holloszy & Booth, 1976). If muscle is inactive, lipid oxidation becomes stagnant and diminishes insulin sensitivity (Goodpaster & Brown, 2005). The association between insulin sensitivity and mitochondrial capacity has become recently popular due to the realization that decreases in both mitochondrial oxidative activity and mitochondrial ATP synthesis play a critical role in insulin resistance, a prominent feature of type 2 diabetes (Lowell & Shulman, 2005). The stimulus for maintaining mitochondria (Booth & Holloszy, 1977) and oxidative capacity (Henriksson & Reitman, 1977) in skeletal muscle is withdrawn within 1 to 2 d when physical activity ceases. These changes, if maintained for weeks, lead to maladaptive processes that lower the capacity of skeletal muscle to oxidize fatty acids.

These concepts and the current scientific data suggest that the major driver in the pathogenesis of type 2 diabetes is inactivity of the skeletal muscle. When inactivity exists for days, it begets insulin resistance, lowers the capacity to oxidize substrates, particularly fatty acids, and adds to abdominal fat. Therefore, we need to reconsider the nutritional emphasis generally attributed to the development of diabetes.

New Ammunitions

It would be facile to state that the complex pathophysiology of diabetes and its treatment can be reduced to a single factor. Clearly, multifactorial approaches to describing and treating diabetes are required. One such approach is the understanding of the biology of physical inactivity. Some genes in humans and animals were selected to support physical activity for survival tasks such as acquiring food, building shelter, and defending against predators (Booth, Chakravarthy, Gordon & Spangenburg, 2002; Booth, Chakravarthy & Spangenburg, 2002; Chakravarthy et al., 2003; Chakravarthy et al., 2004). Each of these activities needs properly functioning metabolic machinery for the efficient utilization and storage of fuel. However, genes alone are not sufficient to fully explain the causes for common chronic diseases. Francis Collins (Collins, 2006), one of the pioneers of the Human Genome Project, stated the following:

> But genes alone do not tell the whole story. Recent increases in chronic diseases like diabetes, childhood asthma, obesity or autism cannot be due to major shifts in the human gene pool as those changes take much more time to occur. They must be due to changes in the environment, including diet and physical activity, which may produce disease in genetically predisposed persons.

Thus, increases in type 2 diabetes prevalence must be due to changes in the environment, including alterations in diet and physical activity (which indeed have changed dramatically in the last 50 y), that produce disease in genetically predisposed individuals. One of the major challenges of the 21st century is to identify a subpopulation of physical inactivity genes that predispose to chronic diseases such as diabetes.

How an organism adapts to physical inactivity is a fundamental question of biology given that during much of human evolutionary history, human genomes were shaped on the background of obligate physical activity needed for survival. Currently in modern medicine, understanding the molecular basis of disease is considered paramount to preventing disease. Ironically, this canonical view that is applied to every known field from cancer to heart disease often is not employed within the field of exercise. Indeed, some people with minimal understanding or training in exercise biology do not even accept exercise as a valid fundable science. Understanding how physical inactivity affects the primordially programmed human genome would not only yield fruitful avenues for research, but has the potential to result in a paradigm shift in the way scientists view and treat chronic disease.

One way to accomplish biological bases for chronic diseases is to support a search for molecular links between physical inactivity and chronic disease, analogous to the demonstration that carcinogens in tobacco directly cause lung cancer. In the issue of *Science* published on October 18, 1996, Denissenko and colleagues showed that a specific carcinogen in tobacco, benzo[a]pyrene diol epoxide, adducts along exons of the p53

gene. On March 20, 1997, a mere 5 mo later, Liggett Group agreed to settle lawsuits in the United States brought by 22 states seeking to recover the cost of smoking-related illnesses. As a part of the accord, Liggett Group agreed to acknowledge that smoking cigarettes causes cancer, which the industry had never previously conceded (Hunter-Gault, 1997). In a recent review regarding the epidemiologic and economic consequences of the global epidemics of obesity and diabetes, Yach and colleagues (Yach, Struckler & Brownell, 2006) made the following statement:

> Prevention and intervention approaches would ideally be based on a thorough knowledge of causes. In the absence of such knowledge, efforts will be scattered, evaluated too rarely and difficult to assess in terms of impact on public health. There is a crucial need to develop precise roadmaps that define appropriate interventions based on the causes of obesity and diabetes at the macro- and microscopic levels from which a coherent prevention plan can be constructed.

Molecular causes will drive policy, but this weapon against inactivity as a cause of diabetes is both underutilized and lacking support.

Future Battle Plans

In 2000, Zimmet proposed that diabetes cannot be stopped by traditional medical approaches. He contends that we must see dramatic changes in worldwide socioeconomic and cultural status. He does not believe that a single individual can make much of a difference; rather, he recommends that international diabetes and public health communities lobby and mobilize politicians and appropriate international nongovernmental agencies to address the socioeconomic, behavioral, nutritional, and public health issues that have led to the pandemic of diabetes. He suggests that some of these parties should form a multidisciplinary task force to help reverse the underlying socioeconomic causes of diabetes. In a recent publication, Brownell's group (Yach, Struckler, Brownell, 2006) wrote the following:

> A major threat to the health systems of developing countries may not be only the

importation of "Westernized lifestyles," but also the importation of Westernized medical responses. As in the US, healthcare services for developing countries tend to be oriented toward acute, reactionary medical care rather than cost-saving preventive approaches. . . . Research-funding agencies favor medical and surgical solutions over health promotion and health systems interventions and policies.

Our overall strategy must shift from a purely reactive treatment approach to a proactive preventative approach based upon a true biological basis. This type of shift will take time, as highlighted by Holloszy (2005): "Although the postexercise increase in muscle insulin sensitivity has been characterized in considerable detail, the basic mechanisms underlying this phenomenon remain a mystery." He is unsurprised by this lack of progress: "When the duration of delineating signaling pathways is considered, the effort to explain how insulin stimulates glucose transport has taken 60 yrs of enormous expenditures of effort and resources by hundreds of talented investigators." Much less effort has been dedicated to understanding how exercise enhances insulin sensitivity, and even less research has been devoted to how physical inactivity diminishes insulin sensitivity.

Given the ever-increasing rate of diabetes prevalence and the overwhelming data supporting the benefits of combating type 2 diabetes with physical activity and by understanding exercise biology, the costs of inaction may prove to be unnecessarily deadly. Time cannot be used as an excuse. To win the war against diabetes we need to mobilize all of our resources in a multidisciplinary effort that includes not only cellular and molecular but also socioeconomic and cultural ammunitions. The army against diabetes must accept soldiers who bring multifaceted talents in order to have a fighting chance; hopefully, those with expertise in physical activity and inactivity will be drafted.

Concluding Remarks

Type 2 diabetes is a stealth pandemic affecting many people at an enormous economic cost. Remarkably, 9 in 10 cases of diabetes could be prevented by a simple lifestyle modification that some people have erroneously categorized as intense. Whether using caloric expenditure, heart rate, or the Borg scale, the profession of exercise classifies walking as a moderate-intensity physical activity and not as intense physical activity unless the exerciser is frail or with physical disability. Moderate physical activity, such as brisk walking for 2.5 h/wk (22 min/d), is capable of reducing the risk of overt diabetes by at least 66% in individuals with prediabetes, as compared to walking for less than 1 h/wk. Conversely, the potent effects of physical inactivity are exemplified by rapid decreases in insulin sensitivity following training cessation in endurance athletes. Additionally, physical activity can ward off numerous other chronic diseases, thereby improving quality of life and decreasing preventable health care costs. Despite the billions of dollars flowing to pharmaceutical companies, the most efficacious preventative antidiabetic drug is still physical activity. Until health care professionals as well as the general public realize and implement physical activity as the most effective exit strategy, the unnecessary war against type 2 diabetes will continue to rage.

References

Alberti, G., P. Zimmet, J. Shaw, Z. Bloomgarden, F. Kaufman, and M. Silink. 2004. Type 2 diabetes in the young: The evolving epidemic: The international diabetes federation consensus workshop. *Diabetes Care* 27:1798-811.

Booth, F.W., M.V. Chakravarthy, S.E. Gordon, and E.E. Spangenburg. 2002. Waging war on physical inactivity: Using modern molecular ammunition against an ancient enemy. *J Appl Physiol* 93:3-30.

Booth, F.W., M.V. Chakravarthy, and E.E. Spangenburg. 2002. Exercise and gene expression: Physiological regulation of the human genome through physical activity. *J Physiol* 543:399-411.

Booth, F.W., and J.O. Holloszy. 1977. Cytochrome c turnover in rat skeletal muscles. *J Biol Chem* 252:416-9.

Brownson, R.C., T.K. Boehmer, and D.A. Luke. 2005. Declining rates of physical activity in the United States: What are the contributors? *Annu Rev Publ Health* 26:421-43.

Burstein, R., C. Polychronakos, C.J. Toews, J.D. MacDougall, H.J. Guyda, and B.I. Posner. 1985. Acute reversal of the enhanced insulin action in trained athletes. Association with insulin receptor changes. *Diabetes* 34:756-60.

Chakravarthy, M.V., and F.W. Booth. 2003. *Hot topics: Exercise.* Philadelphia: Hanley and Belfus (Elsevier).

Chakravarthy, M.V., and F.W. Booth. 2004. Eating, exercise, and "thrifty" genotypes: Connecting the dots toward an evolutionary understanding of modern chronic diseases. *J Appl Physiol* 96:3-10.

Collins, F. 2006. *Fiscal Year 2007 Budget Request.* National Institutes of Health, Department of Health and Human Services. House Subcommittee on Labor-HHS-Education Appropriations.

Denissenko, M.F., A. Pao, M. Tang, and G.P. Pfeifer. 1996.

Preferential formation of benzo[a]pyrene adducts at lung cancer mutational hotspots in p53. *Science* 274:430-2.

Goodpaster, B.H., and N.F. Brown. 2005. Skeletal muscle lipid and its association with insulin resistance: What is the role for exercise? *Exerc Sport Sci Rev* 33:150-4.

Hamman, R.F., R.R. Wing, S.L. Edelstein, J.M. Lachin, G.A. Bray, L. Delahanty, M. Hoskin, A.M. Kriska, E.J. Mayer-Davis, X. Pi-Sunyer, J. Regensteiner, B. Venditti, and J. Wylie-Rosett. 2006. Effect of weight loss with lifestyle intervention on risk of diabetes. *Diabetes Care* 29:2102-7.

Heath, G.W., J.R. Gavin, J.M. Hinderliter III, J.M. Hagberg, S.A. Bloomfield, and J.O. Holloszy. 1983. Effects of exercise and lack of exercise on glucose tolerance and insulin sensitivity. *J Appl Physiol* 55:512-7.

Henriksson, J., and J.S. Reitman. 1977. Time course of changes in human skeletal muscle succinate dehydrogenase and cytochrome oxidase activities and maximal oxygen uptake with physical activity and inactivity. *Acta Physiol Scand* 99:91-7.

Hogan, P., T. Dall, and P. Nikolov. 2003. Economic costs of diabetes in the US in 2002. *Diabetes Care* 26:917-32.

Holloszy, J.O. 2005. Exercise-induced increase in muscle insulin sensitivity. *J Appl Physiol* 99:338-43.

Holloszy, J.O., and F.W. Booth. 1976. Biochemical adaptations to endurance exercise in muscle. *Annu Rev Physiol* 38:273-91.

Hunter-Gault, C. March 20, 1997. Smoking gun. Online News-Hour: Tobacco Company Admission. Available: www.pbs.org/newshour/bb/health/march97/tobacco_3-20a.html.

King, H., R.E. Aubert, and W.H. Herman. 1998. Global burden of diabetes, 1995-2025: Prevalence, numerical estimates, and projections. *Diabetes Care* 21:1414-31.

Knowler, W.C., E. Barrett-Connor, S.E. Fowler, R.F. Hamman, J.M. Lachin, E.A. Walker, and D.M. Nathan. 2002. Reduction in the incidence of type 2 diabetes with lifestyle intervention or metformin. *New Engl J Med* 346:393-403.

Kohrt, W.M., and J.O. Holloszy. 1995. Loss of skeletal muscle mass with aging: Effect on glucose tolerance. *J Gerontol A Biol Sci Med Sci* 50:68-72.

Kump, D.S., and F.W. Booth. 2005a. Alterations in insulin receptor signalling in the rat epitrochlearis muscle upon cessation of voluntary exercise. *J Physiol* 562:829-38.

Kump, D.S., and F.W. Booth. 2005b. Sustained rise in triacylglycerol synthesis and increased epididymal fat mass when rats cease voluntary wheel running. *J Physiol* 565:911-25.

Laaksonen, D.E., J. Lindstrom, T.A. Lakka, J.G. Eriksson, L. Niskanen, K. Wikstrom, S. Aunola et al. 2005. Physical activity in the prevention of type 2 diabetes: The Finnish diabetes prevention study. *Diabetes* 54:158-65.

Lipman, R.L., P. Raskin, T. Love, J. Triebwasser, F.R. Lecocq, and J.J. Schnure. 1972. Glucose intolerance during decreased physical activity in man. *Diabetes* 21:101-7.

Lowell, B.B., and G.I. Shulman. 2005. Mitochondrial dysfunction and type 2 diabetes. *Science* 307:384-7.

Mokdad, A.H., J.S. Marks, D.F. Stroup, and J.L. Gerberding. 2004. Actual causes of death in the United States, 2000. *JAMA* 291:1238-45.

Oshida, Y., K. Yamanouchi, S. Hayamizu, J. Nagasawa, I. Ohsawa, and Y. Sato. 1991. Effects of training and training cessation on insulin action. *Int J Sports Med* 12:484-6.

Pan, X.R., G.W. Li, Y.H. Hu, J.X. Wang, W.Y. Yang, Z.X. An, Z.X. Hu et al. 1997. Effects of diet and exercise in preventing NIDDM in people with impaired glucose tolerance. The Da Qing IGT and Diabetes Study. *Diabetes Care* 20:537-44.

Schnell, O., and E. Standl. 2006. Impaired glucose tolerance, diabetes, and cardiovascular disease. *Endocr Pract* 12:16-9.

Seider, M.J., W.F. Nicholson, and F.W. Booth. 1982. Insulin resistance for glucose metabolism in disused soleus muscle of mice. *Am J Physiol* 242:E12-8.

Smorawinski, J., H. Kaciuba-Uscilko, K. Nazar, P. Kubala, E. Kaminska, A.W. Ziemba, J. Adrain, and J.E. Greenleaf. 2000. Effects of three-day bed rest on metabolic, hormonal and circulatory responses to an oral glucose load in endurance or strength trained athletes and untrained subjects. *J Physiol Pharmacol* 51:279-89.

Spencer, G., and R. Mackar. 2006. Two NIH initiatives launch intensive efforts to determine genetic and environmental roots of common diseases. Available: www.genome.gov/17516707.

Tuomilehto, J., J. Lindstrom, J.G. Eriksson, T.T. Valle, H. Hamalainen, P. Ilanne-Parikka, S. Keinanen-Kiukaanniemi et al. 2001. Prevention of type 2 diabetes mellitus by changes in lifestyle among subjects with impaired glucose tolerance. *New Engl J Med* 344:1343-50.

Wei, M., L.W. Gibbons, T.L. Mitchell, J.B. Kampert, C.D. Lee, and S.N. Blair. 1999. The association between cardiorespiratory fitness and impaired fasting glucose and type 2 diabetes mellitus in men. *Ann Intern Med* 130:89-96.

Yach, D., D. Stuckler, and K.D. Brownell. 2006. Epidemiologic and economic consequences of the global epidemics of obesity and diabetes. *Nat Med* 12:62-6.

Zimmet, P. 2000. Globalization, coca-colonization and the chronic disease epidemic: Can the doomsday scenario be averted? *J Intern Med* 247:301-10.

Zimmet, P., K.G. Alberti, and J. Shaw. 2001. Global and societal implications of the diabetes epidemic. *Nature* 414:782-7.

Part II

Defects in Metabolism and Insulin Resistance

Chapter 3

Fatty Acid Uptake and Insulin Resistance

Arend Bonen, PhD; Adrian Chabowski, MD, PhD; Jan F.C. Glatz, PhD; and Joost J.F.P. Luiken, PhD

Insulin resistance in skeletal muscle is a cardinal feature of obesity and type 2 diabetes. Fatty acids and their metabolites have been implicated in this pathology, since infusion with fatty acids induces insulin resistance (Belfort et al. 2005; Kashyap et al. 2004; Kruszynska et al. 2002; Yu et al. 2002), and there is a negative relationship between insulin-stimulated glucose uptake and accumulation of intramuscular lipids, including triacylglycerols (TAG; Jacob et al. 1999; Kelley and Goodpaster 2001; Krssak et al. 1999; Pan et al. 1997), diacylglycerols (DAG; Itani et al. 2002; Kelley and Mandarino 2000), long-chain fatty acyl-CoAs (LCFA-CoA; Cooney et al. 2002; Ellis et al. 2000), and ceramides (Bruce et al. 2006; Dobrzyn et al. 2005). These fatty acid and lipid metabolite effects appear to contribute to insulin resistance in muscle by interfering with the postreceptor insulin-stimulated signaling mechanisms, including reducing the insulin-stimulated phosphorylation of phosphoinositide 3-kinase (PI3K) and impairing activations of insulin substrate 1 (IRS1)-associated PI3K (Kashyap et al. 2004; Kruszynska et al. 2002; Yu et al. 2002), protein kinase B (Akt), and atypical protein kinase C (PKC-ζ and PKC-λ) a downstream mediator of PI3K (Kim, Shulman, and Kahn 2002; Storz et al. 1999). It has been speculated that a decrease in fatty acid oxidation in skeletal muscle contributes to the accumulation of intracellular lipid metabolites, thereby inducing insulin resistance (Kelley et al. 2000, 2002). However, while the accumulation of intramuscular fatty acid is associated with impaired fatty oxidation in individuals who are severely obese (BMI = 54 kg/m^2; Hulver et al. 2003; Kim et al. 2000), this finding is not observed in individuals who are less severely obese (BMI = 30-35 kg/m^2; Bonen et al. 2004; Steinberg, Parolin et al. 2002). Thus, it appears that mechanisms other than altered rates of fatty acid oxidation also contribute to the accumulation of intramuscular fatty acid and impaired insulin signaling, thereby leading to insulin resistance in muscle.

The transport of fatty acids across the plasma membrane is mediated by one or more fatty acid binding proteins. These fatty acid transporters are now known to be a key contributor to the accumulation of fatty acids in skeletal muscle in obesity and type 2 diabetes. In this chapter we discuss the identification of a protein-mediated fatty acid transport system as well as the known fatty acid transporters, their acute and chronic regulation, and their recently identified role in insulin resistance in skeletal muscle in animal models and humans.

Studies in our laboratories are supported by the Canadian Institutes of Health Research, the Natural Sciences and Engineering Research Council of Canada (NSERC), the Heart and Stroke Foundation of Ontario, the Netherlands Heart Foundation (D98.012), and the Canada Research Chair program.

A. Bonen is the Canada Research Chair in Metabolism and Health.

J.F.C. Glatz is the Netherlands Heart Foundation Professor of Cardiac Metabolism.

J.J.F.P. Luiken is the recipient of a VIDI-Innovation Research Grant from the Netherlands Organization for Scientific Research (NWO-ZonMw grant 016.036.305).

Long-Chain Fatty Acids and Their Uptake Across the Sarcolemma

Long-chain fatty acids (LCFA), released from adipocytes, bind to albumin in the plasma. The circulation delivers this fatty acid–albumin complex to target tissues, such as muscle and heart. Before fatty acids are taken up by these tissues, they encounter several barriers. On their way from the microvascular compartment to cardiac or skeletal muscle myocytes, fatty acids have to pass the capillary endothelium. In muscle tissue, unlike in liver tissue, the interendothelial clefts do not allow the albumin–LCFA complex to pass at a rate that could explain the observed rates of fatty acid uptake by muscle tissue (van der Vusse, van Bilsen, and Glatz 2000). The corollary is that LCFAs are released from albumin and then traffic through the capillary endothelial cells to bind to albumin present in the interstitial compartment. A similar problem exists at the point where albumin-bound fatty acids move from the interstitial compartment across the plasma membrane into the myocyte, as the fatty acid–albumin complex is too large to be taken up across the sarcolemma. Thus, fatty acids somehow dissociate from albumin and are transferred unbound across the plasma membrane, after which their miscibility in the cytosol is restored by binding to the 15 kDa cytosolic fatty acid binding protein (FABPc).

For a long time it was thought that the uptake of fatty acids into peripheral tissues occurred via simple diffusion (Hamilton and Kamp 1999), such that the circulating concentrations of fatty acids dictated the rate of entry into parenchymal cells. However, this view seems to be at odds with the fact that fatty acids are an important substrate for a diversity of cellular processes, including membrane synthesis, protein modification, transcription regulation, and intracellular signaling (Amri et al. 1995; Distel, Robinson, and Spiegelman 1992; Glatz et al. 1995; Newsholme, Calder, and Yaqoob 1993; Schaffer and Lodish 1994), as well as serving as an important energy source for metabolically active tissues such as skeletal muscle and the heart (van der Vusse and de Groot 1992; van der Vusse and Reneman 1996). These multiple roles suggest that carefully regulating fatty acid disposition, including

its uptake, is highly desirable. Beginning in the early 1980s, support for a protein-mediated fatty acid transport mechanism was proposed for a number of different tissues, including adipocytes (Abumrad et al. 1993; Abumrad et al. 1991; Harmon, Luce, and Abumrad 1992; Schaffer and Lodish 1994; Schwieterman et al. 1988; Storch, Lechene, and Kleinfeld 1991), intestine (Gore and Hoinard 1993; Trotter, Ho, and Storch 1996), kidney (Trimble 1989), liver (Stremmel and Berk 1986; Stremmel, Strohmeyer, and Berk 1986; Stremmel et al. 1985), and cardiac myocytes (DeGrella and Light 1980; Sorrentino et al. 1989; Stremmel 1988, 1989), but these studies were conducted using actively metabolizing cells, in which the cellular uptake and metabolism of LCFA are tightly coupled. These systems therefore precluded a definitive resolution as to whether fatty acids traverse the plasma membrane via a protein-mediated system.

We resolved this dilemma in our laboratory by using giant vesicles prepared from heart or skeletal muscle (Bonen et al. 1998; Luiken, Turcotte, and Bonen 1999) as well as from liver and adipose tissue (Koonen et al. 2002; Luiken, Arumugam et al. 2001). Giant vesicles are derived from the plasma membrane and are large (~10-15 μm in diameter) and spherical in shape. The plasma membranes enveloping the vesicles are oriented 100% right side out. Fortuitously, the lumen of the vesicles contains cytosolic fatty acid binding protein (FABPc), a low molecular weight protein that acts as a fatty acid sink for fatty acids that have fully traversed the plasma membrane and have entered the cytosol compartment (Bonen et al. 1998; Luiken, Turcotte, and Bonen 1999). Mitochondria and the machinery to esterify fatty acids are absent in these vesicles, and thus fatty acid metabolism does not occur (Bonen et al. 1998; Luiken, Turcotte, and Bonen 1999). Detailed characterization studies of this vesicle preparation have provided convincing evidence that fatty acids traverse the plasma membrane by a protein-mediated mechanism that is specific for LCFAs in both heart and skeletal muscle. Fatty acid uptake is inhibited by protein-modifying agents and inhibitors of selected fatty acid transporters. In addition, excess oleate, but not glucose, displaces palmitate uptake (Bonen et al. 1998; Luiken, Turcotte, and Bonen 1999). Moreover, kinetic studies indicated that the K_m for fatty acid transport into the heart and in red and white skeletal muscles was similar (6-9 nM), while the V_{max} differed widely among these tissues (Luiken, Turcotte, and Bonen 1999). The rates of protein-mediated fatty transport (heart >> red muscle > white muscle) are scaled proportionately with the markedly different capacities for fatty acid oxidation among muscle tissues (heart >> red muscle > white muscle; see figure 3.1).

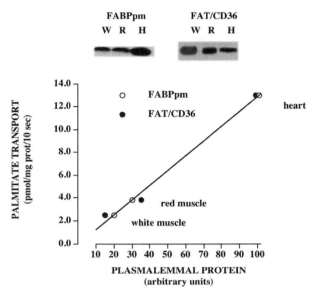

Figure 3.1 Comparison of the fatty acid transporter proteins FABPpm and FAT/CD36 located at the plasma membrane and the rate of palmitate transport into giant sarcolemmal vesicles obtained from white (W) and red (R) skeletal muscle and heart (H).

Data redrawn from J.J.F.P. Luiken, L.P. Turcotte, A. Bonen, 1999, "Protein-mediated palmitate uptake and expression of fatty acid transport proteins in heart giant vesicles," *J Lipid Res* 40(6): 1007-1016.

Fatty Acid Transporters

A number of proteins that can stimulate fatty acid transport have been identified. However, they are not all isoforms with different transport capacities, such as the transport proteins for glucose (Joost et al. 2002) and monocarboxylates (Halestrap and Meredith 2004). Among the known fatty acid transporters are (a) fatty acid translocase (FAT/CD36), the integral membrane protein that is the rat homolog of the human CD36 (the native protein is 53 kDa, but it is heavily glycosylated and is detected at 88 kDa; Abumrad et al. 1993; Ibrahimi et al. 1996); (b) plasma membrane fatty acid binding protein (FABPpm), a 40 kDa peripheral membrane protein (Isola et al. 1995); and (c) a family of integral plasma membrane fatty acid transport proteins

(FATP) ranging in size from 63 to 70 kDa (FATP1-FATP6; Gimeno et al. 2003; Hirsch, Stahl, and Lodish 1998; Schaffer and Lodish 1995). FABPpm and FAT/CD36 have been detected in virtually all tissues examined, while the FATPs exhibit a more restricted pattern of tissue expression (Gimeno et al. 2003; Hirsch, Stahl, and Lodish 1998). In rodent skeletal muscle and heart, FAT/CD36, FABPpm, FATP1, and FATP4 are expressed (Bonen et al. 1998; Luiken, Turcotte, and Bonen 1999; Nickerson and Bonen 2005), while in human muscle FAT/CD36, FABPpm, and FATP1 and 4 are expressed (Bonen, Miskovic, and Kiens 1999; Bonen, unpublished data). Information on whether other FATPs are expressed in human muscle is lacking. The recent identification of FATP6 as an important fatty acid transporter in the murine heart (Gimeno et al. 2003) is contradicted by studies indicating that FATP6 has poor fatty acid transport capacities when overexpressed in yeast cells (DiRusso et al. 2005). Nevertheless, all of the identified fatty acid transporters stimulate the rate of fatty acid transport when overexpressed in cells or muscle (Clarke et al. 2004; DiRusso et al. 2005; Gimeno et al. 2003; Ibrahimi et al. 1996; Isola et al. 1995; see figure 3.2). However, because the membrane topology of these

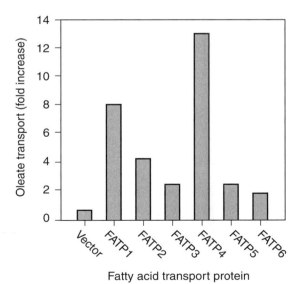

Figure 3.2 Effects of overexpressing FATP1 through FATP6 on the increase in fatty acid (oleate) transport. FATP1 through FATP6 were each overexpressed to the same level in *Saccharomyces cerevisiae*.

Created from data in C.C. DiRusso et al., 2005, "Comparative biochemical studies of the murine fatty acid transport proteins (FATP) expressed in yeast," *J Biol Chem* 280(17): 16829-16837.

proteins is still poorly understood, the exact nature by which they facilitate transmembrane movement of LCFAs is not yet known.

The need for so many different fatty acid transporters is not clear. Presumably, their diversity reflects their different fatty acid transport capacities, congruent with different capacities for fatty acid metabolism in different tissues. It appears that some fatty acid transporters interact with each other to move fatty acids across the plasma membrane (Gimeno et al. 2003; Glatz and Storch 2001; Luiken, Turcotte, and Bonen 1999), since several transporters coimmunoprecipitate, including FAT/CD36 and FATP6 in mouse heart (Gimeno et al. 2003) and FAT/CD36 and FABPpm in rat heart (Chabowski and Bonen, unpublished data). Blocking either FAT/CD36 or FABPpm inhibits fatty acid transport in heart and skeletal muscle (Luiken, Turcotte, and Bonen 1999), further suggesting that these two transporters collaborate at the plasma membrane. We (Luiken et al. 2004; Luiken, Turcotte, and Bonen 1999) have speculated that FABPpm acts as a receptor for LCFAs, facilitating the diffusion of the fatty acid–albumin complex through the unstirred fluid layer, and that it then interacts with FAT/CD36 to mediate the transmembrane passage of LCFAs, possibly by facilitating their flip-flop across the bilayer. Thereafter, FAT/CD36 may interact with FABPc (Spitsberg, Matitashvili, and Gorewit 1995) to facilitate the desorption of fatty acids from the inner leaflet or intracellular site of the transporter to this intracellular carrier protein.

Observations in cell lines have led to the speculation that the existence of different fatty acid transporters allows each to interact with specific intracellular proteins in order to channel fatty acids to different metabolic fates within the cell (Bastie et al. 2004; Hatch et al. 2002; Richards et al. 2006; Zou et al. 2003). However, this hypothesis may apply only to cell lines in which normal physiological regulation is absent, as this concept is not supported by existing evidence in mammalian cells. Specifically, there is already evidence that the plasmalemmal FAT/CD36 and FABPpm do not by themselves channel fatty acids toward specific metabolic fates within skeletal muscle. It has been shown repeatedly that these fatty acid destinations are orchestrated by metabolic signals within the muscle cell (Bonen et al. 2000; Dyck and Bonen 1998; Dyck, Steinberg, and Bonen 2001; Ibrahimi et al. 1999; Luiken et al. 2003; Luiken, Dyck et al. 2002; Luiken, Koonen et al. 2002; Luiken, Willems et al. 2001). Thus, in metabolically active muscle tissue, a given fatty acid transporter does not seem to predestine the delivery of LCFAs to a specific metabolic fate.

Fatty Acid Transport and Fatty Acid Transporter Expression

The correlation between the mRNA expression of fatty acid transporters and their protein products or rates of fatty acid transport has proved to be poor (Berk et al. 1997, 1999). We have observed that the expression of FAT/CD36 and FABPpm appears to be regulated at the posttranscriptional level, and hence a direct correlation between a fatty acid transporter transcript and its protein product is frequently not observed (Chabowski, Chatham et al. 2006; Luiken, Arumugam et al. 2001, 2002). Moreover, FABPpm is identical to mitochondrial aspartate aminotransferase (mAspAT; Berk et al. 1990; Bradbury and Berk 2000; Stump, Zhou, and Berk 1993), and these two proteins have different functions depending on their subcellular location. Clearly, mRNA transcripts cannot convey this type of information. Even the expression of fatty acid transporter protein does not always correlate with rates of fatty acid transport, as in several instances these proteins traffic between an intracellular endosomal compartment and the plasma membrane without altering the total pool of fatty acid transporters (see next section below). Thus, it is only the fatty acid transporters present at the plasma membrane that facilitate the fatty acid movement across the plasma membrane. In heart and skeletal muscle, the maximal rates of fatty acid transport correlate highly with either plasma membrane FAT/CD36 or FABPpm (Luiken, Turcotte, and Bonen 1999; see figure 3.1).

Regulating Fatty Acid Transport by Altering the Expression and Subcellular Location of Fatty Acid Transporters

Altered rates of fatty acid transport are attributable to (a) the changes in the expression of selected fatty acid transporter proteins and (b) the subcellular distribution of fatty acid transporter proteins.

Acute Regulation of Fatty Acid Transport by Muscle Contraction, Insulin, and Leptin

When muscle transitions from rest to contraction, it rapidly changes its rate of fatty acid uptake to meet its increased energetic demands (Dyck and Bonen 1998; Turcotte, Petry, and Richter 1998). However, this upregulation can occur without any accompanying change in the concentration of fatty acid being supplied to the muscle (Dyck and Bonen 1998; Turcotte, Petry, and Richter 1998). Therefore, we examined the rates of fatty acid transport in giant vesicles prepared from rat lower-leg muscles after short-term (30 min) electrical stimulation via the sciatic nerve. Compared to transport in resting muscle, the transport rate of palmitate was markedly increased in giant vesicles obtained from contracting muscles (see figure 3.3, *a* and *b;* Bonen et al. 2000). Concomitantly, the plasma membrane FAT/CD36 content was also increased, due to the translocation of FAT/CD36 from an intracellular depot to the plasma (figure 3.3*c;* Bonen et al. 2000). After a 45 min recovery from muscle contraction, both palmitate transport and vesicular FAT/CD36

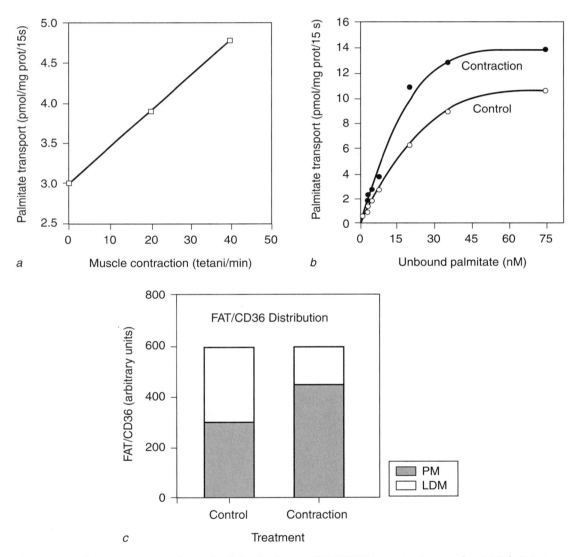

Figure 3.3 Rates of palmitate transport and the subcellular distribution of FAT/CD36 in contracting muscles. *(a)* Palmitate transport at different rates of muscle contraction (0 = resting muscle). *(b)* Kinetics of palmitate transport in resting (control) and contracting muscle (20 tetani/min). *(c)* Subcellular distribution of FAT/CD36 in resting (control) and contracting muscle (20 tetani/min). LDM is low-density membrane (intracellular FAT/CD36 depot) and PM is plasma membrane.

Data redrawn from A. Bonen et al., 2000, "Acute regulation of fatty acid uptake involves the cellular redistribution of fatty acid translocase," *J Biol Chem.* 275(19): 14501-14508.

content were restored to control (Bonen et al. 2000). These studies were the first to demonstrate that the transport of fatty acid into muscle is regulated acutely (i.e., within minutes) by inducing the translocation of FAT/CD36.

In subsequent studies in cardiac myocytes and perfused hind-limb muscles we found that the contraction-induced translocation of FAT/CD36 involves the activation of AMP-activated protein kinase (AMPK; Chabowski et al. 2005; Luiken et al. 2003). In addition, we observed that FABPpm but not FATP1 can be induced to move from an intracellular depot to the plasma membrane when AMPK in cardiac myocytes is activated by 5'-phosphoribosyl-5-aminoimidazole-4-carboxamide (AICAR; figure 3.4; Chabowski et al. 2005).

Our group was also the first to demonstrate that insulin rapidly upregulates the rate of fatty acid transport in muscle (Luiken, Dyck et al. 2002) and in heart (Chabowski et al. 2005; Luiken, Koonen et al. 2002; figure 3.4). Perfusion of rat hind-limb muscles with insulin promoted a rapid uptake of fatty acids, most of which were esterified (Luiken, Dyck et al. 2002). In both heart and skeletal muscle, insulin induced the translocation of FAT/CD36 from an endosomal pool to the plasma membrane (Luiken, Dyck et al. 2002; Luiken, Koonen et al. 2002). This effect

was inhibited when the insulin-signaling protein PI3K was inhibited (Luiken, Dyck et al. 2002). Thus insulin induces the translocation of not only the glucose transporter GLUT4 but also the fatty acid transporter FAT/CD36. More recent work has shown that insulin fails to induce the translocation of FABPpm and FATP1 in the heart (figure 3.4; Chabowski et al. 2005). In contrast, others have observed that insulin may perhaps induce FATP1 translocation in skeletal muscle (Wu et al. 2006). Only minimal insulin-induced translocation of FABPpm (<20%) is observed in skeletal muscle (unpublished data).

Recent evidence from our laboratory indicates that leptin, which stimulates fatty acid oxidation, also regulates fatty acid transport. Activation of AMPK by leptin induced the translocation of both FAT/CD36 and FABPpm in heart, thereby increasing the rate of palmitate uptake (unpublished data). Collectively, the results of these studies indicate that the acute regulation of FAT/CD36 by both insulin and AMPK resembles the regulation of GLUT4, as we have reviewed in detail elsewhere (Luiken et al. 2004). This may mean that there are separate insulin-sensitive and AMPK-sensitive pools of FAT/CD36, while there is only an AMPK-sensitive pool for FABPpm. The acute regulation of FAT/CD36 is shown schematically in figure 3.5.

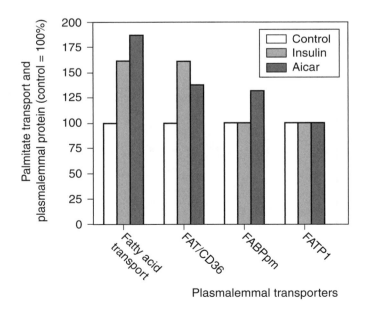

Figure 3.4 Acute (30 min) insulin- and AICAR-stimulated rates of palmitate transport and plasmalemmal fatty acid transporters in cardiac myocytes.

Data redrawn from A. Chabowski et al., 2005, "The subcellular compartmentation of fatty acid transporters is regulated differently by insulin and by aicar," *FEBS Lett.* 579(11): 2428-2432.

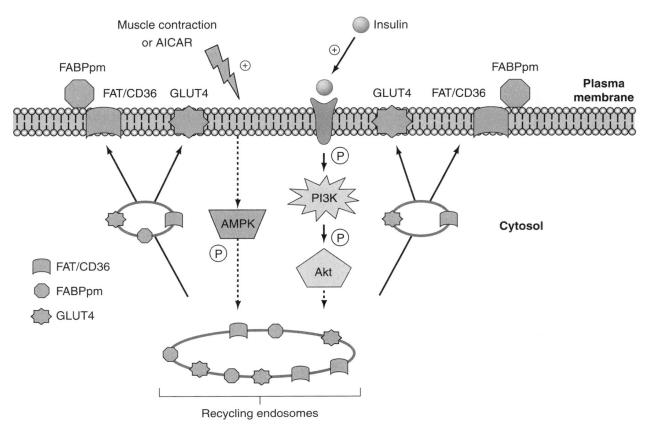

Figure 3.5 Schematic representation of stimuli that induce the translocation of FAT/CD36 and FABPpm from intracellular depots to the plasma membrane.

Created from data in Bonen et al., 2000, Chabowski et al., 2005, Luiken et al., 2003, Luiken et al, 2002b and Luiken et al., 2002c.

Chronic Regulation of Fatty Acid Transport by Muscle Contraction and Leptin

Increased muscle activity lasting weeks or days can upregulate substrate transport by changing the expression of the substrate transporters such as GLUT4 and monocarboxylate transporter 1 (MCT1; Johannsson et al. 1996; McCullagh et al. 1996; Megeney et al. 1993). Similarly, increasing muscle activity with chronic muscle stimulation (5-7 d) upregulated protein expression and plasmalemmal content of FAT/CD36 as well as enhanced the rates of fatty acid transport (figure 3.6; Benton et al. 2006; Koonen et al. 2004). Exercise training also upregulated FABPpm and FAT/CD36 (Kiens et al. 1997; Roepstorff et al. 2004; Tunstall et al. 2002; Turcotte et al. 1999) and increased palmitate utilization (Turcotte et al. 1999). Conversely, when muscle activity was eliminated (denervation, 7 d), the rate of fatty acid transport decreased due to a reduction in plasmalemmal FAT/CD36 and FABPpm, the expression of which was not altered

(figure 3.6; Koonen et al. 2004). These alterations in fatty acid transporters may be mediated via AMPK, since activation of this kinase by AICAR upregulates the expression and plasmalemmal content of FAT/CD36 and FABPpm (Chabowski, Momken et al. 2006) and blocking AMPK activity during prolonged (5-7 d) muscle stimulation prevents the contraction-induced upregulation of fatty acid transporter expression (unpublished data).

Endocrine regulation of fatty acid transport and transporters also occurs. In cardiac myocytes, insulin did not alter the expression of FABPpm, but it did upregulate, in a dose-dependent manner, the expression of FAT/CD36 protein. Concurrently, plasmalemmal FAT/CD36 and fatty acid transport also increased (Chabowski et al. 2004). In L6 muscle cells, prolonged exposure (24 h) to resistin reduced the uptake of palmitate, which correlated with reduced FAT/CD36 content at the cell surface and lower expression of FATP1 (Palanivel and Sweeney 2005). Finally, prolonged (2 wk) infusion

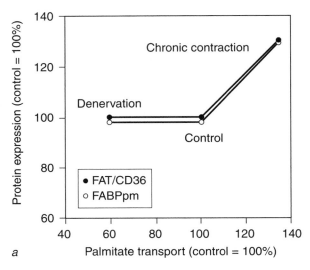
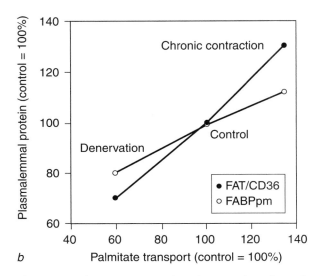

Figure 3.6 Comparison of palmitate transport in relation to expression of FABPpm and FAT/CD36 *(a)* and in relation to plasmalemmal changes in FABPpm and FAT/CD36 *(b)* in skeletal muscle in which contractile activity has been upregulated (chronic contraction for 7 d) or eliminated (denervation for 7 d). These figures demonstrate that plasmalemmal FABPpm and FAT/CD36 content can be reduced (see denervation in *b)* while their expression is not altered (see denervation in *a).*

Data redrawn from D.P.Y. Koonen et al., 2004, "Different mechanism can alter fatty acid transport when muscle contractile activity is chronically altered," *Am J Physiol Endocrinol Metab* 286(6): 1042-1049.

with leptin, as opposed to acute (15 min) leptin effects, repressed the protein expression of FAT/CD36 and FABPpm. This repression contributed to their reduced plasmalemmal content and concomitantly lowered rates of fatty acid transport (Steinberg, Dyck et al. 2002). It is possible that these leptin-induced reductions in fatty acid transporter expression and the transport of fatty acids are compensatory mechanisms, designed to limit the loss of circulating fatty acids.

Thus, more chronic perturbations affect fatty acid transport in two ways: (1) by altering the protein expression of fatty acid transporters, which results in their change at the plasma membrane, and/or (2) by permanently relocating the fatty acid transporters to the plasma membrane without altering their expression (figure 3.6).

LCFA Transport and Transporters in Obesity and Type 2 Diabetes

Fatty acid metabolism in skeletal muscle is dysregulated in obesity and type 2 diabetes, resulting in the accumulation of lipids within the muscle cell. These intramuscular lipid products interfere with insulin signaling. Given that reduced oxida-

tion of fatty acids in insulin-resistant muscle does not necessarily account for the lipid accumulation (Bonen et al. 2004; Steinberg, Parolin et al. 2002), it became of interest to determine whether fatty acid transport and transporters are upregulated in insulin-resistant skeletal muscle.

Initial studies in insulin-resistant animal models (Zucker obese rats and Zucker diabetic fatty rats) failed to establish a link between impaired insulin-stimulated glucose transport and increased fatty acid transport (Berk et al. 1997, 1999). However, neither of these studies examined skeletal muscle, which is now recognized as one of the key tissues contributing to insulin resistance.

The regulation of FAT/CD36 trafficking (Chabowski et al. 2005; Luiken, Dyck et al. 2002; Luiken, Koonen et al. 2002) and expression by insulin (Chabowski et al. 2004) suggested that FAT/CD36 could play a central role in insulin-resistant states such as obesity and type 2 diabetes. Therefore we examined fatty acid transport and transporters in muscle of the lean rat and the obese Zucker rat, a well-known model of insulin resistance in which insulin fails to induce the translocation of GLUT4 from its intracellular depot in muscle to the sarcolemma (Brozinick Jr. et al. 1992, 1993). In the skeletal muscle of obese Zucker rats, the rates of fatty acid transport were markedly increased

(Luiken, Arumugam et al. 2001). Unexpectedly, this finding was not attributable to changes in protein expression of FAT/CD36 or FABPpm, as expression did not differ in lean and obese muscle (Luiken, Arumugam et al. 2001; figure 3.7). However, examination of plasmalemmal fatty acid transporters showed that sarcolemmal FAT/CD36 was markedly increased while sarcolemmal FABPpm remained unaltered (figure 3.7). This revealed an important new mechanism: namely, that it is possible to permanently relocate FAT/CD36 to the plasma membrane without altering its expression in insulin-resistant muscle. Hence, the subcellular trafficking machinery regulating the subcellular localization of FAT/CD36 (Luiken, Arumugam et al. 2001) as well as that of GLUT4 (Brozinick Jr. et al. 1992, 1993) is impaired in obesity-associated insulin-resistant skeletal muscle. However, the sequestering of these transport proteins is juxtaposed, with GLUT4 being retained within its intracellular depot and FAT/CD36 being retained at the plasma membrane.

The higher levels of circulating insulin in obese Zucker rats may contribute to the permanent relocation of FAT/CD36 to the plasma membrane. This suggestion is based on our recent work showing that insulin fails to induce FAT/CD36 translocation in heart (Coort et al. 2004) and skeletal muscle

(unpublished data) of obese Zucker rats in which plasmalemmal FAT/CD36 is already upregulated, while normal insulin-induced FAT/CD36 translocation does occur in heart (Coort et al. 2004) and skeletal muscle (unpublished data) of lean animals. Interestingly, under conditions of insulin stimulation, plasmalemmal FAT/CD36 and fatty acid uptake were similar in lean and obese animals, leading to similar rates of palmitate esterification in muscle of lean and obese Zucker rats (unpublished data). Thus, with respect to fatty acid transport and FAT/CD36 as well as with respect to fatty acid esterification, insulin induces a temporary obesity-like effect in muscle.

Although the obese Zucker rat is a model of insulin resistance, it does not display frank type 2 diabetes. However, the Zucker diabetic fatty (ZDF) rat is a good model of type 2 diabetes. In this animal model the disease progresses rapidly within weeks, and the animal exhibits many of the changes in the circulating milieu that are observed in human type 2 diabetes (Chatham and Seymour 2002; Friedman et al. 1991). At 12 wk of age the ZDF animals are diabetic. At that point we found that protein expression and plasmalemmal content of FAT/CD36 as well as rates of fatty acid transport were markedly upregulated in ZDF animals compared to lean animals (figure 3.8; Chabowski, Chatham et al. 2006).

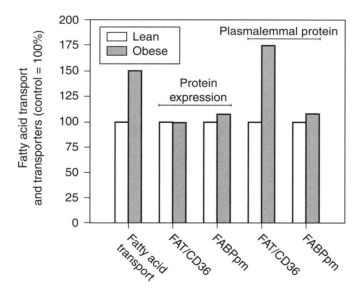

Figure 3.7 Comparison of fatty acid transport in relation to expression of FABPpm and FAT/CD36 and in relation to plasmalemmal changes in FABPpm and FAT/CD36 in skeletal muscle of obese Zucker rats. Plasmalemmal FAT/CD36 can be increased while its expression is not altered in obese rat muscle.

Data redrawn from J.J.F.P. Luiken et al., 2001, "Increased rates of fatty acid uptake and plasmalemmal fatty acid transporters in obese Zucker rats," *J Biol Chem.* 276(44): 40567-40573.

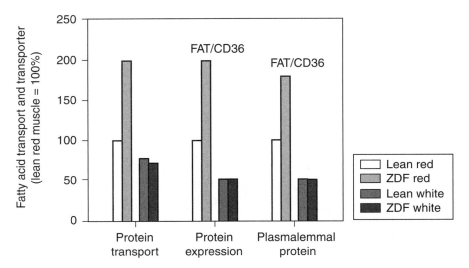

Figure 3.8 Rates of palmitate transport, FAT/CD36 protein expression, and plasmalemmal FAT/CD36 content in red and white muscle of lean and ZDF rats at 12 wk of age, when type 2 diabetes has been established in ZDF rats.

Data redrawn from A. Chabowski et al., 2006, "Fatty acid transport and FAT/CD36 are increased in red but not in white muscle skeletal muscle of Zucker diabetic fatty (ZDF) rats," *Am J Physiol Endocrinol Metab* 291: E675-E682.

Remarkably, as the ZDF animals progressed from insulin resistance (week 6) to severe type 2 diabetes (week 24), the increases observed in FAT/CD36 protein expression, FAT/CD36 plasmalemmal content, and rates of fatty acid transport occurred only in oxidative (red) skeletal muscle, while no changes of any kind were observed in glycolytic (white) skeletal muscle (Chabowski, Chatham et

al. 2006). It had been observed previously that GLUT4 mRNA and protein are repressed primarily in red muscle of ZDF rats (Etgen and Oldham 2000; Friedman et al. 1991; Henriksen et al. 2003; Slieker et al. 1992). When comparing protein expression as well as plasmalemmal content of FAT/CD36 and GLUT4, we also observed a large inverse relationship between these two transporters in red skeletal muscle of lean and ZDF rats (figure 3.9; Chabowski, Chatham et al. 2006).

Improved Insulin Sensitivity With Reduced Fatty Acid Transport

As a converse to insulin resistance mediated by fatty acid, improved insulin sensitivity might occur when LCFA transport and transporters are reduced. Indeed, such improved insulin sensitivity has been observed in FAT/CD36 null mice (Febbraio et al. 1999; Goudriaan et al. 2003) as well as in FATP1 null mice, in which a high-fat diet failed to induce insulin resistance (Kim et al. 2004; Wu et al. 2006). Along similar lines, leptin-mediated improvement of insulin sensitivity in muscle (Muoio et al. 1997, 1999; Yaspelkis III et al. 1999) could be explained, in part, by the repression of FAT/CD36 and FABPpm protein expression and the resulting reduction in their plasmalemmal content, which reduced the rate of fatty acid transport into the muscle cell (Steinberg, Dyck et al. 2002). This reduction would lower the

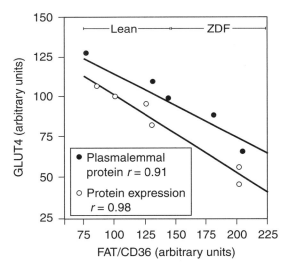

Figure 3.9 Comparison of FAT/CD36 and GLUT4 protein expression and plasmalemmal content in red muscle of lean and ZDF rats at selected ages (weeks 6-24).

Data redrawn from A. Chabowski et al., 2006, "Fatty acid transport and FAT/CD36 are increased in red but not in white muscle skeletal muscle of Zucker diabetic fatty (ZDF) rats," *Am J Physiol Endocrinol Metab* 291: E675-E682.

interference of fatty acids and their metabolites on the insulin signaling pathway, further suggesting that fatty acid transport and glucose transport are reciprocally related.

Fatty Acid Transport and Transporters in Human Obesity and Type 2 Diabetes

Given the foregoing evidence that fatty acid transport rates and plasmalemmal FAT/CD36 are upregulated in animal models of obesity (Luiken, Arumugam et al. 2001) and type 2 diabetes (Chabowski, Chatham et al. 2006), it was of considerable interest to determine whether similar mechanisms occur in human obesity and type 2 diabetes.

Studies in cultured muscle cells obtained from people with type 2 diabetes appeared to indicate that protein-mediated fatty acid uptake was reduced (Wilmsen et al. 2003). Similarly, studies using stable isotopes indicated that LCFA transport into muscle of subjects with type 2 diabetes is also reduced (Blaak and Wagenmakers 2002). In both of these studies, however, there are methodological concerns. The estimates of fatty acid uptake in cultured muscle cells were based on phloretin-mediated inhibition of palmitate uptake (Wilmsen et al. 2003), which is a nonspecific measure of fatty acid transport. In the studies of Blaak and Wagenmakers (2002), there also seems to be a methodological problem, as, inexplicably, the investigators were not able to observe any oxidation of fatty acids. Moreover, it is difficult to reconcile how reduced fatty acid uptake in people with type 2 diabetes (Blaak and Wagenmakers 2002; Wilmsen et al. 2003) can result in the well-known increase in intramuscular triacylglycerol concentration observed in muscle of individuals who are obese or who have type 2 diabetes (Bonen et al. 2004; Perseghin et al. 2003; van Loon et al. 2004). In addition, recent studies in insulin-resistant obese muscle (Bonen et al. 2004; Steinberg, Parolin et al. 2002) do not concur with the results of Blaak and Wagenmakers (2002) and Wilmsen and colleagues (2003). These recent studies of insulin-resistant obese muscle strongly suggest that fatty acid transport is upregulated in obese human muscle (the sum of fatty acid esterification into tri-, di-, and monoacylgycerols and fatty acid oxidation) increased by 72% while fatty acid oxidation is not impaired (Bonen et al. 2004; Steinberg, Parolin et al. 2002).

We have now gathered direct evidence that fatty acid transport is upregulated in insulin-resistant human muscle (Bonen et al. 2004). We compared individuals who were lean ($n = 11$; BMI = 21.8 ± 0.6 kg/m^2), overweight ($n = 10$; BMI = 26.9 ± 0.5 kg/m^2), or obese ($n = 7$; BMI = 34.7 ± 1.7 kg/m^2) or who had been diagnosed with type 2 diabetes ($n = 5$; BMI = 25.6 ± 1.1 kg/m^2). From a small sample of the rectus abdominis muscle obtained during surgery, we measured protein expression of fatty acid transporter. We also prepared giant sarcolemmal vesicles for fatty acid transport studies and determination of plasma membrane fatty acid transporters. In some individuals who were lean or obese, sufficient muscle tissue was also obtained for preparing intact muscle strips to determine rates of palmitate oxidation and esterification in vitro (Bonen et al. 2004).

Rates of palmitate transport into giant sarcolemmal vesicles did not differ in muscle obtained from individuals who were lean or who were overweight. However, in comparison to subjects who were lean, the rates of palmitate transport were greatly increased in individuals who were obese (3.8-fold) and individuals who were diagnosed with type 2 diabetes (4.3-fold). In addition, there was a good correlation between BMI and rates of fatty acid transport among the individuals who were lean, overweight, or obese, although this correlation did not apply to the subjects with type 2 diabetes (figure 3.10; Bonen et al. 2004). This upregulation in fatty acid transport, however, was not attributable to changes in protein expression or in plasmalemmal content of FABPpm in muscles from individuals who were obese or who were diagnosed with type 2 diabetes (Bonen et al. 2004). Previously, Simoneau and colleagues (1999) had observed that expression of FABPpm was increased in muscle obtained from humans who were obese. However, our present results in human muscle (Bonen et al. 2004) parallel a similar lack of change in FABPpm that is seen in muscles from obese Zucker rats (Luiken, Arumugam et al. 2001) and ZDF rats (Chabowski, Chatham et al. 2006).

In contrast to the results for FABPpm, the change we observed in FAT/CD36 was marked. Notably, the plasmalemmal content of FAT/CD36 was increased in individuals who were obese or who were diagnosed with type 2 diabetes, but the protein expression of FAT/CD36 was not altered. Thus, the marked increase in fatty acid transport

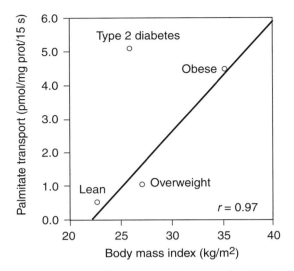

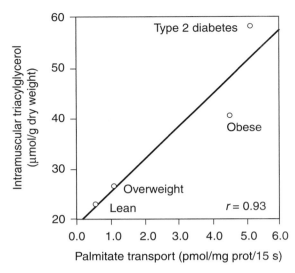

Figure 3.10 Relationship between body mass index (BMI) and the rate of palmitate transport into giant sarcolemmal vesicles prepared from skeletal muscle obtained from individuals who were lean, overweight, or obese, or who had been diagnosed with type 2 diabetes. (Correlation is based on individuals who were lean, overweight, or obese.)

Data drawn from results published by A. Bonen et al., 2004, "Triacylglycerol accumulation in human obesity and type 2 diabetes is associated with increased rates of skeletal muscle fatty acid transport and increased sarcolemmal FAT/CD36," *FASEB J* 18(10): 1144-1146.

Figure 3.11 Comparison between intramuscular triacylglycerol concentrations and rates of palmitate transport into giant sarcolemmal vesicles prepared from skeletal muscle obtained from individuals who were lean, overweight, or obese, or who had been diagnosed with with type 2 diabetes.

Data redrawn from results published by A. Bonen et al., 2004, "Triacylglycerol accumulation in human obesity and type 2 diabetes is associated with increased rates of skeletal muscle fatty acid transport and increased sarcolemmal FAT/CD36," *FASEB J* 18(10): 1144-1146.

found in muscle from individuals who were obese or diagnosed with type 2 diabetes was associated with an increase in plasmalemmal FAT/CD36. Remarkably, this finding parallels our previous observations in muscle of obese Zucker rats, in which increased rates of fatty acid transport were also associated with increased plasmalemmal FAT/CD36 and unaltered FAT/CD36 protein expression (Luiken, Arumugam et al. 2001). Thus, it appears that both in animal models of insulin resistance and in humans with insulin resistance, rates of fatty acid transport are increased in skeletal muscle, due in large part to the permanent relocation of FAT/CD36 to the plasma membrane.

In these same studies the intramuscular triacylglycerol content did not differ in individuals who were lean or overweight. When compared to control subjects, the intramuscular triacylglycerol concentrations were twofold greater in the obese subjects and threefold greater in the type 2 diabetic subjects. Interestingly, the rates of palmitate transport and plasmalemmal FAT/CD36 were positively associated with intramuscular triacylglycerol concentrations (figure 3.11).

Thus, it is likely that the increased rate of fatty acid esterification is secondary to a greater rate of

fatty acid transport into the muscle. Hence, it can be concluded that a key perturbation in human obesity and type 2 diabetes is the permanent subcellular relocation of FAT/CD36 to the sarcolemma, a move that increases the rate of LCFA transport into skeletal muscle and thus leads to an excess accumulation of intramuscular lipids, as demonstrated by the increased concentration of intramuscular triacylglycerol. Presumably, the FAT/CD36-mediated increase in intramuscular lipids (Bonen et al. 2004) contributes substantially to the well-known lipid-mediated impairment in insulin signaling (Shulman 2000). The likely involvement of FAT/CD36 in this process is shown in figure 3.12.

In skeletal muscle of individuals who are obese or who have type 2 diabetes, a permanent redistribution of FAT/CD36 to the plasma membrane, and not altered expression of FAT/CD36, contributes to an increased influx of fatty acids that results in the accumulation of intramuscular triacylglycerol and other lipid metabolites that interfere with insulin signaling, thereby contributing to insulin resistance. In this scheme fatty acid oxidation is not altered as has been reported (see Bonen et al. 2004).

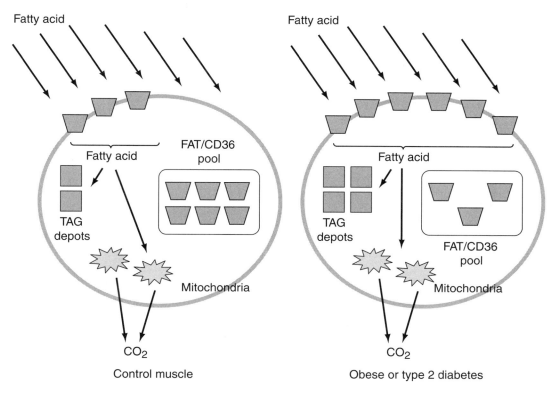

Figure 3.12   Schematic representation of how FAT/CD36 contributes to increased fatty acid transport and intramuscular lipid accumulation, which has been associated with the development of insulin resistance. In control muscle the normal situation is depicted, in which circulating fatty acids are largely taken up into the muscle cell by FAT/CD36 located at the plasma membrane. Once inside the muscle cells fatty acids are oxidized and stored as TAG. In the muscle of obese or type 2 diabetics, more of the FAT/CD36 protein is permanently present at the plasma membrane (note, the total pool of FAT/CD36 is not altered). This allows more fatty acids to be taken up into the muscle cell. In our studies rates of fatty acid oxidation are not altered, but the additional fatty acids are stored as TAG, a marker of insulin resistance. Presumably, fatty acid metabolites (diacylglycerol, ceramide) are also increased and these interfere with insulin signaling to GLUT4. (TAG = triacylglycerol)

Unsuspected Role of FAT/CD36 in Mitochondrial Fatty Acid Oxidation

In individuals exhibiting severe obesity, fatty acid oxidation is impaired (Hulver et al. 2003; Kim et al. 2000). There may be a role for FAT/CD36 in this pathology. Recent work from our laboratory has shown that FAT/CD36 is located at the mitochondrion in rodent (Campbell et al. 2004) and human muscle (Bezaire et al. 2006; Holloway et al. 2006). Blocking FAT/CD36 in isolated mitochondria greatly reduces long-chain, but not short-chain, fatty acid oxidation (Bezaire et al. 2006; Campbell et al. 2004; Holloway et al. 2006). In addition, we also observed that during muscle contraction FAT/CD36 translocates to the mitochondrion (Campbell et al. 2004; Holloway et al. 2006). This finding suggests that this protein may play a role in facilitating fatty acid oxidation. In humans there is a good correla-

tion between the FAT/CD36 at the mitochondrion and the rate of fatty acid oxidation during exercise (Holloway et al. 2006). How FAT/CD36 supports fatty acid oxidation at the mitochondrion is not clear. It appears to interact with carnitine palmitoyltransferase I (CPT1) to deliver fatty acyl moieties across the mitochondrial membrane (Bezaire et al. 2006), although the exact mechanisms remain to be determined. Whether impaired mitochondrial FAT/CD36 availability would impair fatty acid oxidation as is observed in severe obesity is under investigation. Nevertheless, FAT/CD36 appears to be involved with transporting LCFAs not only into the muscle cell but also into the mitochondrion.

Concluding Remarks

Physiological, biochemical, and molecular evidence has shown that fatty acid movement across the plasma membrane is regulated acutely by muscle contraction, AMPK activation and selected hormones (insulin, leptin, resistin) as well as chronically by altered muscle activity and prolonged leptin infusion. Acutely altered up- or downregulation of fatty acid transport is mediated by the subcellular redistribution of one or more fatty acid transporters. Chronically altered fatty acid transport can occur either by up- or downregulation of the expression of selected fatty acid transporters, with concomitant changes at the plasma membrane, or by relocation of selected fatty acid transporters to or away from the plasma membrane, without change in their protein expression.

In insulin-resistant muscle (muscle from Zucker obese and ZDF rats and from humans with obesity or type 2 diabetes), rates of LCFA transport are increased. This increase is associated with an increase in plasmalemmal LCFA transporters that is due to either an increased presence of FAT/CD36 at the plasma membrane (i.e., when the total availability of the LCFA transporters is not altered, as is seen in Zucker obese rats and in humans who are obese or who have type 2 diabetes) or an increased expression of FAT/CD36 that occurs with a concomitant increase in plasmalemmal FAT/CD36 (ZDF rats). Clearly, the increased rate of fatty acid transport into muscle is a mechanism that contributes to insulin resistance mediated by fatty acids.

References

Abumrad, N.A., M.R. El-Maghrabi, E.-Z. Amri, E. Lopez, and P. Grimaldi. 1993. Cloning of a rat adipocyte membrane protein implicated in binding or transport of long chain fatty acids that is induced during preadipocyte differentiation. Homology with human CD36. *J Biol Chem* 268:17665-8.

Abumrad, N.A., C.C. Forest, D.M. Regen, and S. Sanders. 1991. Increase in membrane uptake of long-chain fatty acids early during preadipocyte differentiation. *Proc Natl Acad Sci USA* 88:6008-12.

Amri, E.Z., F. Bonino, G. Ailhaud, N.A. Abumrad, and P.A. Grimaldi. 1995. Cloning of a protein that mediates transcriptional effects of fatty acids in preadipocytes. Homology to peroxisome proliferator-activated receptors. *J Biol Chem* 270:2367-71.

Bastie, C.C., T. Hajri, V.A. Drover, P.A. Grimaldi, and N.A. Abumrad. 2004. CD36 in myocytes channels fatty acids to a lipase-accessible triglyceride pool that is related to cell lipid and insulin responsiveness. *Diabetes* 53:2209-16.

Belfort, R., L. Mandarino, S. Kashyap, K. Wirfel, T. Pratipanawatr, R. Berria, R.A. Defronzo, and K. Cusi. 2005. Dose-response effect of elevated plasma free fatty acid on insulin signaling. *Diabetes* 54:1640-8.

Benton, C.R., D.P. Koonen, J. Calles-Escandon, N.N. Tandon, J.F.C. Glatz, J.J.F.P. Luiken, J.J. Heikkila, and A. Bonen. 2006. Differential effects of muscle contraction and PPAR agonists on the expression of fatty acid transporters in rat skeletal muscle. *J Physiol* 573:199-210.

Berk, P.D., H. Wada, Y. Horio, B.J. Potter, D. Sorrentino, S.-L. Zhou, L.M. Isola, D. Stump, C.-L. Kiang, and S. Thung. 1990. Plasma membrane fatty acid-binding protein and mitochondrial glutamic-oxalocetic transaminase of rat liver are related. *Proc Natl Acad Sc USA* 87:3484-8.

Berk, P.D., S.-L. Zhou, C.-L. Kiang, D. Stump, M. Bradbury, and L. Isola. 1997. Uptake of long chain fatty acids is selectively up-regulated in adipocytes of Zucker rats with genetic obesity and non-insulin-dependent diabetes mellitus. *J Biol Chem* 272:8830-5.

Berk, P.D., S.-L. Zhou, C.-L. Kiang, D. Stump, X. Fan, and M. Bradbury. 1999. Selective upregulation of fatty acid uptake by adipocytes characterizes both genetic and diet-induced obesity in rodents. *J Biol Chem* 274:28626-31.

Bezaire, V., C.R. Bruce, G.J. Heigenhauser, N.N. Tandon, J.F. Glatz, J.J. Luiken, A. Bonen, and L.L. Spriet. 2006. Identification of fatty acid translocase on human skeletal muscle mitochondrial membranes: Essential role in fatty acid oxidation. *Am J Physiol Endocrinol Metab* 290:E509-15.

Blaak, E.E., and A.J. Wagenmakers. 2002. The fate of [U-^{13}C] palmitate extracted by skeletal muscle in subjects with type 2 diabetes and control subjects. *Diabetes* 51:784-9.

Bonen, A., D.J. Dyck, A. Ibrahimi, and N.A. Abumrad. 1999. Muscle contractile activity increases fatty acid metabolism and transport and FAT/CD36. *Am J Physiol Endocrinol Metab* 276:E642-9.

Bonen, A., J.J.F.P. Luiken, Y. Arumugam, J.F.C. Glatz, and N.N. Tandon. 2000. Acute regulation of fatty acid uptake involves the cellular redistribution of fatty acid translocase. *J Biol Chem* 275:14501-8.

Bonen, A., J.J.F.P. Luiken, S. Lui, D.J. Dyck, B. Kiens, S. Kristiansen, L. Turcotte, G.J. van der Vusse, and J.F.C. Glatz. 1998. Palmitate transport and fatty acid transporters in red and white muscles. *Am J Physiol Endocrinol Metab* 275: E471-8.

Bonen, A., D. Miskovic, and B. Kiens. 1999. Fatty acid transporters (FABPpm, FAT, FATP) in human muscle. *Can J Appl Physiol* 24:515-23.

Bonen, A., M.L. Parolin, G.R. Steinberg, J. Calles-Escandon, N.N. Tandon, J.F.C. Glatz, J.J.F.P. Luiken, G.J.F. Heigenhauser, and D.J. Dyck. 2004. Triacylglycerol accumulation in human obesity and type 2 diabetes is associated with increased rates of skeletal muscle fatty acid transport and increased sarcolemmal FAT/CD36. *FASEB J* 18:1144-6.

Bradbury, M.W., and P.D. Berk. 2000. Mitochondrial aspartate aminotransferase: Direction of a single protein with two distinct functions to two subcellular sites does not require alternative splicing of the mRNA. *Biochem J* 345:423-7.

Brozinick Jr., J.T., G.J. Etgen Jr., B.B. Yaspelkis III, and J.L. Ivy. 1992. Contraction-activated glucose uptake is normal in insulin-resistant muscle of the obese Zucker rat. *J Appl Physiol* 73:382-7.

Brozinick Jr., J.T., G.J. Etgen, B.B. Yaspelkis III, H.Y. Kang, and J.L. Ivy. 1993. Effects of exercise training on muscle GLUT-4 protein content and translocation in obese Zucker rats. *Am J Physiol Endocrinol Metab* 265:E419-27.

Bruce, C.R., A.B. Thrush, V.A. Mertz, V. Bezaire, A. Chabowski, G.J. Heigenhauser, and D.J. Dyck. 2006. Endurance training in obese humans improves glucose tolerance, mitochondrial fatty acid oxidation and alters muscle lipid content. *Am J Physiol Endocrinol Metab* 291:E99-E107.

Campbell, S.E., N.N. Tandon, G. Woldegiorgis, J.J.F.P. Luiken, J.F.C. Glatz, and A. Bonen. 2004. A novel function for FAT/CD36: Involvement in long chain fatty acid transfer into the mitochondria. *J Biol Chem* 279:36325-41.

Chabowski, A., J.C. Chatham, N.N. Tandon, J. Calles-Escandon, J.F.C. Glatz, J.J.F.P. Luiken, and A. Bonen. 2006. Fatty acid transport and FAT/CD36 are increased in red but not in white muscle skeletal muscle of Zucker diabetic fatty (ZDF) rats. *Am J Physiol Endocrinol Metab* 291: E675-E682.

Chabowski, A., S.L. Coort, J. Calles-Escandon, N.N. Tandon, J.F. Glatz, J.J. Luiken, and A. Bonen. 2004. Insulin stimulates fatty acid transport by regulating expression of FAT/CD36 but not FABPpm. *Am J Physiol Endocrinol Metab* 287: E781-9.

Chabowski, A., S.L.M. Coort, J. Calles-Escandon, N.N. Tandon, J.F.C. Glatz, J.J.F.P. Luiken, and A. Bonen. 2005. The subcellular compartmentation of fatty acid transporters is regulated differently by insulin and by AICAR. *FEBS Lett* 579:2428-32.

Chabowski, A., I. Momken, S.L.M. Coort, J. Calles-Escandon, N.N. Tandon, J.F.C. Glatz, J.J.F.P. Luiken, and A. Bonen. 2006. Prolonged AMPK activation increases the expression of fatty acid transporters in cardiac myocytes and perfused hearts. *Mol Cell Biochem* 288:201-212.

Chatham, J.C., and A.M. Seymour. 2002. Cardiac carbohydrate metabolism in Zucker diabetic fatty rats. *Cardiovasc Res* 55:104-12.

Clarke, D.C., D. Miskovic, X.-X. Han, J. Calles-Escandon, J.F.C. Glatz, J.J.F.P. Luiken, J.J. Heikkila, and A. Bonen. 2004. Overexpression of membrane associated fatty acid binding protein (FABPpm) in vivo increases fatty acid sarcolemmal transport and metabolism. *Physiol Genomics* 17:31-7.

Cooney, G.J., A.L. Thompson, S.M. Furler, J. Ye, and E.W. Kraegen. 2002. Muscle long-chain acyl coa esters and insulin resistance. *Ann N Y Acad Sci* 967:196-207.

Coort, S.L., J.J. Luiken, G.J. Van Der Vusse, A. Bonen, and J.F. Glatz. 2004. Increased fat (fatty acid translocase)/CD36-mediated long-chain fatty acid uptake in cardiac myocytes from obese Zucker rats. *Biochem Soc Trans* 32:83-5.

DeGrella, R.F., and R.J. Light. 1980. Uptake and metabolism of fatty acids by dispersed adult rat heart myocytes. *J Biol Chem* 255:9731-8.

DiRusso, C.C., H. Li, D. Darwis, P.A. Watkins, J. Berger, and P.N. Black. 2005. Comparative biochemical studies of the murine fatty acid transport proteins (FATP) expressed in yeast. *J Biol Chem* 280:16829-37.

Distel, R.J., G.S. Robinson, and B.M. Spiegelman. 1992. Fatty acid regulation of gene expression. *J Biol Chem* 267:5937-41.

Dobrzyn, A., P. Dobrzyn, S.H. Lee, M. Miyazaki, P. Cohen, E. Asilmaz, D.G. Hardie, J.M. Friedman, and J.M. Ntambi. 2005. Stearoyl-CoA desaturase-1 deficiency reduces ceramide synthesis by downregulating serine palmitoyltransferase and increasing beta-oxidation in skeletal muscle. *Am J Physiol Endocrinol Metab* 288:E599-607.

Dyck, D.J., and A. Bonen. 1998. Muscle contraction increases palmitate esterification and oxidation, and triacylglycerol oxidation. *Am J Physiol Endocrinol Metab* 275:E888-96.

Dyck, D.J., G. Steinberg, and A. Bonen. 2001. Insulin increases FFA uptake and esterification but reduces lipid utilization in isolated contracting muscles. *Am J Physiol Endocrinol Metab* 281:E600-7.

Ellis, B.A., A. Poynten, A.J. Lowy, S.M. Furler, D.J. Chisholm, E.W. Kraegen, and G.J. Cooney. 2000. Long chain acyl-CoA esters as indicators of lipid availability and insulin sensitivity in rat and human muscle. *Am J Physiol Endocrinol Metab* 279:E554-60.

Etgen, G.J., and B.A. Oldham. 2000. Profiling of Zucker diabetic fatty rats in their progression to the overt diabetic state. *Metabolism* 49:684-8.

Febbraio, M., N.A. Abumrad, D.P. Hajjar, K. Sharma, W. Cheng, S. Frieda, A. Pearce, and R.L. Silverstein. 1999. A null mutation in murine CD36 reveals an important role in fatty acid and lipoprotein metabolism. *J Biol Chem* 274:19055-062.

Friedman, J.E., J.E. De Vente, R.G. Peteson, and G.L. Dohm. 1991. Altered expression of muscle glucose transporter GLUT-4 in diabetic fatty Zucker rats (zdf/drt-fa). *Am J Physiol Endocrinol Metab* 261:E782-8.

Gimeno, R.E., A.M. Ortegon, S. Patel, S. Punreddy, P. Ge, Y. Sun, H.F. Lodish, and A. Stahl. 2003. Characterization of a heart-specific fatty acid transport protein. *J Biol Chem* 278:16039-44.

Glatz, J.F., T. Borchers, F. Spener, and G. van der Vusse. 1995. Fatty acids in cell signalling: Modulation by lipid binding proteins. *Prostaglandins Leukot Essent Fatty Acids* 52:121-7.

Glatz, J.F.C., and J. Storch. 2001. Unraveling the significance of cellular fatty acid binding proteins. *Curr Opin Lipidol* 12:267-74.

Gore, L., and C. Hoinard. 1993. Linolenic acid transport in hamster intestinal cells is carrier-mediated. *J Nutr* 123:66-73.

Goudriaan, J.R., V.E. Dahlmans, B. Teusink, D.M. Ouwens, M. Febbraio, J.A. Maassen, J.A. Romijn, L.M. Havekes, and P.J. Voshol. 2003. CD36 deficiency increases insulin sensitivity in muscle, but induces insulin resistance in the liver in mice. *J Lipid Res* 44:2270-7.

Halestrap, A.P., and D. Meredith. 2004. The slc16 gene family—from monocarboxylate transporters (MCTs) to aromatic amino acid transporters and beyond. *Pflugers Arch* 447:619-28.

Hamilton, J.A., and F. Kamp. 1999. How are free fatty acids transported in membranes? Is it by proteins or by free diffusion through the lipids? *Diabetes* 48:2255-69.

Harmon, C.M., P. Luce, and N.A. Abumrad. 1992. Labeling of an 88 kDa adipocyte membrane protein by sulpho-n-succinimidyl long-chain fatty acids: Inhibition of fatty acid transport. *Biochem Soc Trans* 20:811-3.

Hatch, G.M., A.J. Smith, F.Y. Xu, A.M. Hall, and D.A. Bernlohr. 2002. FATP1 channels exogenous FA into 1,2,3-triacyl-sn-glycerol and down-regulates sphingomyelin and cholesterol metabolism in growing 293 cells. *J Lipid Res* 43:1380-9.

Henriksen, E.J., T.R. Kinnick, M.K. Teachey, M.P. O'Keefe, D. Ring, K.W. Johnson, and S.D. Harrison. 2003. Modulation of muscle insulin resistance by selective inhibition of GSK-3 in Zucker diabetic fatty rats. *Am J Physiol Endocrinol Metab* 284:E892-900.

Hirsch, D., A. Stahl, and H.F. Lodish. 1998. A family of fatty acid transporters conserved from mycobacterium to man. *Proc Natl Acad Sci USA* 95:8625-9.

Holloway, G.P., V. Bezaire, G.J. Heigenhauser, N.N. Tandon, J.F. Glatz, J.J. Luiken, A. Bonen, and L.L. Spriet. 2006. Mitochondrial long chain fatty acid oxidation, fatty acid translocase/cd36 content and carnitine palmitoyltransferase activity in human skeletal muscle during aerobic exercise. *J Physiol* 571:201-10.

Hulver, M.W., J.R. Berggren, R.N. Cortright, R.W. Dudek, R.P. Thompson, W.J. Pories, K.G. MacDonald, G.W. Cline, G.I. Shulman, G.L. Dohm, and J.A. Houmard. 2003. Skeletal muscle lipid metabolism with obesity. *Am J Physiol Endocrinol Metab* 284:E741-7.

Ibrahimi, A., A. Bonen, W.D. Blinn, T. Hajri, X. Li, K. Zhong, R. Cameron, and N.A. Abumrad. 1999. Muscle-specific overexpression of FAT/CD36 enhances fatty acid oxidation by contracting muscles, reduces plasma triglycerides and fatty acids, and increases plasma glucose and insulin. *J Biol Chem* 274:26761-6.

Ibrahimi, A., Z. Sfeir, H. Magharaine, E.Z. Amri, P. Grimaldi, and N.A. Abumrad. 1996. Expression of the CD36 homolog (FAT) in fibroblast cells: Effects on fatty acid transport. *Proc Natl Acad Sci USA* 93:2646-51.

Isola, L.M., S.L. Zhou, C.L. Kiang, D.D. Stump, M.W. Bradbury, and P.D. Berk. 1995. 3t3 fibroblasts transfected with a cDNA for mitochondrial aspartate aminotransferase express plasma membrane fatty acid-binding protein and saturable fatty acid uptake. *Proc Natl Acad Sci USA* 92:9866-70.

Itani, S.I., N.B. Ruderman, F. Schmieder, and G. Boden. 2002. Lipid-induced insulin resistance in human muscle is associated with changes in diacylglycerol, protein kinase C, and iKappab-a. *Diabetes* 51:2005-11.

Jacob, S., J. Machann, K. Rett, K. Brechtel, A. Volk, W. Renn, E. Maerker, S. Matthaei, F. Schick, C.D. Claussen, and H.U. Haring. 1999. Association of increased intramyocellular lipid content with insulin resistance in lean nondiabetic offspring of type 2 diabetic subjects. *Diabetes* 48:1113-9.

Johannsson, E., K.J.A. McCullagh, X. Han, P.K. Fernando, J. Jensen, H.A. Dahl, and A. Bonen. 1996. Effect of overexpressing glut-1 and glut-4 on insulin- and contraction stimulated glucose transport in muscle. *Am J Physiol Endocrinol Metab* 271:E547-55.

Joost, H.G., G.I. Bell, J.D. Best, M.J. Birnbaum, M.J. Charron, Y.T. Chen, H. Doege, D.E. James, H.F. Lodish, K.H. Moley, J.F. Moley, M. Mueckler, S. Rogers, A. Schurmann, S. Seino, and B. Thorens. 2002. Nomenclature of the GLUT/SLC2a family of sugar/polyol transport facilitators. *Am J Physiol Endocrinol Metab* 282:E974-6.

Kashyap, S.R., R. Belfort, R. Berria, S. Suraamornkul, T. Pratipranawatr, J. Finlayson, A. Barrentine, M. Bajaj, L. Mandarino, R. DeFronzo, and K. Cusi. 2004. Discordant effects of a chronic physiological increase in plasma FFA on insulin signaling in healthy subjects with or without a family history of type 2 diabetes. *Am J Physiol Endocrinol Metab* 287:E537-46.

Kelley, D.A., and L.J. Mandarino. 2000. Fuel selection in human skeletal muscle in insulin resistance. A reexamination. *Diabetes* 49:677-83.

Kelley, D.E., and B.H. Goodpaster. 2001. Skeletal muscle triglyceride: An aspect of regional adiposity and insulin resistance. *Diabetes Care* 24:933-41.

Kelley, D.E., J. He, E.V. Menshikova, and V.B. Ritov. 2002. Dysfunction of mitochondria in human skeletal muscle in type 2 diabetes. *Diabetes* 51:2944-50.

Kelley, D.E., F.L. Thaete, F. Troost, T. Huwe, and B.H. Goodpaster. 2000. Subdivisions of subcutaneous abdominal adipose tissue and insulin resistance. *Am J Physiol Endocrinol Metab* 278:E941-8.

Kiens, B., S. Kristiansen, P. Jensen, E.A. Richter, and L.P. Turcotte. 1997. Membrane associated fatty acid binding protein (FABPpm) in human skeletal muscle is increased by endurance training. *Biochem Biophys Res Commun* 231:463-5.

Kim, J.K., R.E. Gimeno, T. Higashimori, H.-J. Kim, H. Choi, S. Punreddy, R.L. Mozell, G. Tan, A. Stricker-Krongrad, D.J. Hirsch, J.J. Fillmore, Z.-X. Liu, J. Dong, G. Cline, A. Stahl, H.F. Lodish, and G.I. Shulman. 2004. Inactivation of fatty acid transport protein 1 prevents fat-induced insulin resistance in skeletal muscle. *J Clin Invest* 113:756-63.

Kim, J.Y., R.L. Hickner, R.N. Cortright, G.L. Dohm, and J.A. Houmard. 2000. Lipid oxidation is reduced in obese human skeletal muscle. *Am J Physiol Endocrinol Metab* 279:E1039-44.

Kim, Y.-B., G.I. Shulman, and B.B. Kahn. 2002. Fatty acid infusion selectively impairs insulin action on akt1 and protein kinase C l/z but not on glycogen synthase kinase-3. *J Biol Chem* 277:32915-22.

Koonen, D.P.Y., C.R. Benton, Y. Arumugam, N.N. Tandon, J. Calles-Escandon, J.F.C. Glatz, J.J.F.P. Luiken, and A. Bonen. 2004. Different mechanism can alter fatty acid transport when muscle contractile activity is chronically altered. *Am J Physiol Endocrinol Metab* 286:1042-9.

Koonen, D.P.Y., W.A. Coumans, Y. Arumugam, A. Bonen, J.F.C. Glatz, and J.J.F.P. Luiken. 2002. Giant membrane vesicles as a model to study cellular substrate uptake dissected from metabolism. *Mol Cell Biochem* 239:121-30.

Krssak, M., K. Falk Petersen, A. Dresner, L. DiPietro, S.M. Vogel, D.L. Rothman, M. Roden, and G.I. Shulman. 1999. Intramyocellular lipid concentrations are correlated with insulin sensitivity in humans: A 1H NMR spectroscopy study. *Diabetologia* 42:113-6.

Kruszynska, Y.T., D. Sears Worrall, J. Ofrecio, J.P. Frias, G. Macaraeg, and J.M. Olefsky. 2002. Fatty acid-induced insulin resistance: Decreased muscle PI3K activation but unchanged Akt phosphorylation. *J Clin Endocrinol Metab* 87:226-34.

Luiken, J.J., S.L. Coort, D.P. Koonen, D.J. van der Horst, A. Bonen, A. Zorzano, and J.F. Glatz. 2004. Regulation of cardiac long-chain fatty acid and glucose uptake by translocation of substrate transporters. *Pflugers Arch* 448:1-15.

Luiken, J.J.F.P., Y. Arumugam, R.C. Bell, J. Calles-Escandon, N.N. Tandon, J.F.C. Glatz, and A. Bonen. 2002. Changes in fatty acid transport and transporters are related to the severity of insulin deficiency. *Am J Physiol Endocrinol Metab* 282:612-21.

Luiken, J.J.F.P., Y. Arumugam, D.J. Dyck, R.C. Bell, M.L. Pelsers, L.P. Turcotte, N.N. Tandon, J.F.C. Glatz, and A. Bonen. 2001. Increased rates of fatty acid uptake and plasmalemmal fatty acid transporters in obese Zucker rats. *J Biol Chem* 276:40567-73.

Luiken, J.J.F.P., S.M.L. Coort, J. Willems, W.A. Coumans, A. Bonen, G.J. van der Vusse, and J.F.C. Glatz. 2003. Contraction-induced fatty acid translocase/CD36 translocation in rat cardiac myocytes is mediated through AMP-activated protein kinase signaling. *Diabetes* 52:1627-34.

Luiken, J.J.F.P., D.J. Dyck, X.-X. Han, N.N. Tandon, Y. Arumugam, J.F.C. Glatz, and A. Bonen. 2002. Insulin induces the translocation of the fatty acid transporter FAT/CD36 to the plasma membrane. *Am J Physiol Endocrinol Metab* 282:E491-5.

Luiken, J.J.F.P., D.P.Y. Koonen, J. Willems, A. Zorzano, Y. Fischer, G.J. van der Vusse, A. Bonen, and J.F.C. Glatz. 2002. Insulin stimulates long-chain fatty acid utilization by rat cardiac myocytes through cellular redistribution of FAT/CD36. *Diabetes* 51:3113-9.

Luiken, J.J.F.P., L.P. Turcotte, and A. Bonen. 1999. Protein-mediated palmitate uptake and expression of fatty acid transport proteins in heart giant vesicles. *J Lipid Res* 40:1007-16.

Luiken, J.J.F.P., J. Willems, G.J. van der Vusse, and J.F.C. Glatz. 2001. Electrostimulation enhances FAT/CD36-mediated long-chain fatty acid uptake by isolated rat cardiac myocytes. *Am J Physiol Endocrinol Metab* 281:E704-12.

McCullagh, K.J.A., C. Juel, M. O'Brien, and A. Bonen. 1996. Chronic muscle stimulation increases lactate transport in rat skeletal muscle. *Mol Cell Biochem* 156:51-7.

Megeney, L.A., P.D. Neufer, G.L. Dohm, M.H. Tan, C.A. Blewett, G.C.B. Elder, and A. Bonen. 1993. Effects of muscle activity and fiber composition on glucose transport and GLUT-4. *Am J Physiol Endocrinol Metab* 264:E583-93.

Muoio, D.M., G.L. Dohm, F.T. Fiedorek, E.B. Tapscott Jr., and R.A. Coleman. 1997. Leptin directly alters lipid partitioning in skeletal muscle. *Diabetes* 46:1360-3.

Muoio, D.M., G.L. Dohm, E.B. Tapscott, and R.A. Coleman. 1999. Leptin opposes insulin's effects on fatty acid partitioning in muscles isolated from obese ob/ob mice. *Am J Physiol Endocrinol Metab* 276:E913-21.

Newsholme, E.A., P. Calder, and P. Yaqoob. 1993. The regulatory, informational, and immunomodulator roles of fat fuels. *Am J Clin Nutr Suppl.* no. 57:738S-51S.

Nickerson, J., and A. Bonen. 2005. Defining a role for skeletal muscle fatty acid transport proteins. Paper presented at the 2nd Northern Lights Conference. Canadian Federation of Biological Societies, Guelph, Ontario, Canada.

Palanivel, R., and G. Sweeney. 2005. Regulation of fatty acid uptake and metabolism in L6 skeletal muscle cells by resistin. *FEBS Lett* 579:5049-54.

Pan, D.A., S. Lillioja, A.D. Kriketos, M.R. Milner, L.A. Baur, A.B. Jenkins, and L.H. Storlien. 1997. Skeletal muscle triglyceride levels are inversely related to insulin action. *Diabetes* 46:983-8.

Perseghin, G., G. Lattuada, M. Danna, L.P. Sereni, P. Maffi, F. De Cobelli, A. Battezzati, A. Secchi, A. Del Maschio, and L. Luzi. 2003. Insulin resistance, intramyocellular lipid content, and plasma adiponectin in patients with type 1 diabetes. *Am J Physiol Endocrinol Metab* 285:E1174-81.

Richards, M.R., J.D. Harp, D.S. Ory, and J.E. Schaffer. 2006. Fatty acid transport protein 1 and long-chain acyl coenzyme A synthetase 1 interact in adipocytes. *J Lipid Res* 47:665-72.

Roepstorff, C., B. Vistisen, K. Roepstorff, and B. Kiens. 2004. Regulation of plasma long-chain fatty acid oxidation in relation to uptake in human skeletal muscle during exercise. *Am J Physiol Endocrinol Metab* 287:E696-705.

Schaffer, J.E., and H.F. Lodish. 1994. Expression cloning and characterization of a novel adipocyte long chain fatty acid transport protein. *Cell* 79:427-36.

Schaffer, J.E., and H.F. Lodish. 1995. Molecular mechanism of long-chain fatty acid uptake. *Trends Cardiovasc Med* 5:218-24.

Schwieterman, W., D. Sorrentino, B.J. Potter, J. Rand, C.-L. Kiang, D. Stump, and P.D. Berk. 1988. Uptake of oleate by isolated rat adipocytes is mediated by a 40-kDa plasma membrane fatty acid binding protein closely related to that in liver and gut. *Proc Natl Acad Sci USA.* 85:359-63.

Shulman, G.I. 2000. Cellular mechanisms of insulin resistance. *J Clin Invest* 106:171-6.

Simoneau, J.-A., J.-A. Veerkamp, L.P. Turcotte, and D.E. Kelley. 1999. Markers of capacity to utilize fatty acids in human skeletal muscle: Relation to insulin resistance and obesity and effects of weight loss. *FASEB J* 13:2051-60.

Slieker, L.J., K.L. Sundell, W.F. Heath, H.E. Osborne, J. Bue, J. Manetta, and J.R. Sportsman. 1992. Glucose transporter levels in tissues of spontaneously diabetic Zucker fa/fa rat (zdf/drt) and viable yellow mouse (avy/a). *Diabetes* 41:187-93.

Sorrentino, D., R.B. Robinson, C.-L. Kiang, and P. D. Berk, 1989. At physiologic albumin/oleate concentrations oleate uptake by isolated hepatocytes, cardiac myocytes, and adipocytes is a saturable function of the unbound oleate concentration. Uptake kinetics are consistent with the conventional theory. *J Clin Invest* 84:1325-33.

Spitsberg, V.L., E. Matitashvili, and R.C. Gorewit. 1995. Association and coexpression of fatty-acid-binding protein and glycoprotein cd36 in the bovine mammary gland. *Eur J Biochem* 230:872-8.

Steinberg, G.R., D.J. Dyck, J. Calles-Escandon, N.N. Tandon, J.J.F.P. Luiken, J.F. Glatz, and A. Bonen. 2002. Chronic leptin administration decreases fatty acid uptake and fatty acid transporters in rat skeletal muscle. *J Biol Chem* 277:8854-60.

Steinberg, G.R., M.L. Parolin, G.J.F. Heigenhauser, and D.J. Dyck. 2002. Leptin increases fatty acid oxidation in lean but not obese human skeletal muscle: Evidence of peripheral leptin resistance. *Am J Physiol Endocrinol Metab* 283: E187-92.

Storch, J., C. Lechene, and A.M. Kleinfeld. 1991. Direct determination of free fatty acid transport across the adipocyte plasma membrane using quantitative fluorescence microscopy. *J Biol Chem* 266:13473-6.

Storz, P., H. Doppler, A. Wernig, K. Pfizenmaier, and G. Muller. 1999. Cross-talk mechanisms in the development of insulin resistance of skeletal muscle cells. Palmitate rather than tumor necrosis factor inhibits insulin-dependent protein kinase B (PKB)/Akt stimulation and glucose uptake. *Eur J Biochem* 266:17-25.

Stremmel, W. 1988. Fatty acid uptake by isolated heart myocytes represents a carrier-mediated transport process. *J Clin Invest* 81:844-52.

Stremmel, W. 1989. Transmembrane transport of fatty acids in the heart. *Mol Cell Biochem* 88:23-9.

Stremmel, W., and P.D. Berk. 1986. Hepatocellular influx of [^{14}C]oleate reflects membrane transport rather than intracellular metabolism or binding. *Proc Natl Acad Sci USA* 83:3086-90.

Stremmel, W., G. Strohmeyer, and P.D. Berk. 1986. Hepatocellular uptake of oleate is energy dependent, sodium linked, and inhibited by an antibody to a hepatocyte plasma membrane fatty acid binding protein. *Proc Natl Acad Sci USA* 83:3584-8.

Stremmel, W., G. Strohmeyer, F. Borchard, S. Kochwa, and P.D. Berk. 1985. Isolation and partial characterization of a fatty acid binding protein in rat liver plasma membranes. *Proc Natl Acad Sci USA* 82:4-8.

Stump, D.D., S.-L. Zhou, and P.D. Berk. 1993. Comparison of plasma membrane fabp and mitochondrial isoform of aspartate aminotransferase from rat liver. *Am J Physiol Gastrointest Liver Physiol* 265:G894-902.

Trimble, M.E. 1989. Mediated transport of long-chain fatty acids by rat renal basolateral membranes. *Am J Physiol* 257:F539-46.

Trotter, P.J., S.Y. Ho, and J. Storch. 1996. Fatty acid uptake by caco-2 human intestinal cells. *J Lipid Res* 7:336-46.

Tunstall, R.J., K.A. Mehan, G.D. Wadley, G.R. Collier, A. Bonen, M. Hargreaves, and D. Cameron-Smith. 2002. Exercise training increases lipid metabolism gene expression in human skeletal muscle. *Am J Physiol Endocrinol Metab* 283:E66-72.

Turcotte, L.P., C. Petry, and E.A. Richter. 1998. Contraction-induced increase in Vmax of palmitate uptake and oxidation in perfused skeletal muscle. *J Appl Physiol* 84:1788-94.

Turcotte, L.P., J.R. Swenberger, M.Z. Tucker, and A.J. Yee. 1999. Training-induced elevation in FABPpm is associated with increased palmitate use in contracting muscle. *J Appl Physiol* 87:285-93.

van der Vusse, G.J., and M.J.M. de Groot. 1992. Interrelationship between lactate and cardiac fatty acid metabolism. *Mol Cell Biochem* 116:11-17.

van der Vusse, G.J., and R.S. Reneman. 1996. Lipid metabolism in muscle. In *Handbook of physiology. Exercise: Regulation and integration of multiple systems,* ed. L.B. Rowell and J.T. Shepherd, 952-994. Bethesda, MD: American Physiological Society.

van der Vusse, G.J., M. van Bilsen, and J.F.C. Glatz. 2000. Cardiac fatty acid uptake and transport in health and disease. *Cardiovasc Res* 45:279-93.

van Loon, L.J., R. Koopman, R. Manders, W. van der Weegen, G.P. van Kranenburg, and H.A. Keizer. 2004. Intramyocellular lipid content in type 2 diabetes patients compared with overweight sedentary men and highly trained endurance athletes. *Am J Physiol Endocrinol Metab* 287:E558-65.

Wilmsen, H.M., T.P. Ciaraldi, L. Carter, N. Reehman, S.R. Mudaliar, and R.R. Henry. 2003. Thiazolidinediones upregulate impaired fatty acid uptake in skeletal muscle of type 2 diabetics. *Am J Physiol Endocrinol Metab* 285:E354-62.

Wu, Q., A.M. Ortegon, B. Tsang, H. Doege, K.R. Feingold, and A. Stahl. 2006. FATP1 is an insulin-sensitive fatty acid transporter involved in diet-induced obesity. *Mol Cell Biol* 26:3455-67.

Yaspelkis III, B.B., L. Ansari, E.A. Ramey, and S.F. Loy. 1999. Chronic leptin administration increases insulin-stimulated skeletal muscle glucose uptake. *Metabolism* 48:671-6.

Yu, C., Y. Chen, G.W. Cline, D. Zhang, H. Zong, Y. Wang, R. Bergeron, J.K. Kim, S.W. Cushman, G.W. Cooney, B. Atcheson, M.F. White, E.W. Kraegen, and G.I. Shulman. 2002. Mechanism by which fatty acids inhibit insulin activation of insulin receptor substrate-1 (irs-1)-associated phosphatidylinositol 3-kinase activity in muscle. *J Biol Chem* 277:50230-6.

Zou, Z., F. Tong, N.J. Faergeman, C. Borsting, P.N. Black, and C.C. DiRusso. 2003. Vectorial acylation in Saccharomyces cerevisiae. Fat1p and fatty acyl-CoA synthetase are interacting components of a fatty acid import complex. *J Biol Chem* 278:16414-22.

Chapter 4

Lipid Metabolism and Insulin Signaling

Jason R. Berggren, PhD; Leslie A. Consitt, PhD;
and Joseph A. Houmard, PhD

Whole-body insulin resistance is linked to a range of detrimental conditions such as obesity, type 2 diabetes, and cardiovascular disease; the combination of these disorders is often referred to as the *metabolic syndrome* (Reaven 1988). Skeletal muscle is the primary site for insulin-mediated glucose uptake, and it is estimated to dispose of 70% to 80% of an ingested glucose load (DeFronzo et al. 1981). Thus, discerning whether defects in skeletal muscle contribute to insulin resistance is fundamentally important when studying obesity and diabetes. Accordingly, adaptations in skeletal muscle may be responsible, at least in part, for the dramatic improvements in insulin action evident with interventions such as physical activity or weight loss.

The effects of insulin on the muscle cell begin with insulin binding to its receptor on the cell membrane. This binding initiates a complex signaling cascade that ultimately facilitates the diffusion of glucose into the cell via the insulin-sensitive glucose transporters (GLUT4). The insulin signaling cascade in skeletal muscle can be impaired by obesity and type 2 diabetes and subsequently enhanced with physical activity (Zierath and Wallberg-Henriksson 2002). However, the biological mechanisms by which insulin signal transduction is impaired in the insulin-resistant state are not clear. This chapter provides evidence that demonstrates the importance of intracellular lipid accumulation in regulating insulin action and discusses how interventions recommended for the treatment and prevention of type 2 diabetes and obesity (physical activity, weight loss) affect this system.

Lipid Metabolism in Skeletal Muscle

Lipid metabolism in skeletal muscle is a dynamic process that has several sites of regulation (Rasmussen and Wolfe 1999). The first step governing the utilization of fatty acids involves transporting fatty acids into the cell. Once transported, lipids can be partitioned toward either storage or oxidation. In this section we briefly address components of lipid metabolism in order to provide a basis for understanding defects that occur with diabetes and obesity. For a comprehensive review of lipid metabolism and fatty acid uptake in muscle, see chapter 3.

Fatty Acid Transport

During times of increased energy demand or excess nutrient supply (i.e., high-fat feeding), fatty acids are transported to skeletal muscle, where fatty acid uptake occurs via simple diffusion and protein-mediated transport (Schaffer 2002). FATP, FABPpm, and FAT/CD36 are expressed in skeletal muscle, and each has distinct actions (Hajri and Abumrad 2002). FATP is located primarily in the cytosol and may serve as an acyl-CoA synthetase for LCFA esterification. Moreover, it has been suggested that FATP channels fatty acids away from oxidation and toward synthesis. In response to high-fat diet or acute lipid infusion, FATP knockout (KO) mice accumulate a lesser amount of fatty acyl-CoA in skeletal muscle, a finding suggesting that FATP plays an important role in fatty acid uptake and acyl-CoA synthesis. However, other work suggests that FATP is important but not indispensable for fatty acid uptake. Insulin stimulation and muscle contraction induce the translocation of FAT/CD36 from an intracellular compartment to the plasma membrane. Concomitantly, insulin increases fatty acid uptake by 51%, while both effects are negated by inhibition of insulin signal transduction (Bonen et al. 2003; Luiken et al. 2002). FATP and FAT/CD36 are involved in the transport of fatty acids and may be an important metabolic step in regulating lipid oxidation. Therefore, the potential role of the fatty acid transporters is worth considering when either diabetes or obesity induces lipid accumulation in the cell.

Mitochondrial Fatty Acid Oxidation

Mitochondria are the organelles principally responsible for fatty acid catabolism. For mitochondrial fatty acid uptake to occur, the fatty acid must be esterified to a CoA derivate. This reaction is catalyzed by fatty acyl-CoA synthetase (ACS). LCFA-CoA cross the mitochondrial membrane by a carnitine-mediated system involving CPT1. Regulation of fatty acid transport into the mitochondria occurs primarily at CPT1, which can be allosterically inhibited by malonyl-CoA (Rasmussen et al. 2002; Ruderman et al. 1999). Malonyl-CoA is derived from the carboxylation of acetyl-CoA, a process that is regulated by acetyl-CoA carboxylase (ACC). Both glucose and insulin enhance ACC activity, while excess LCFA-CoA inhibits ACC activity. When glucose availability is low (e.g., during fasting), malonyl-CoA levels decrease, while reliance on fatty acid oxidation increases. Thus, metabolic permutations that favor fatty acid catabolism, such as exercise and fasting, are associated with decreased levels of malonyl-CoA. A study examining the effects of chronic (12 wk) endurance exercise training determined that malonyl-CoA decreased in human skeletal muscle. This reduction may underlie the increase in fatty acid oxidation observed at rest or during exercise (Kuhl et al. 2006).

Though CPT1 is the classical rate-limiting step of fatty acid uptake in mitochondria, recent evidence indicates that FAT/CD36 is also present in mitochondria (Campbell et al. 2004). Inhibiting FAT/CD36 resulted in a dose-dependent decrease in fatty acid uptake and caused a 95% decrease at the maximum inhibitor concentration (Campbell

et al. 2004). Further research is needed to clarify the role of FAT/CD36 in mitochondrial fatty acid uptake.

Once fatty acids enter the mitochondria, they undergo a series of regulated processes that oxidize the fatty acid to carbon dioxide (CO_2). First the chains of the fatty acids are shortened through beta-oxidation, in which β-hydroxyacyl-CoA dehydrogenase (β-HAD) is believed to be the rate-limiting enzyme, providing acetyl-CoA for the tricarboxylic acid (TCA) cycle. Citrate synthase (CS) is often used as a marker of TCA cycle activity and mitochondrial content, and it is the first step in preparing acetyl-CoA for oxidation to CO_2. Both the beta-oxidation and TCA cycle provide reducing equivalents for the electron transport chain, where ATP is synthesized.

Fatty Acid Partitioning: Storage or Oxidation?

Upon entering the cell, fatty acids are channeled either to oxidation in mitochondria or to storage. As metabolic demands increase (i.e., during exercise and starvation), fatty acids are channeled toward oxidation to meet energy demands. The complex interplay of metabolic control that regulates substrate selection and partitioning is not fully known. Two factors that may play important roles in the skeletal muscle of individuals who are obese or diabetic are pyruvate dehydrogenase kinase 4 (PDK4) and stearoyl-CoA desaturase (SCD).

Glucose and fatty acids serve as competing substrates for meeting metabolic energy demands. In the postprandial state, glucose is the preferred energy supply, while after fasting overnight the oxidation of fatty acids accounts for about 80% of resting energy expenditure. The shift in fuel utilization is achieved, at least in part, by an increase in PDK4 activity (Spriet et al. 2004). PDK4 phosphorylates and inactivates pyruvate dehydrogenase, which converts pyruvate to acetyl-CoA. Inhibiting pyruvate dehydrogenase thus decreases glycolytic flux (acetyl-CoA derived from glucose) and shifts fuel utilization to fatty acid metabolism. The expression and activity of PDK4 are increased by the same metabolic alterations that decrease malonyl-CoA levels and favor fat oxidation. During fasting or starvation, an inhibitor of fatty acid uptake (malonyl-CoA) is lowered and pyruvate catabolism is inhibited, and thus glucose metabolism decreases and fatty acids are channeled to oxidation.

The content of intramuscular fatty acids is determined not only by substrate delivery but also by fatty acid utilization. Several enzymes, including SCD1, have been implicated in the regulation of fatty acid partitioning (figure 4.1). SCD1 is a delta-9-desaturase that catalyzes the desaturation of palmitoyl-CoA (C16) and stearoyl-CoA (C18) to palmitoleoyl- and oleoyl-CoA, respectively. SCD1 KO animals have increased beta-oxidation, have decreased lipid storage, and are resistant to dietary-induced obesity (Dobrzyn et al. 2005; Dobrzyn and Ntambi 2005). Overexpression of SCD1 in human skeletal muscle cells decreases fat oxidation and increases esterification of fatty acids into lipid storage pools (Hulver et al. 2005). Although PDK4 and SCD are not the only ways to selectively partition lipid in the muscle cell, they do indicate that whether a fatty acid is diverted to oxidation or to storage depends on tightly controlled mechanisms. If SCD, PDK4, or another step in lipid metabolism is altered, lipid could easily be preferentially partitioned toward storage. This case may be what occurs in obesity and type 2 diabetes (figure 4.1).

Although adipose tissue is the primary site of lipid storage, skeletal muscle also stores fatty acids as triglycerides packaged in lipid droplets. Glycerol-3-phosphate acyltransferase (GPAT) catalyzes the first and committed step in glycerolipid synthesis in a variety of tissues, including skeletal muscle, while diacylglycerol acyltransferase (DGAT) is an important regulator of TAG synthesis. At least one lipase, hormone-sensitive lipase (HSL), is involved in TAG degradation. In addition to increasing DAG and TAG synthesis, excess saturated fatty acids are likely to increase ceramide synthesis via serine palmitoyl transferase and ceramide synthase (Chavez et al. 2003; Lee et al. 2006; Powell et al. 2004). However, the regulation of these enzymes in TAG synthesis, lipid droplet formation, and lipid degradation are not well characterized. It does appear that the lipid droplets are often positioned near mitochondria, potentially for ease of mobilization during physical activity, and represent a significant source of energy during low-intensity exercise (Hoppeler et al. 1985). The cycling of lipid synthesis and degradation represents normal fluxes in skeletal muscle metabolism. Moreover, it has been proposed that substrate cycling (degradation followed by resynthesis) is imperative for maintaining normal lipid homeostasis (Dulloo et al. 2004).

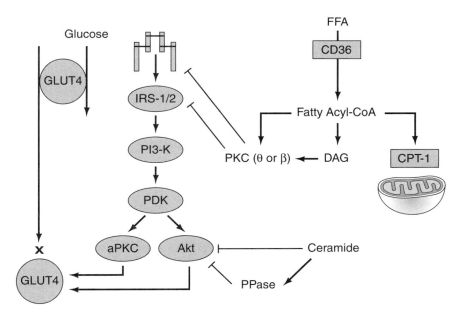

Figure 4.1   Proposed mechanism of lipid accumulation that occurs in cases of obesity or diabetes. Increased fatty acid uptake and decreased fatty acid catabolism contribute to the increased lipid accumulation that is observed with obesity and type 2 diabetes. PPARs, peroxisome proliferator-activated receptors.

Transcriptional Control of Lipid Metabolism

Lipid metabolism is also controlled by a variety of transcription factors, including the family of peroxisome proliferator-activated receptors (PPARs), which are ligand-inducible nuclear receptors. The PPARs control the expression of several lipid regulatory genes, including PDK4 and CPT1. PPARs are expressed in liver, skeletal muscle, and adipose tissue and are activated by endogenous free fatty acid (FFA), fatty acid metabolites, and synthetic agonists such as fibrates and thiazolidinediones (TZDs; Barish, Narkar, and Evans 2006; Glass 2006; Lefebvre et al. 2006; Semple, Chatterjee, and O'Rahilly 2006). Three isoforms have been identified (alpha, delta, and gamma), and while all of them are expressed in oxidative and lipogenic tissues, their relative distribution and contribution to fuel partitioning are not completely elucidated. For optimal activity, PPARs must bind an endogenous ligand (FFA) and complex with the 9-cis-retinoic acid activated retinoid X receptor (RXR) before binding to the specific PPAR response elements (PPRE) of DNA and initiating the transcription of target genes (Evans, Barish, and Wang 2004; Ferre 2004; Schoonjans, Staels, and Auwerx 1996). Activation of these transcription factors results in global expression patterns of lipid regulatory genes that help maintain systemic lipid homeostasis. Both PPARδ (Fredenrich and Grimaldi 2005; Grimaldi 2005) and PPARα agonists increase fatty acid oxidation in skeletal (Muoio et al. 2002) and cardiac (Cheng et al. 2004) muscle. Skeletal muscle PPARδ gene expression and protein content increase with 24 and 48 h of fasting and decrease upon refeeding, suggesting nutritional control (Holst et al. 2003). Skeletal muscle was the only tissue that demonstrated an increase in PPARδ expression as a result of fasting. Transgenic mice with targeted expression of an activated form of PPARδ exhibit elevated oxidative enzyme expression, mitochondrial biogenesis, and a transformation of fiber types to more slow-twitch oxidative fibers (Wang et al. 2003, 2004). This muscle remodeling elicits a metabolic phenotype that protects against dietary-induced metabolic disturbances and insulin resistance. Additionally, PPARδ overexpression increases CS and β-HAD while reducing fat pad weight (Grimaldi 2003). Whether PPARδ is the primary transcription factor regulating lipid metabolism in skeletal muscle or is working concomitantly with PPARα is not known. Additionally, other regulators such as the PPAR gamma coactivator 1 (PGC-1) have been implicated in regulating fiber type switching, mitochondrial biogenesis, and adaptive responses to metabolic permutations (Finck and Kelly 2006; Koves et al. 2005). Thus, PPAR regulation may be important in governing the oxidative capacity of muscle (figure 4.1).

Substrate demands and availability may dictate metabolic flux. In support of this contention, a high-fat diet (>65% calories from fat) compared to a high-carbohydrate diet (70%-75% calories from carbohydrate) significantly increases fat oxidation during exercise (Cameron-Smith et al. 2003). This adaptation appears to be mediated through increased expression and protein content of FAT/CD36 as well as through increased expression of β-HAD (Cameron-Smith et al. 2003). Therefore, a greater FFA availability appears to increase the capacity for fatty acid uptake and beta-oxidation. This finding suggests that nutrients have a potent effect on gene transcription.

The complexity of lipid metabolism in skeletal muscle amplifies the difficulty of identifying causative factors of metabolic disease. Alterations in lipid oxidation, in lipid storage, and in the regulation of these metabolic pathways may have deleterious effects on insulin action in skeletal muscle (discussed subsequently). Dysregulation of any of these steps may lead to intramuscular lipid accumulation and insulin resistance.

Insulin Signaling

Insulin is a potent anabolic hormone that regulates numerous cellular processes, including glucose transport, fatty acid metabolism, protein synthesis, and gene expression. This section briefly addresses insulin signaling and its role in glucose transport in skeletal muscle. Readers are also referred to several reviews of this topic (Bjornholm and Zierath 2005; Chang, Chiang, and Saltiel 2004; Czech and Corvera 1999; Krook, Wallberg-Henriksson, and Zierath 2004; Watson and Pessin 2006).

Insulin has a site-specific heterotetrameric receptor that consists of two extracellular alpha subunits and two intracellular beta subunits. Insulin binding to the alpha subunits results in a cascade of signaling events that can initiate glucose transport or a host of other anabolic processes. In glucose transport, insulin binding stimulates the autophosphorylation of the beta subunits and thus increases kinase activity. Phosphorylation on tyrosine (Tyr-960) creates a recognition site that binds several intracellular signal molecules, including the insulin receptor substrates (IRS1-IRS12; figure 4.2; Sun et al. 1991; White et al. 1988). IRS1 is phosphorylated on tyrosine residues and activates its downstream target PI3K. PI3K then associates with the plasma membrane, where it phosphorylates inositol phospholipids to generate phosphatidylinositol (3,4,5)-trisphosphate (PIP_3). PIP_3 activates 3-phosphoinositide dependent protein kinase-1 (PDK1). Adding to the complexity of insulin signaling, PDK1 also phosphorylates several kinases and thus activates numerous insulin responsive pathways. In terms of insulin-stimulated glucose transport, PDK1 phosphorylates Akt proteins (Akt1-Akt3) as well as aPKC. Before PDK1 activation, Akt translocates to the cell membrane in response to PIP_3 binding in the Akt PH domain. Of the three isoforms, Akt1 and Akt2 appear to regulate glucose transport in skeletal muscle (Jiang et al. 2003; Zhou et al. 2004).

In addition to Akt, which must be activated for GLUT4 translocation, several substrates of Akt are now believed to be important in regulating GLUT4 translocation. Of these, the Akt substrate 160 (AS160) is perhaps the best candidate for involvement in insulin-mediated glucose transport (Larance et al. 2005). Physiological insulin concentrations increase AS160 phosphorylation in human skeletal muscle (Karlsson et al. 2005). AS160 has at least five phosphorylation sites responsive to insulin, and mutation of these sites impairs insulin-mediated GLUT4 translocation (Sano et al. 2003).

In addition to Akt, PDK is responsible for activating atypical PKCξ and PKCλ. Although the precise signaling mechanisms are not known, it appears that PKCξ and PKCλ are signaling molecules mediating GLUT4 translocation (Farese, Sajan, and Standaert 2005). Targeted disruption of PKCξ or PKCλ impairs insulin-mediated glucose transport in both adipose tissue and skeletal muscle.

Although the insulin signaling cascade mediated by PI3K is vital to insulin-mediated glucose transport, there is evidence that activation of PI3K alone does not induce GLUT4 translocation. A novel insulin signaling pathway commonly termed the *Cbl pathway* may also be involved (Chiang, Chang, and Saltiel 2006). Briefly, when insulin binds to its receptor, it stimulates tyrosine phosphorylation of APS, recruitment and tyrosine phosphorylation of Cbl proteins, and Cbl-associated protein (CAP) corecruitment. This activated complex subsequently recruits CrkII and C3G to lipid rafts, where GTP binding activates TC10. Several TC10 downstream targets have been identified, including Exo70, CIP4, and PKCλ. Inhibition of this pathway has yielded inconsistent results regarding insulin-stimulated GLUT4 translocation. While the PI3K and Cbl/TC10 pathways are distinct, collectively they may regulate GLUT4 translocation. Although extensive research has attempted to elucidate the mechanism by which insulin stimulates GLUT4 translocation, more work is needed to discern the interaction of these two pathways.

Glucose uptake in skeletal muscle occurs primarily by facilitated diffusion involving the glucose transporter proteins (GLUT1-GLUT12; figure 4.2). Predominantly expressed in skeletal muscle (Klip and Marette 1992; Stuart et al. 2000), GLUT4 is an insulin-responsive transporter responsible for postprandial glucose uptake (Bell et al. 1993; James, Strube, and Mueckler 1989). Upon insulin stimulation, vesicles containing GLUT4 translocate from an intracellular location to the sarcolemma or to the T-tubules (Wang et al. 1996). The vesicles fuse with the cell membrane, allowing glucose to enter the cell via facilitated diffusion (Klip and Marette 1992). GLUT4 is recycled to intracellular vesicles when the insulin stimulation is removed. The trafficking of GLUT4 is a dynamic process involving several protein-mediated events that include GLUT4 movement along actin filaments and fusion at the plasma membrane. While research is providing new insight into the mechanisms of GLUT4 vesicle formation, translocation, and fusion, more investigation is needed to discern these processes.

In summary, insulin-mediated glucose transport is initiated when insulin binds to its receptor and triggers a series of signaling events. While more than one pathway may be involved, the PI3K pathway is most accepted as the signaling mechanism leading to GLUT4 translocation.

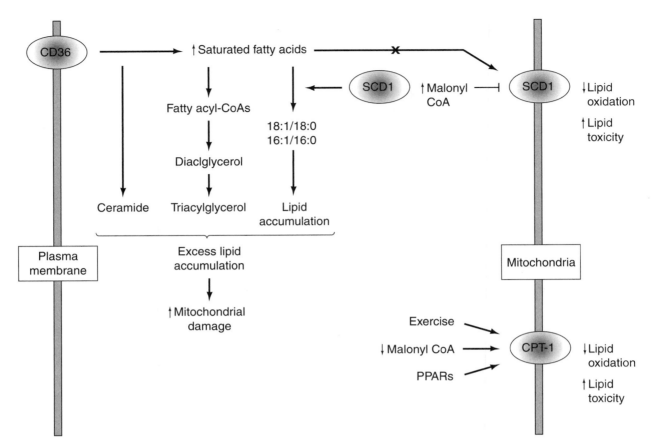

Figure 4.2   Proposed mechanisms of the lipid-induced insulin resistance that occurs with obesity and type 2 diabetes. Increased intracellular accumulation of lipids can inhibit the insulin signaling pathway at several steps. PPase, protein phosphatase.

Does Lipid Exposure Impair Insulin Action?

Conditions that promote intracellular lipid accumulation via lipid oversupply can induce whole-body insulin resistance. For example, fasting can produce whole-body insulin resistance by elevating plasma concentrations of FFA, glycerol, and palmitic acid (Klein et al. 1993; Romijn et al. 1990). Belfort and colleagues (2005) observed an inverse dose–response relationship between plasma FFA concentrations and insulin action when adults who were lean and healthy were given an intravenous infusion of a lipid emulsion. Muscle samples indicated that FFA exposure impaired numerous steps of the insulin signaling process; this decrement in signaling was hypothesized to contribute to insulin resistance (Belfort et al. 2005). These and other data from numerous studies utilizing lipid infusion (Bachmann et al. 2001; Boden 1997; Roden et al. 1996) indicate that increasing the concentration of

plasma lipid, which in turn increases intracellular lipid, impairs insulin-mediated glucose uptake in muscle. Similarly, a high-fat diet induces insulin resistance via the accumulation of intramuscular lipid stores (Bachmann et al. 2001; Stettler et al. 2005). For example, insulin-mediated glucose transport was depressed in muscle strips isolated from rodents fed high-fat diets (compared to glucose transport in their chow-fed counterparts), indicating a defect specific to skeletal muscle (Thompson et al. 2000).

Some of the more convincing evidence for lipid-induced insulin resistance comes from studies utilizing cell culture systems in which a single variable can be manipulated. Schmitz-Peiffer, Craig, and Biden (1999) reported that including fatty acids of various saturations and chain lengths in the incubation media for C2C12 muscle cells impaired insulin-stimulated glucose uptake and glycogen synthesis. Studies have also demonstrated impaired insulin signaling in muscle cell cultures incubated with fatty acids (Chavez et al. 2003; Schmitz-Peiffer, Craig, and Biden 1999). Similarly, insulin action was decreased in rodent muscle strips preincubated with various fatty acids (Thompson et al. 2000). All of these experiments, ranging from cell cultures to rodents to intact human beings, consistently demonstrate that lipid oversupply and accumulation induce an insulin-resistant state in skeletal muscle.

While compelling evidence indicates an interaction between intramuscular lipid accumulation and insulin resistance, the precise cellular mechanism of this interaction is not evident, particularly in relation to obesity and type 2 diabetes. One of the earliest mechanistic explanations was provided by Randle and coworkers (1963) and was based on the premise that lipid oversupply leads to a preferential oxidation of fat that decreases glucose oxidation, which accordingly reduces glucose uptake. However, findings by several independent research groups have cast doubt upon the Randle hypothesis (Hegarty et al. 2003).

A current viable hypothesis explaining the link between intramuscular lipid content and insulin resistance suggests that the accumulation of metabolites from intracellular lipid metabolism interferes with insulin signal transduction (figure 4.2). The initial basis for this theory was the observation of an inverse relationship between insulin action and intramuscular triglyceride content (Hegarty et al. 2003; Kelley, Goodpaster, and Storlien 2002). However, triglyceride is largely inert in mediating intracellular processes; in addition, findings such as elevated intramuscular triglyceride content in endurance athletes, who are very insulin sensitive, imply that intramuscular triglyceride in and of itself is not likely to regulate insulin action (Hegarty et al. 2003; Kelley, Goodpaster, and Storlien 2002).

Such findings have prompted a search for molecules that increase with intramuscular lipid accumulation or lipid oversupply and are more directly responsible for insulin resistance. One candidate includes the long-chain acyl-CoAs (LCACoAs), which are intracellular fatty acids activated by ACS. LCACoA accumulation can directly impair insulin action by inhibiting hexokinase, an event that in turn decreases glucose transport (Cooney et al. 2002). The intracellular presence of LCACoA also activates specific PKC isozymes that

in turn interfere with insulin signal transduction; LCACoAs can also activate or impair transcription factors that control lipid metabolism (Cooney et al. 2002).

Emerging evidence points to other lipid metabolites as the link between intramuscular lipid accumulation and insulin resistance. Ceramide is produced from palmitoyl-CoA or from sphingomyelin, a phospholipid component of cell membranes (Hegarty et al. 2003). It can function as a signaling molecule to control the phosphorylation state and activity of various transcription factors; intramuscular ceramide also increases in states of insulin resistance (Hegarty et al. 2003; Schmitz-Peiffer 2000) and in skeletal muscle from individuals who are obese and insulin resistant (Adams II et al. 2004). Incubation of C2C12 muscle cells with fatty acids increased intracellular ceramide content and concomitantly induced insulin resistance by inhibiting Akt; direct incubation with ceramide produced the same inhibitory effect on insulin action (Chavez et al. 2003; Schmitz-Peiffer, Craig, and Biden 1999).

Another candidate that may be responsible for insulin resistance is DAG, which is a product of both triglyceride and phospholipid metabolism. DAG can activate PKC isozymes that in turn can interfere with insulin signaling (Cortright et al. 2000; figure 4.2). DAG content in skeletal muscle increases with lipid oversupply or insulin resistance, an observation that indicates a potential role in these conditions (Hegarty et al. 2003; Itani et al. 2002; Schmitz-Peiffer 2000). Taken together, these findings indicate that intracellular products that increase with intramuscular lipid accumulation can be linked to insulin resistance. However, the specific interactions of LCACoA, ceramide, DAG, and perhaps even other molecules in relation to insulin action have yet to be clearly defined.

In summary, convincing evidence suggests that intracellular lipid accumulation induces an insulin-resistant state in skeletal muscle. This observation is fundamentally important, as intracellular lipid accumulation occurs with obesity and type 2 diabetes, suggesting a link between intramuscular lipid content and insulin resistance. These findings imply that intramuscular lipid accumulation is a critical contributor to insulin resistance, although the exact relationships remain to be defined.

Perturbations in Substrate Utilization With Type 2 Diabetes and Obesity

A key component of obesity and diabetes is insulin resistance in skeletal muscle (Petersen et al. 1998; Zierath and Wallberg-Henriksson 2002). As discussed earlier, it is thought that insulin resistance may be due to lipid accumulation in skeletal muscle. In this section, we present examples supporting the lipid accumulation hypothesis in individuals who are obese or diabetic.

Diminished Insulin Signal Transduction

Insulin-resistant conditions such as obesity and diabetes are not associated with a decrease in GLUT4 content. Rather, the ability of insulin to stimulate GLUT4 translocation decreases, resulting in a reduced GLUT4 content at the plasma membrane. Such observations suggest that defects in insulin signaling contribute to insulin resistance in the muscle of individuals who are overweight or diabetic.

Numerous steps in the insulin signal cascade are impaired in skeletal muscle from individuals who are obese and insulin resistant. For example, Goodyear and colleagues (1995) showed that insulin receptor phosphorylation, IRS1 phosphorylation, and PI3K activity all decreased with insulin stimulation in muscle taken from subjects who were obese. AS160 phosphorylation decreased in skeletal muscle of subjects with type 2 diabetes compared to skeletal muscle of subjects without diabetes despite similar levels of AS160 content (Karlsson et al. 2005). Akt phosphorylation of threonine (Thr-308) was also reduced in skeletal muscle from people with diabetes, while phosphorylation of serine (Ser-473) was not significantly different (Karlsson et al. 2005). Conflicting reports exist as to whether Akt activation decreases with diabetes. However, this disparity may be explained by differences in the phosphorylation sites and the specific Akt isoform examined. While insulin stimulated the activation of all three Akt isoforms in muscle from controls who were lean, muscle from subjects who were morbidly obese and insulin resistant demonstrated depressed activation of Akt2 and Akt3 (Brozinick, Roberts, and Dohm 2003). These results were supported by the observation that

Akt2 phosphorylation is depressed while Akt1 phosphorylation is normal in muscle from subjects who were obese and type 2 diabetic (Gosmanov et al. 2004). As lipid accretion (lipid infusion and high-fat feeding) in skeletal muscle is linked to adverse insulin signal transduction (figure 4.2), perturbations in lipid metabolism (figure 4.1) likely contribute to insulin resistance.

Impaired Fatty Acid Utilization

Convincing evidence links perturbations in muscle lipid homeostasis to the development of insulin resistance, obesity, and type 2 diabetes. Both diabetes and obesity are associated with diminished insulin-stimulated glucose disposal, with lipid accrual (including DAG, TAG, LCACoA, and ceramide), and with reduced ability to oxidize fatty acids (Adams II et al. 2004; Blaak et al. 2000; Hulver et al. 2003; Kelley and Simoneau 1994; Pan et al. 1997). One candidate contributing to lipid accretion in skeletal muscle could be a disconnect between fatty acid supply and fatty acid utilization (Muoio and Newgard 2006). Both obesity and type 2 diabetes are associated with lowered rates of fat oxidation at rest and during physical activity (Guesbeck et al. 2001; Hickner et al. 2001). The mRNA and activity of proteins responsible for lipid catabolism, as well as the activity of the electron transport system, are decreased in skeletal muscle from subjects who are obese (Hittel et al. 2005; Hulver et al. 2003, 2005; Kelley et al. 1999, 2002; Kim et al. 2000; Lowell and Shulman 2005; Mootha et al. 2003; Patti et al. 2003; Ritov et al. 2005; Simoneau et al. 1999). At least one report indicates that mitochondria from subjects with type 2 diabetes and from individuals who are obese are smaller and less functional (Kelley et al. 2002). Both in vivo and in vitro fatty acid oxidations also decrease with obesity (Berggren et al. 2004). This decline in muscle oxidative capacity partitions lipid toward storage and induces insulin resistance by impairing insulin signal transduction. In vitro models using intact human muscle strips to examine lipid degradation and deposition revealed preferential partitioning of lipids toward synthesis as opposed to oxidation in tissue from individuals who were obese (Hulver et al. 2003). Moreover, cell cultures derived from these individuals displayed this same phenotype of lipid channeling toward storage and diminished capacity for fatty acid oxidation

(Hulver et al. 2005), while offspring of people with type 2 diabetes have increased lipid accumulation in skeletal muscle and decreased mitochondrial activity (Petersen et al. 2004). Thus, lipid oxidation in skeletal muscle can be decreased with diabetes or obesity and lead to lipid accretion.

Augmented Fatty Acid Uptake and Partitioning

In addition to involving a decreased reliance on fatty acid usage, obesity and diabetes may be associated with increased rates of fatty acid uptake. Bonen and coworkers (2004) reported that although total FAT/CD36 content was not different in subjects who were obese or diabetic, these subjects had significantly elevated levels of FAT/CD36 at the plasma membrane. This arrangement promotes increased fatty acid movement into the cell and possibly increases lipid deposition. Human myotubes engineered to overexpress FAT/CD36 display increased fatty acid uptake along with decreased fatty acid oxidation (Garcia-Martinez et al. 2005). In these myotubes, fatty acids were partitioned toward lipid deposition, mimicking an obese phenotype. As insulin has been shown to signal the redistribution of FAT/CD36 to the plasma membrane, it is likely that the hyperinsulinemic environment mediates this process. Thus, skeletal muscle from people who are obese or who have diabetes displays a phenotype of increased capacity for fatty acid uptake accompanied by decreased utilization of the same substrate. This metabolic perturbation may be an important factor causing lipid accretion (Berggren et al. 2004; Blaak 2005).

Other factors may also increase the intramuscular lipid accretion that is evident in insulin-resistant conditions. The mRNA expression of SCD1 is increased in skeletal muscle from subjects who are obese, is positively correlated with BMI and intramuscular triglyceride synthesis, and is negatively correlated with fatty acid oxidation in skeletal muscle (Hulver et al. 2005). Moreover, the metabolic phenotype is maintained in culture, as myotubes derived from obese tissue also display increased mRNA expression and protein content of SCD1 in conjunction with decreased fatty acid oxidation and increased fatty acid esterification. Overexpression of SCD1 resulted in this same obese phenotype of increased channeling of fatty acids

to esterification (Hulver et al. 2005). Deficiency of SCD1 is associated with reduced content of ceramide and LCFA-CoA in skeletal muscle and increased fatty acid oxidation (Dobrzyn et al. 2005). While cell cultures derived from subjects with type 2 diabetes also display decreased fat oxidation, it has not been determined if they demonstrate altered SCD1 regulation (Gaster et al. 2004).

It is not clear if the decreased oxidative capacity observed with obesity and diabetes is a direct cause of lipid accretion or a secondary manifestation of excess lipid influx. The mitochondria, which encode several key proteins of lipid metabolism, are particularly prone to the toxic effects of lipid accumulation, reactive oxygen species (ROS) formation, and lipid peroxidation (figure 4.1). Increasing evidence suggests that obesity and diabetes are associated with a chronic state of low-level inflammation (Houstis, Rosen, and Lander 2006). Regardless of the mechanism, an imbalance between fatty acid supply and fatty acid utilization has detrimental consequences on insulin action. Manipulations that increase mitochondrial fatty acid oxidation affect insulin action favorably, even in the presence of elevated lipid content (Perdomo et al. 2004). Therefore, interventions that improve muscle oxidative capacity, such as exercise training, hold tremendous promise in treating obesity and diabetes.

In summary, skeletal muscle from both subjects who were obese and subjects with diabetes is plagued by lipid accretion that contributes to diminished insulin signal transduction and glucose uptake (figure 4.2). While considerable work is needed to define the causative factors of lipid deposition and insulin resistance, mounting evidence suggests a defect in fatty acid oxidation combined with an increased capacity for fatty acid uptake (figure 4.1). Therapeutic interventions that alleviate this disconnect between fatty acid influx and fatty acid oxidation are thus of critical importance.

The Exercise Paradox

The association between increased intramuscular lipid content and incidence of obesity and type 2 diabetes is well documented (Goodpaster et al. 2000; Jacob et al. 1999). Previous speculation suggested that elevated intramyocellular lipid (IMCL) content was a predictor of insulin resistance in skeletal muscle and a risk factor for type 2 diabetes.

However, researchers (Goodpaster et al. 2001; Bruce et al. 2004) later reported that the association between insulin resistance and IMCL content disappeared in studies of endurance-trained athletes. Thus, an apparent paradox exists in which endurance-trained individuals have enhanced insulin sensitivity despite IMCL levels similar to (Goodpaster et al. 2001) or greater than (van Loon et al. 2004) those observed in individuals with type 2 diabetes.

Van Loon and colleagues (2004) reported that the increased proportion of Type I fibers in endurance-trained individuals could not fully account for the increased IMCL storage observed in these subjects. In addition, lipid droplet size and IMCL distribution within the skeletal muscle fibers were comparable among subjects who were endurance-trained, obese, or type 2 diabetic (van Loon et al. 2004). These similarities suggest that IMCL itself may not be a direct cause of insulin resistance.

A possible explanation for this paradox may lie in the differing ability of skeletal muscle to oxidize intramuscular lipid stores. It is well known that endurance-trained individuals exhibit increased rates of fat oxidation at rest and during submaximal exercise (Goodpaster, Katsiaras, and Kelley 2003; Pruchnic et al. 2004). Also, due to the increased energy demands of repeated bouts of exercise, endurance-trained individuals rely on IMCL as a substrate for energy (Zehnder et al. 2005). Therefore, despite increased storage of IMCL, individuals who are endurance trained appear to have a greater oxidative capacity for lipid and thus a greater IMCL turnover rate compared to their sedentary counterparts.

As a consequence of the longer residence time of IMCL in populations who are obese or diabetic, potentially damaging lipid species such as DAG and ceramide may accumulate. It is hypothesized that the rapid turnover of the lipid stores in individuals who are endurance trained prevents the potentially damaging buildup of fatty acid metabolites that induce insulin resistance. In support of this theory, Bruce and coworkers (2006) reported that endurance training in individuals who were obese reduced ceramide and DAG content and increased fat oxidation, even though skeletal muscle TAG content did not change. Others have reported no change in muscular triglyceride (Gan et al. 2003) or LCACoA content despite improved insulin sensitivity with endurance training (Bruce et al. 2004; Gan et al. 2003).

While the specific explanation of the exercise paradox is still under debate, it appears that IMCL alone does not produce insulin resistance. Instead, the ratio of IMCL to muscle oxidative capacity has been suggested as a more accurate predictor of insulin resistance (van Loon and Goodpaster 2006). In conclusion, endurance exercise training is an effective intervention for insulin resistance. Although not definitive, it appears that the enhanced oxidative capacity of muscle that occurs with physical activity decreases the levels of lipid-associated metabolites. Consequently, individuals who are endurance trained appear to preserve insulin signaling by lowering their levels of damaging lipid metabolites, despite maintained or increased intramuscular triglyceride content.

Effect of Weight Loss on Muscle Lipid Accumulation and Insulin Signaling

Weight loss is commonly recommended for treating diabetes and overweight and can improve insulin action (Goodpaster and Wolf 2004; Houmard et al. 2002). Weight loss can enhance the insulin signaling pathway leading to glucose transport in skeletal muscle, and this effect offers a potential mechanism for improving insulin action (Musi and Goodyear 2006). In conjunction with enhancing insulin action, weight loss also decreases intramuscular triglyceride (Goodpaster and Wolf 2004; Gray et al. 2003) and saturated LCACoA (palmityl- and stearate-CoA) content. Although neither of these molecules may be directly responsible for controlling insulin action, reductions in triglyceride and LCACoA would also decrease active by-products linked with the metabolism of these molecules (i.e., ceramide, DAG). This decrease would concomitantly enhance insulin action. The potentially complex interrelationship of muscle lipid content and metabolism, insulin action, and weight loss, however, remains to be defined.

It is not evident how weight loss decreases intramuscular lipid content. As mentioned earlier in this chapter, the skeletal muscle of individuals who are overweight and insulin resistant can exhibit a decrement in its ability to oxidize lipid. This decrement partitions lipid entering the muscle cell toward storage and leads to insulin resistance (Hulver et al. 2005; Kelley 2005). Weight loss does not reverse this defect in muscular oxidative capacity (Berggren et al. 2004; Goodpaster and Wolf 2004; Gray et al. 2003). Therefore, a mechanism other than change in muscular oxidative capacity is likely responsible for the reduction in intramuscular lipid content that occurs with weight loss; possibilities could be a lowered lipid uptake by the muscle cells (Bonen et al. 2004) or a reduced dietary fat intake with caloric restriction.

In conclusion, weight loss via caloric restriction can improve insulin action. Weight loss is associated with enhanced insulin signaling in skeletal muscle, an effect that may result from a decrease in intramuscular lipid content. This reduction in intramuscular lipid concentration is not likely to be associated with enhanced oxidative capacity of the muscle and may involve other alterations occurring with the intervention.

Concluding Remarks

Mounting evidence suggests that an accumulation of intracellular lipid induces insulin resistance in the skeletal muscle cell. Recent research indicates that this insulin-resistant state is due to lipid accumulation interfering with the insulin signaling pathway within the muscle cell. However, the mechanistic links between lipid accumulation and insulin resistance, particularly in human disease, have yet to be completely elucidated. In the skeletal muscle of individuals who are insulin resistant and overweight, lipid may be preferentially partitioned toward storage via a reduced ability to effectively oxidize lipid, via increased fatty acid transport into the cell, or via other mechanisms. With exercise training, the ability to oxidize lipid in skeletal muscle improves and may contribute to the accompanying enhancement in insulin action, despite elevated intramuscular triglyceride content. Weight loss improves insulin action, an effect that may be linked to a reduction in intramuscular lipid accumulation. Complex and important interrelationships exist among intramuscular lipid metabolism, insulin signal transduction, insulin action, and insulin resistance present with diabetes and obesity.

References

Adams II, J.M., T. Pratipanawatr, R. Berria, E. Wang, R.A. DeFronzo, M.C. Sullards, and L.J. Mandarino. 2004.

Ceramide content is increased in skeletal muscle from obese insulin-resistant humans. *Diabetes* 53:25-31.

Bachmann, O.P., D.B. Dahl, K. Brechtel, J. Machann, M. Haap, T. Maier, M. Loviscach, M. Stumvoll, C.D. Claussen, F. Schick, H.U. Haring, and S. Jacob. 2001. Effects of intravenous and dietary lipid challenge on intramyocellular lipid content and the relation with insulin sensitivity in humans. *Diabetes* 50:2579-84.

Barish, G.D., V.A. Narkar, and R.M. Evans. 2006. PPAR delta: A dagger in the heart of the metabolic syndrome. *J Clin Invest* 116:590-7.

Belfort, R., L. Mandarino, S. Kashyap, K. Wirfel, T. Pratipanawatr, R. Berria, R.A. Defronzo, and K. Cusi. 2005. Dose-response effect of elevated plasma free fatty acid on insulin signaling. *Diabetes* 54:1640-8.

Bell, G.I., C.F. Burant, J. Takeda, and G.W. Gould. 1993. Structure and function of mammalian facilitative sugar transporters. *J Biol Chem* 268:19161-4.

Berggren, J.R., M.W. Hulver, G.L. Dohm, and J.A. Houmard. 2004. Weight loss and exercise: Implications for muscle lipid metabolism and insulin action. *Med Sci Sports Exerc* 36:1191-5.

Bjornholm, M., and J.R. Zierath. 2005. Insulin signal transduction in human skeletal muscle: Identifying the defects in type II diabetes. *Biochem Soc Trans* 33:354-7.

Blaak, E.E. 2005. Metabolic fluxes in skeletal muscle in relation to obesity and insulin resistance. *Best Pract Res Clin Endocrinol Metab* 19:391-403.

Blaak, E.E., A.J. Wagenmakers, J.F. Glatz, B.H. Wolffenbuttel, G.J. Kemerink, C.J. Langenberg, G.A. Heidendal, and W.H. Saris. 2000. Plasma FFA utilization and fatty acid-binding protein content are diminished in type 2 diabetic muscle. *Am J Physiol Endocrinol Metab* 279:E146-54.

Boden, G. 1997. Role of fatty acids in the pathogenesis of insulin resistance and NIDDM. *Diabetes* 46:3-10.

Bonen, A., C.R. Benton, S.E. Campbell, A. Chabowski, D.C. Clarke, X.X. Han, J.F. Glatz, and J.J. Luiken. 2003. Plasmalemmal fatty acid transport is regulated in heart and skeletal muscle by contraction, insulin and leptin, and in obesity and diabetes. *Acta Physiol Scand* 178:347-56.

Bonen, A., M.L. Parolin, G.R. Steinberg, J. Calles-Escandon, N.N. Tandon, J.F. Glatz, J.J. Luiken, G.J. Heigenhauser, and D.J. Dyck. 2004. Triacylglycerol accumulation in human obesity and type 2 diabetes is associated with increased rates of skeletal muscle fatty acid transport and increased sarcolemmal FAT/CD36. *FASEB J* 18:1144-6.

Brozinick Jr., J.T., B.R. Roberts, and G.L. Dohm. 2003. Defective signaling through Akt-2 and -3 but not Akt-1 in insulin-resistant human skeletal muscle: Potential role in insulin resistance. *Diabetes* 52:935-41.

Bruce, C.R., A.D. Kriketos, G.J. Cooney, and J.A. Hawley. 2004. Disassociation of muscle triglyceride content and insulin sensitivity after exercise training in patients with type 2 diabetes. *Diabetologia* 47:23-30.

Bruce, C.R., A.B. Thrush, V.A. Mertz, V. Bezaire, A. Chabowski, G.J.F. Heigenhauser, and D.J. Dyck. 2006. Endurance training in obese humans improves glucose tolerance, mitochondrial fatty acid oxidation and alters muscle lipid content. *Am J Physiol Endocrinol Metab* 291(1): E99-E107.

Cameron-Smith, D., L.M. Burke, D.J. Angus, R.J. Tunstall, G.R. Cox, A. Bonen, J.A. Hawley, and M. Hargreaves. 2003. A short-term, high-fat diet up-regulates lipid metabolism and gene expression in human skeletal muscle. *Am J Clin Nutr* 77:313-8.

Campbell, S.E., N.N. Tandon, G. Woldegiorgis, J.J.F.P. Luiken, J.F.C. Glatz, and A. Bonen. 2004. A novel function for fatty acid translocase (FAT)/CD36: Involvement in long chain fatty acid transfer into the mitochondria. *J Biol Chem* 279:36235-41.

Chang, L., S.H. Chiang, and A.R. Saltiel. 2004. Insulin signaling and the regulation of glucose transport. *Mol Med* 10:65-71.

Chavez, J.A., T.A. Knotts, L.P. Wang, G. Li, R.T. Dobrowsky, G.L. Florant, and S.A. Summers. 2003. A role for ceramide, but not diacylglycerol, in the antagonism of insulin signal transduction by saturated fatty acids. *J Biol Chem* 278:10297-303.

Cheng, L., G. Ding, Q. Qin, Y. Huang, W. Lewis, N. He, R.M. Evans, M.D. Schneider, F.A. Brako, Y. Xiao, Y.E. Chen, and Q. Yang. 2004. Cardiomyocyte-restricted peroxisome proliferator-activated receptor-delta deletion perturbs myocardial fatty acid oxidation and leads to cardiomyopathy. *Nat Med* 10:1245-50.

Chiang, S.H., L. Chang, and A.R. Saltiel. 2006. TC10 and insulin-stimulated glucose transport. *Methods Enzymol* 406:701-14.

Cooney, G.J., A.L. Thompson, S.M. Furler, J. Ye, and E.W. Kraegen. 2002. Muscle long-chain acyl CoA esters and insulin resistance. *Ann NY Acad Sci* 967:196-207.

Cortright, R.N., J.L. Azevedo Jr., Q. Zhou, M. Sinha, W.J. Pories, S.I. Itani, and G.L. Dohm. 2000. Protein kinase C modulates insulin action in human skeletal muscle. *Am J Physiol Endocrinol Metab* 278:E553-62.

Czech, M.P., and S. Corvera. 1999. Signaling mechanisms that regulate glucose transport. *J Biol Chem* 274:1865-8.

DeFronzo, R.A., E. Jacot, E. Jequier, E. Maeder, J. Wahren, and J.P. Felber. 1981. The effect of insulin on the disposal of intravenous glucose. Results from indirect calorimetry and hepatic and femoral venous catheterization. *Diabetes* 30:1000-7.

Dobrzyn, A., P. Dobrzyn, S.H. Lee, M. Miyazaki, P. Cohen, E. Asilmaz, D.G. Hardie, J.M. Friedman, and J.M. Ntambi. 2005. Stearoyl-CoA desaturase-1 deficiency reduces ceramide synthesis by downregulating serine palmitoyltransferase and increasing beta-oxidation in skeletal muscle. *Am J Physiol Endocrinol Metab* 288:E599-607.

Dobrzyn, A., and J.M. Ntambi. 2005. The role of stearoyl-CoA desaturase in the control of metabolism. *Prostaglandins Leukot Essent Fatty Acids* 73:35-41.

Dulloo, A.G., M. Gubler, J.P. Montani, J. Seydoux, and G. Solinas. 2004. Substrate cycling between de novo lipogenesis and lipid oxidation: A thermogenic mechanism against skeletal muscle lipotoxicity and glucolipotoxicity. *Int J Obes Relat Metab Disord* 28:S29-37.

Evans, R.M., G.D. Barish, and Y.X. Wang. 2004. PPARs and the complex journey to obesity. *Nat Med* 10:355-61.

Farese, R.V., M.P. Sajan, and M.L. Standaert. 2005. Atypical protein kinase C in insulin action and insulin resistance. *Biochem Soc Trans* 33:350-3.

Ferre, P. 2004. The biology of peroxisome proliferator-activated receptors: Relationship with lipid metabolism and insulin sensitivity. *Diabetes* 53:S43-50.

Finck, B.N., and D.P. Kelly. 2006. PGC-1 coactivators: Inducible regulators of energy metabolism in health and disease. *J Clin Invest* 116:615-22.

Fredenrich, A., and P.A. Grimaldi. 2005. PPAR delta: An uncompletely known nuclear receptor. *Diabetes Metab* 31:23-27.

Gan, S.K., A.D. Kriketos, B.A. Ellis, C.H. Thompson, E.W. Kraegen, and D.J. Chisholm. 2003. Changes in aerobic capacity and visceral fat but not myocyte lipid levels predict increased insulin action after exercise in overweight and obese men. *Diabetes Care* 26:1706-13.

Garcia-Martinez, C., M. Marotta, R. Moore-Carrasco, M. Guitart, M. Camps, S. Busquets, E. Montell, and A.M. Gomez-Foix. 2005. Impact on fatty acid metabolism and differential localization of FATP1 and FAT/CD36 proteins delivered in cultured human muscle cells. *Am J Physiol Cell Physiol* 288:C1264-72.

Gaster, M., A.C. Rustan, V. Aas, and H. Beck-Nielsen. 2004. Reduced lipid oxidation in skeletal muscle from type 2 diabetic subjects may be of genetic origin: Evidence from cultured myotubes. *Diabetes* 53:542-8.

Glass, C.K. 2006. Going nuclear in metabolic and cardiovascular disease. *J Clin Invest* 116:556-60.

Goodpaster, B.H., J. He, S. Watkins, and D.E. Kelley. 2001. Skeletal muscle lipid content and insulin resistance: Evidence for a paradox in endurance-trained athletes. *J Clin Endocrinol Metab* 86:5755-61.

Goodpaster, B.H., A. Katsiaras, and D.E. Kelley. 2003. Enhanced fat oxidation through physical activity is associated with improvements in insulin sensitivity in obesity. *Diabetes* 52:2191-7.

Goodpaster, B.H., R. Theriault, S.C. Watkins, and D.E. Kelley. 2000. Intramuscular lipid content is increased in obesity and decreased by weight loss. *Metabolism* 49:467-72.

Goodpaster, B.H., and D. Wolf. 2004. Skeletal muscle lipid accumulation in obesity, insulin resistance, and type 2 diabetes. *Pediatr Diabetes* 5:219-26.

Goodyear, L.J., F. Giorgino, L.A. Sherman, J. Carey, R.J. Smith, and G.L. Dohm. 1995. Insulin receptor phosphorylation, insulin receptor substrate-1 phosphorylation, and phosphatidylinositol 3-kinase activity are decreased in intact skeletal muscle strips from obese subjects. *J Clin Invest* 95:2195-204.

Gosmanov, A.R., G.E. Umpierrez, A.H. Karabell, R. Cuervo, and D.B. Thomason. 2004. Impaired expression and insulin-stimulated phosphorylation of Akt-2 in muscle of obese patients with atypical diabetes. *Am J Physiol Endocrinol Metab* 287:E8-15.

Gray, R.E., C.J. Tanner, W.J. Pories, K.G. MacDonald, and J.A. Houmard. 2003. Effect of weight loss on muscle lipid content in morbidly obese subjects. *Am J Physiol Endocrinol Metab* 284:E726-32.

Grimaldi, P.A. 2003. Roles of PPARdelta in the control of muscle development and metabolism. *Biochem Soc Trans* 31:1130-2.

Grimaldi, P.A. 2005. Regulatory role of peroxisome proliferator-activated receptor delta (PPAR delta) in muscle metabolism. A new target for metabolic syndrome treatment? *Biochimie* 87:5-8.

Guesbeck, N.R., M.S. Hickey, K.G. MacDonald, W.J. Pories, I. Harper, E. Ravussin, G.L. Dohm, and J.A. Houmard. 2001. Substrate utilization during exercise in formerly morbidly obese women. *J Appl Physiol* 90:1007-12.

Hajri, T., and N.A. Abumrad. 2002. Fatty acid transport across membranes: Relevance to nutrition and metabolic pathology. *Annu Rev Nutr* 22:383-415.

Hegarty, B.D., S.M. Furler, J. Ye, G.J. Cooney, and E.W. Kraegen. 2003. The role of intramuscular lipid in insulin resistance. *Acta Physiol Scand* 178:373-83.

Hickner, R.C., J. Privette, K. McIver, and H. Barakat. 2001. Fatty acid oxidation in African-American and Caucasian women during physical activity. *J Appl Physiol* 90:2319-24.

Hittel, D.S., Y. Hathout, E.P. Hoffman, and J.A. Houmard. 2005. Proteome analysis of skeletal muscle from obese and morbidly obese women. *Diabetes* 54:1283-8.

Holst, D., S. Luquet, V. Nogueira, K. Kristiansen, X. Leverve, and P.A. Grimaldi. 2003. Nutritional regulation and role of peroxisome proliferator-activated receptor delta in fatty acid catabolism in skeletal muscle. *Biochim Biophys Acta* 1633:43-50.

Hoppeler, H., H. Howald, K. Conley, S.L. Lindstedt, H. Claassen, P. Vock, and E.R. Weibel. 1985. Endurance training in humans: Aerobic capacity and structure of skeletal muscle. *J Appl Physiol* 59:320-7.

Houmard, J.A., C.J. Tanner, C. Yu, P.G. Cunningham, W.J. Pories, K.G. MacDonald, and G.I. Shulman. 2002. Effect of weight loss on insulin sensitivity and intramuscular long-chain fatty acyl-CoAs in morbidly obese subjects. *Diabetes* 51:2959-63.

Houstis, N., E.D. Rosen, and E.S. Lander. 2006. Reactive oxygen species have a causal role in multiple forms of insulin resistance. *Nature* 440:944-8.

Hulver, M.W., J.R. Berggren, M.J. Carper, M. Miyazaki, J.M. Ntambi, E.P. Hoffman, J.P. Thyfault, R. Stevens, G.L. Dohm, J.A. Houmard, and D.M. Muoio. 2005. Elevated stearoyl-CoA desaturase-1 expression in skeletal muscle contributes to abnormal fatty acid partitioning in obese humans. *Cell Metab* 2:251-61.

Hulver, M.W., J.R. Berggren, R.N. Cortright, R.W. Dudek, R.P. Thompson, W.J. Pories, K.G. MacDonald, G.W. Cline, G.I. Shulman, G.L. Dohm, and J.A. Houmard. 2003. Skeletal muscle lipid metabolism with obesity. *Am J Physiol Endocrinol Metab* 284:E741-7.

Itani, S.I., N.B. Ruderman, F. Schmieder, and G. Boden. 2002. Lipid-induced insulin resistance in human muscle is associated with changes in diacylglycerol, protein kinase C, and IkappaB-alpha. *Diabetes* 51:2005-11.

Jacob, S., J. Machann, K. Rett, K. Brechtel, A. Volk, W. Renn, E. Maerker, S. Matthaei, F. Schick, C.D. Claussen, and H.U. Haring. 1999. Association of increased intramyocellular lipid content with insulin resistance in lean nondiabetic offspring of type 2 diabetic subjects. *Diabetes* 48:1113-9.

James, D.E., M. Strube, and M. Mueckler. 1989. Molecular cloning and characterization of an insulin-regulatable glucose transporter. *Nature* 338:83-7.

Jiang, Z.Y., Q.L. Zhou, K.A. Coleman, M. Chouinard, Q. Boese, and M.P. Czech. 2003. Insulin signaling through Akt/protein kinase B analyzed by small interfering RNA-mediated gene silencing. *Proc Natl Acad Sci USA* 100:7569-74.

Karlsson, H.K., J.R. Zierath, S. Kane, A. Krook, G.E. Lienhard, and H. Wallberg-Henriksson. 2005. Insulin-stimulated phosphorylation of the Akt substrate AS160 is impaired in skeletal muscle of type 2 diabetic subjects. *Diabetes* 54:1692-7.

Kelley, D.E. 2005. Skeletal muscle fat oxidation: Timing and flexibility are everything, *J Clin Invest* 115:1699-1702.

Kelley, D.E., B. Goodpaster, R.R. Wing, and J.A. Simoneau. 1999. Skeletal muscle fatty acid metabolism in association with insulin resistance, obesity, and weight loss. *Am J Physiol* 277:E1130-41.

Kelley, D.E., B.H. Goodpaster, and L. Storlien. 2002. Muscle triglyceride and insulin resistance. *Annu Rev Nutr* 22:325-46.

Kelley, D.E., J. He, E.V. Menshikova, and V.B. Ritov. 2002. Dysfunction of mitochondria in human skeletal muscle in type 2 diabetes. *Diabetes* 51:2944-50.

Kelley, D.E., and J.A. Simoneau. 1994. Impaired free fatty acid utilization by skeletal muscle in non-insulin-dependent diabetes mellitus. *J Clin Invest* 94:2349-56.

Kim, J.Y., R.C. Hickner, R.L. Cortright, G.L. Dohm, and J.A. Houmard. 2000. Lipid oxidation is reduced in obese human skeletal muscle. *Am J Physiol Endocrinol Metab* 279:E1039-44.

Klein, S., Y. Sakurai, J.A. Romijn, and R.M. Carroll. 1993. Progressive alterations in lipid and glucose metabolism during short-term fasting in young adult men. *Am J Physiol* 265:E801-6.

Klip, A., and A. Marette. 1992. Acute and chronic signals controlling glucose transport in skeletal muscle. *J Cell Biochem* 48:51-60.

Koves, T.R., P. Li, J. An, T. Akimoto, D. Slentz, O. Ilkayeva, G.L. Dohm, Z. Yan, C.B. Newgard, and D.M. Muoio. 2005. Peroxisome proliferator-activated receptor-gamma co-activator 1alpha-mediated metabolic remodeling of skeletal myocytes mimics exercise training and reverses lipid-induced mitochondrial inefficiency. *J Biol Chem* 280:33588-98.

Krook, A., H. Wallberg-Henriksson, and J.R. Zierath. 2004. Sending the signal: Molecular mechanisms regulating glucose uptake. *Med Sci Sports Exerc* 36:1212-7.

Kuhl, J.E., N.B. Ruderman, N. Musi, L.J. Goodyear, M.E. Patti, S. Crunkhorn, D. Dronamraju, A. Thorell, J. Nygren, O. Ljungkvist, M. Degerblad, A. Stahle, T.B. Brismar, K.L. Andersen, A.K. Saha, S. Efendic, and P.N. Bavenholm. 2006. Exercise training decreases the concentration of malonyl-CoA and increases the expression and activity of malonyl-CoA decarboxylase in human muscle. *Am J Physiol Endocrinol Metab* 290:E1296-303.

Larance, M., G. Ramm, J. Stockli, E.M. van Dam, S. Winata, V. Wasinger, F. Simpson, M. Graham, J.R. Junutula, M. Guilhaus, and D.E. James. 2005. Characterization of the role of the Rab GTPase-activating protein AS160 in insulin-regulated GLUT4 trafficking. *J Biol Chem* 280:37803-13.

Lee, J.S., S.K. Pinnamaneni, S.J. Eo, I.H. Cho, J.H. Pyo, C.K. Kim, A.J. Sinclair, M.A. Febbraio, and M.J. Watt. 2006. Saturated, but not n-6 polyunsaturated, fatty acids induce insulin resistance: Role of intramuscular accumulation of lipid metabolites. *J Appl Physiol* 100:1467-74.

Lefebvre, P., G. Chinetti, J.C. Fruchart, and B. Staels. 2006. Sorting out the roles of PPAR alpha in energy metabolism and vascular homeostasis. *J Clin Invest* 116:571-80.

Lowell, B.B., and G.I. Shulman. 2005. Mitochondrial dysfunction and type 2 diabetes. *Science* 307:384-7.

Luiken, J.J., D.J. Dyck, X.X. Han, N.N. Tandon, Y. Arumugam, J.F. Glatz, and A. Bonen. 2002. Insulin induces the translocation of the fatty acid transporter FAT/CD36 to the plasma membrane. *Am J Physiol Endocrinol Metab* 282:E491-5.

Mootha, V.K., Lindgren, C.M., Eriksson, K.F., Subramanian, A., Sihag, S., Lehar, J., Puigserver, P., Carlsson, E., Ridderstrale, M., Laurila, E., Houstis, N., Daly, M.J., Patterson, N., Mesirov, J.P., Golub, T.R., Tamayo, P., Spiegelman, B., Lander, E.S., Hirschhorn, J.N., Altshuler, D., and Groop, L.C. 2003. PGC-1alpha-responsive genes involved in oxidative phosphorylation are coordinately downregulated in human diabetes. *Nat Genet* 34(3): 267-73.

Muoio, D.M., P.S. MacLean, D.B. Lang, S. Li, J.A. Houmard, J.M. Way, D.A. Winegar, J.C. Corton, G.L. Dohm, and W.E. Kraus. 2002. Fatty acid homeostasis and induction of lipid regulatory genes in skeletal muscles of peroxisome proliferator-activated receptor (PPAR) alpha knock-out mice. Evidence for compensatory regulation by *PPAR delta*. *J Biol Chem* 277:26089-97.

Muoio, D.M., and C.B. Newgard. 2006. Obesity-related derangements in metabolic regulation. *Annu Rev Biochem* 75: 367-401.

Musi, N., and L.J. Goodyear. 2006. Insulin resistance and improvements in signal transduction, *Endocrine* 29:73-80.

Pan, D.A., S. Lillioja, A.D. Kriketos, M.R. Milner, L.A. Baur, C. Bogardus, A.B. Jenkins, and L.H. Storlien. 1997. Skeletal muscle triglyceride levels are inversely related to insulin action. *Diabetes* 46:983-8.

Patti, M.E., A.J. Butte, S. Crunkhorn, K. Cusi, R. Berria, S. Kashyap, Y. Miyazaki, I. Kohane, M. Costello, R. Saccone, E.J. Landaker, A.B. Goldfine, E. Mun, R. DeFronzo, J. Finlayson, C.R. Kahn, and L.J. Mandarino. 2003. Coordinated reduction of genes of oxidative metabolism in humans with insulin resistance and diabetes: Potential role of PGC1 and NRF1. *Proc Natl Acad Sci USA* 100:8466-71.

Perdomo, G., S.R. Commerford, A.M. Richard, S.H. Adams, B.E. Corkey, R.M. O'Doherty, and N.F. Brown. 2004. Increased beta-oxidation in muscle cells enhances insulin-stimulated glucose metabolism and protects against fatty acid-induced insulin resistance despite intramyocellular lipid accumulation. *J Biol Chem* 279:27177-86.

Petersen, K.F., S. Dufour, D. Befroy, R. Garcia, and G.I. Shulman. 2004. Impaired mitochondrial activity in the insulin-resistant offspring of patients with type 2 diabetes. *New Engl J Med* 350:664-71.

Petersen, K.F., R. Hendler, T. Price, G. Perseghin, D.L. Rothman, N. Held, J.M. Amatruda, and G.I. Shulman. 1998. 13C/31P NMR studies on the mechanism of insulin resistance in obesity. *Diabetes* 47:381-6.

Powell, D.J., S. Turban, A. Gray, E. Hajduch, and H.S. Hundal. 2004. Intracellular ceramide synthesis and protein kinase C zeta activation play an essential role in palmitate-induced insulin resistance in rat L6 skeletal muscle cells. *Biochem J* 382:619-29.

Pruchnic, R., A. Katsiaras, J. He, D.E. Kelley, C. Winters, and B.H. Goodpaster. 2004. Exercise training increases intramyocellular lipid and oxidative capacity in older adults. *Am J Physiol Endocrinol Metab* 287:E857-62.

Randle, P.J., P.B. Garland, C.N. Hales, and E.A. Newsholme. 1963. The glucose fatty-acid cycle. Its role in insulin sensitivity and the metabolic disturbances of diabetes mellitus. *Lancet* 1:785-9.

Rasmussen, B.B., U.C. Holmback, E. Volpi, B. Morio-Liondore, D. Paddon-Jones, and R.R. Wolfe. 2002. Malonyl coenzyme A and the regulation of functional carnitine palmitoyltransferase-1 activity and fat oxidation in human skeletal muscle. *J Clin Invest* 110:1687-93.

Rasmussen, B.B., and R.R. Wolfe. 1999. Regulation of fatty acid oxidation in skeletal muscle. *Annu Rev Nutr* 19:463-84.

Reaven, G.M. 1988. Banting lecture 1988. Role of insulin resistance in human disease. *Diabetes* 37:1595-607.

Ritov, V.B., E.V. Menshikova, J. He, R.E. Ferrell, B.H. Goodpaster, and D.E. Kelley. 2005. Deficiency of subsarcolemmal mitochondria in obesity and type 2 diabetes. *Diabetes* 54:8-14.

Roden, M., T.B. Price, G. Perseghin, K.F. Petersen, D.L. Rothman, G.W. Cline, and G.I. Shulman. 1996. Mechanism of free fatty acid-induced insulin resistance in humans. *J Clin Invest* 97:2859-65.

Romijn, J.A., M.H. Godfried, M.J. Hommes, E. Endert, and H.P. Sauerwein. 1990. Decreased glucose oxidation during short-term starvation. *Metabolism* 39:525-30.

Ruderman, N.B., A.K. Saha, D. Vavvas, and L.A. Witters. 1999. Malonyl-CoA, fuel sensing, and insulin resistance. *Am J Physiol* 276:E1-18.

Sano, H., S. Kane, E. Sano, C.P. Miinea, J.M. Asara, W.S. Lane, C.W. Garner, and G.E. Lienhard. 2003. Insulin-stimulated phosphorylation of a Rab GTPase-activating protein regulates GLUT4 translocation. *J Biol Chem* 278:14599-602.

Schaffer, J.E. 2002. Fatty acid transport: The roads taken. *Am J Physiol Endocrinol Metab* 282:E239-46.

Schmitz-Peiffer, C. 2000. Signalling aspects of insulin resistance in skeletal muscle: Mechanisms induced by lipid oversupply. *Cell Signal* 12:583-94.

Schmitz-Peiffer, C., D.L. Craig, and T.J. Biden. 1999. Ceramide generation is sufficient to account for the inhibition of the insulin-stimulated PKB pathway in C2C12 skeletal muscle cells pretreated with palmitate. *J Biol Chem* 274:24202-10.

Schoonjans, K., B. Staels, and J. Auwerx. 1996. Role of the peroxisome proliferator-activated receptor (PPAR) in mediating the effects of fibrates and fatty acids on gene expression. *J Lipid Res* 37:907-25.

Semple, R.K., V.K. Chatterjee, and S. O'Rahilly. 2006. PPAR gamma and human metabolic disease. *J Clin Invest* 116:581-9.

Simoneau, J.A., J.H. Veerkamp, L.P. Turcotte, and D.E. Kelley. 1999. Markers of capacity to utilize fatty acids in human skeletal muscle: Relation to insulin resistance and obesity and effects of weight loss. *FASEB J* 13:2051-60.

Spriet, L.L., R.J. Tunstall, M.J. Watt, K.A. Mehan, M. Hargreaves, and D. Cameron-Smith. 2004. Pyruvate dehydrogenase activation and kinase expression in human skeletal muscle during fasting. *J Appl Physiol* 96:2082-7.

Stettler, R., M. Ith, K.J. Acheson, J. Decombaz, C. Boesch, L. Tappy, and C. Binnert. 2005. Interaction between dietary lipids and physical inactivity on insulin sensitivity and on intramyocellular lipids in healthy men. *Diabetes Care* 28:1404-9.

Stuart, C.A., G. Wen, W.C. Gustafson, and E.A. Thompson. 2000. Comparison of GLUT1, GLUT3, and GLUT4 mRNA and the subcellular distribution of their proteins in normal human muscle. *Metabolism* 49:1604-9.

Sun, X.J., P. Rothenberg, C.R. Kahn, J.M. Backer, E. Araki, P.A. Wilden, D.A. Cahill, B.J. Goldstein, and M.F. White. 1991. Structure of the insulin receptor substrate IRS-1 defines a unique signal transduction protein. *Nature* 352:73-7.

Thompson, A.L., M.Y. Lim-Fraser, E.W. Kraegen, and G.J. Cooney. 2000. Effects of individual fatty acids on glucose uptake and glycogen synthesis in soleus muscle in vitro. *Am J Physiol Endocrinol Metab* 279:E577-84.

van Loon, L.J., and B.H. Goodpaster. 2006. Increased intramuscular lipid storage in the insulin-resistant and endurance-trained state. *Pflugers Arch* 451:606-16.

van Loon, L.J., R. Koopman, R. Manders, W. van der Weegen, G.P. van Kranenburg, and H.A. Keizer. 2004. Intramyocellular lipid content in type 2 diabetes patients compared with overweight sedentary men and highly trained endurance athletes. *Am J Physiol Endocrinol Metab* 287:E558-65.

Wang, W., P.A. Hansen, B.A. Marshall, J.O. Holloszy, and M. Mueckler. 1996. Insulin unmasks a COOH-terminal Glut4 epitope and increases glucose transport across T-tubules in skeletal muscle. *J Cell Biol* 135:415-30.

Wang, Y.X., C.H. Lee, S. Tiep, R.T. Yu, J. Ham, H. Kang, and R.M. Evans. 2003. Peroxisome-proliferator-activated receptor delta activates fat metabolism to prevent obesity. *Cell* 113:159-70.

Wang, Y.X., C.L. Zhang, R.T. Yu, H.K. Cho, M.C. Nelson, C.R. Bayuga-Ocampo, J. Ham, H. Kang, and R.M. Evans. 2004. Regulation of muscle fiber type and running endurance by PPARdelta. *PLoS Biol* 2:e294.

Watson, R.T., and J.E. Pessin. 2006. Bridging the GAP between insulin signaling and GLUT4 translocation. *Trends Biochem Sci* 31:215-22.

White, M.F., J.N. Livingston, J.M. Backer, V. Lauris, T.J. Dull, A. Ullrich, and C.R. Kahn. 1988. Mutation of the insulin receptor at tyrosine 960 inhibits signal transmission but does not affect its tyrosine kinase activity. *Cell* 54:641-9.

Zehnder, M., M. Ith, R. Kreis, W. Saris, U. Boutellier, and C. Boesch. 2005. Gender-specific usage of intramyocellular lipids and glycogen during exercise. *Med Sci Sports Exerc* 37:1517-24.

Zhou, Q.L., J.G. Park, Z.Y. Jiang, J.J. Holik, P. Mitra, S. Semiz, A. Guilherme, A.M. Powelka, X. Tang, J. Virbasius, and M.P. Czech. 2004. Analysis of insulin signalling by RNAi-based gene silencing. *Biochem Soc Trans* 32:817-21.

Zierath, J.R., and H. Wallberg-Henriksson. 2002. From receptor to effector: Insulin signal transduction in skeletal muscle from type II diabetic patients. *Ann N Y Acad Sci* 967:120-34.

Chapter 5

Metabolic Inflexibility and Insulin Resistance in Skeletal Muscle

Bret H. Goodpaster, PhD; and David E. Kelley, MD

Skeletal muscle utilizes both fatty acids and carbohydrate under normal physiological conditions. The extent to which healthy muscle relies relatively more or less on fatty acids or carbohydrate depends on a variety of factors. These include metabolic demands, substrate availability, hormonal milieu, and capacity for efficient delivery, transport, and utilization of substrate. Skeletal muscle adapts to two quite different physiological conditions—reduced energy intake during fasting and increased energy expenditure during exercise—by increasing its reliance on fat oxidation (Henriksson 1995). Greater reliance on fat oxidation preserves plasma glucose for brain utilization during fasting and delays consumption of muscle glycogen during exercise. These are well-recognized homeostatic adaptations. Indeed, over 50 y ago, Andres, Cader, and Zierler (1956) reported that in volunteers who are lean and healthy, skeletal muscle chiefly relies on fat oxidation after just an overnight fast.

This ability to adapt to a variety of physiological conditions, termed *metabolic flexibility*, is not a new concept in normal physiology but has recently received worthy attention due to its link with the etiology, prevention, and treatment of insulin resistance in obesity and type 2 diabetes (Kelley and Mandarino 2000; Storlien, Oakes, and Kelley 2004). This chapter focuses on the role of flexibility in fuel utilization in response to increased metabolic demands in healthy individuals and, conversely, looks at how metabolic *inflexibility* has been implicated in obesity, insulin resistance, and type 2 diabetes. In addition, it discusses how weight loss and exercise, two first-line therapies for these pathophysiological states, alter various aspects of metabolic inflexibility.

Substrate Utilization During Rest in Individuals Who Are Lean and Healthy

Before we can discuss aspects of impaired substrate metabolism related to the pathophysiology of obesity and type 2 diabetes, we must first describe normal fuel metabolism under conditions of various metabolic demands. Skeletal muscle can oxidize either lipid or carbohydrate to yield energy. During postabsorptive conditions, as occur after an overnight fast, skeletal muscle of lean persons predominantly relies on lipid oxidation (Andres, Cadar, and Zierler 1956; Baltzan et al. 1962). During fasting conditions, there is also a high rate of extraction of plasma FFA by skeletal muscle of approximately 40% (Dagenais, Tancredi, and Zierler 1976). Oxidation of plasma FFA taken up by muscle, if these fatty acids were to be completely oxidized, would account for nearly 80% of resting oxygen consumption by muscle. Even in isolated muscle cell cultures derived from biopsy of human muscle, the capacity for fat oxidation is positively related to body leanness and aerobic fitness (Ukropcova et al. 2005). Thus, it is clear that skeletal muscle can play an important role in systemic patterns of fatty acid utilization, especially during postabsorptive metabolism.

The metabolic flexibility characteristic of individuals who are lean and healthy essentially has two facets. One aspect is the ability to respond to conditions favoring increased glucose utilization or storage, such as those after a meal in which insulin levels are high and in which glucose storage would be beneficial. The other side of metabolic flexibility is

The authors would like to thank the National Institutes of Health and the American Diabetes Association for supporting our research.

the ability to increase reliance on fat oxidation. Insulin normally suppresses fat oxidation and stimulates glucose oxidation, while fasting stimulates a reliance on fat oxidation. This ability to shift toward greater fat oxidation is useful to preserve valuable glucose for brain functioning during fasting conditions. Another condition that can be implicated with a metabolic flexibility is that of exercise, in which both fat and carbohydrate oxidation are increased.

Response to Acute Exercise

Total energy expenditure can increase nearly 20-fold during intense exercise. Both the intensity and duration of exercise affect the mix of substrate utilization. With increasing exercise intensity there is normally a dramatic shift from predominantly fat to exclusively glucose utilization (Friedlander et al. 1998; Brooks 1997). This ability to shift from fat to carbohydrate is metabolic flexibility, which is influenced by a variety of factors, including the hormonal milieu, substrate availability, pH of muscle and blood, and capacity for substrate delivery and oxidation. While this shift is common in individuals who are lean and healthy, little is known regarding whether states of insulin resistance such as obesity and type 2 diabetes exhibit as robust a shift in fuel use with increasing exercise intensity.

Insulin Resistance

The normal response to physiological levels of insulin is to stimulate the uptake and utilization of plasma glucose in tissues. A person with this response intact is insulin sensitive. The converse of insulin sensitivity is insulin resistance, which is defined clinically as the relative inability of insulin to increase glucose uptake and utilization. Insulin acts by binding to its plasma membrane receptor, initiating a cascade of intracellular postreceptor protein interactions known as *insulin signaling*. It is within this intracellular signaling pathway that defects leading to insulin resistance and type 2 diabetes occur. Several mechanisms have been proposed as possible causes of insulin resistance. These include (a) obesity, particularly increases in abdominal obesity and nonadipose tissue lipids within liver and skeletal muscle; (b) physical inactivity; (c) genetic defects affecting proteins involved in insulin action; and (d) fetal malnutrition.

Substrate Utilization in Individuals Who Are Insulin Resistant

Several studies have begun to address whether classical states of insulin resistance such as obesity and type 2 diabetes are characterized and perhaps even mechanistically linked by patterns of altered lipid utilization. In 1963, Randle spawned a series of subsequent investigations over the following four decades by examining the biochemical mechanisms involved in the switch from carbohydrate to fat oxidation (Randle et al. 1963).

The key aspects of Randle's model were that increased fat oxidation in muscle results in the accumulation of acetyl-CoA and citrate, thereby inhibiting pyruvate dehydrogenase (PDH) and phosphofructokinase, respectively. This inhibition of the glycolytic pathway then increases glucose-6-phosphate concentrations, inhibiting hexokinase and resulting in reduced glucose uptake and oxidation. This homeostatic mechanism became known as the *glucose–fatty acid cycle* or the *Randle cycle*.

The link between high rates of fat oxidation and insulin resistance was supported by evidence that people with type 1 diabetes have elevated plasma FFA and triglyceride concentrations as well as high rates of fat oxidation. However, some studies cast doubt on whether this glucose–fatty acid cycle was operative in all tissues under all circumstances (Rennie and Holloszy 1977; Zorzano et al. 1985). Their findings led to a substantial effort to determine whether the glucose–fatty acid cycle is responsible for human insulin resistance and type 2 diabetes.

The advent of the euglycemic-hyperinsulinemic glucose clamp enabled a number of studies over the next 30 y to demonstrate that maintaining or increasing plasma FFA concentrations during an insulin infusion inhibits insulin-stimulated glucose uptake, as predicted by the glucose–fatty acid cycle. Given that the original glucose–fatty acid cycle also predicted that excess fatty acid inhibits glucose oxidation, other studies were conducted to determine whether excess glucose availability could also inhibit lipid oxidation.

Kelley and Mandarino (1990) found that, in contrast to the predictions of the original glucose–fatty acid cycle, under conditions of postabsorptive hyperglycemia glucose oxidation is higher and fat oxidation is concomitantly lower in muscle of people with type 2 diabetes. Induction of hyperglycemia in people without diabetes produced a similar pattern, and this effect was exacerbated by obesity (Kelley et al. 1993a; Mandarino et al. 1996). Several other studies indicated that glucose inhibits fat oxidation (Sidossis et al. 1996; Wolfe et al. 1988), which is the reverse of the Randle cycle. This finding could be pertinent to the observation that insulin-resistant skeletal muscle in animal models of obesity has increased malonyl-CoA content (Winder et al. 1990), which would inhibit CPT and thus fatty acid oxidation (McGarry 1995).

Fatty Acids Induce Insulin Resistance

Skeletal muscle plays a key role in determining systemic insulin sensitivity because under insulin-stimulated conditions, a major proportion of glucose utilization occurs in muscle. Impaired glucose utilization in muscle determines the severity of systemic insulin resistance in common metabolic diseases such as type 2 diabetes and obesity. In order to evaluate whether the impaired capacity to increase reliance on fat oxidation relates to the pathogenesis of insulin resistance, we must first address the apparently contradictory observation that insulin resistance in muscle can be induced by elevated plasma fatty acid levels (Boden et al. 1991, 1994, 2001) and that such elevations are associated with increased fat oxidation. Even the experimental maintenance of fasting levels of plasma fatty acids, which prevent the suppression that normally occurs during elevated insulin levels, is sufficient to significantly lower insulin-stimulated glucose uptake by skeletal muscle, impair insulin suppression of lipid oxidation, and blunt stimulation of glucose oxidation in muscle (Kelley et al. 1993a).

These experimental conditions, which induce insulin resistance in skeletal muscle by elevating levels of plasma fatty acids, have a strong congruence with traits found in obesity and type 2 diabetes (Lillioja et al. 1985). The strength and consistency of such observations make it challenging to reconcile them with the notion that insulin-resistant skeletal muscle can also manifest reduced efficiency of fat oxidation during fasting metabolism despite the elevated levels of plasma FFA usually present in obesity and type 2 diabetes.

Fat Oxidation by Muscle as a Component of Insulin Sensitivity

One of the first links between insulin resistance and impaired fat oxidation in skeletal muscle was observed by Colberg and colleagues (1995), who found that postabsorptive rates of FFA utilization by muscle were diminished in relation to visceral obesity despite elevated circulating concentrations of plasma fatty acids. This study, together with studies by Blaak and coworkers (Blaak et al. 2000; Mensink et al. 2001), provided some of the first evidence that defects of lipid utilization and defects of insulin-stimulated glucose utilization might occur together. These observations are consistent with previous studies by Ravussin and colleagues, who found that obesity is associated with an impaired capacity for oxidizing fat calories (Zurlo et al. 1990). These studies included data indicating that lower rates of fat oxidation during fasting conditions predict subsequent weight gain over several years (Ravussin and Gautier 1999).

Recently, Kelley and colleagues examined fasting patterns of lipid metabolism and insulin-stimulated conditions in obesity (Kelley et al. 1999). In subjects who were lean, insulin significantly stimulated glucose oxidation, a classic insulin-sensitive response. In subjects who were obese, however, the response to insulin was severely blunted, representing insulin resistance (figure 5.1). During insulin-stimulated conditions, subjects who were obese also failed to suppress fat oxidation, such that the rate of lipid oxidation was unchanged from that of fasting conditions. During fasting conditions, persons who are obese have a reduced capacity for fatty acid oxidation compared with individuals who are lean, as illustrated by their higher respiratory exchange quotient (RQ). In response to insulin, individuals who are obese have an impaired ability to stimulate glucose oxidation (lower RQ), which is classic insulin resistance. Thus in obesity there is a diminished capacity to increase both fatty acid and glucose metabolism according to metabolic demand. These findings are not disparate but are interconnected pieces of the puzzle of how insulin resistance manifests within skeletal muscle in obesity. The concept that links these two findings is one of metabolic flexibility as a component of insulin sensitivity in lean individuals and of metabolic

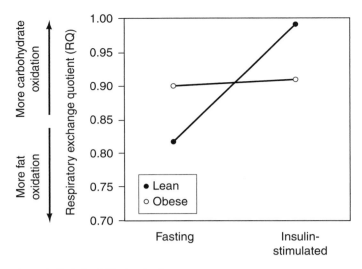

Figure 5.1 Two components of metabolic inflexibility in obesity.

inflexibility as a component of insulin resistance in obesity.

Potential Cellular Mechanisms of Metabolic Flexibility in Fat Oxidation

Characteristics of muscle insulin resistance are retained in myocyte culture (Henry, Abrams, and Nikoulina 1995; Henry et al. 1996), and it has been reported that muscle obtained from patients with type 2 diabetes has a reduced capacity for fat oxidation in culture (Gaster et al. 2004). What are the cellular characteristics of myocytes that determine metabolic flexibility? Recent studies performed in human muscle cell cultures indicate that decreased mitochondrial function is related to reduced fatty acid oxidation by muscle (Ukropcova et al. 2005). This finding provides additional insight why properties of metabolic flexibility in the transition between fat and glucose oxidation that are manifested in vivo are retained in myocyte cultures (Ukropcova et al. 2005). These studies serve as an impetus for further research into these properties even as they suggest that mitochondria may play a critical role. It will be of great interest to more fully examine mitochondrial metabolism as well as to undertake broader analyses of known metabolic pathways in order to elucidate the mechanisms that permit or constrain the metabolic adaptation of skeletal muscle toward a high rate of fat oxidation.

Skeletal Muscle Lipid Accumulation and Insulin Resistance of Obesity and Type 2 Diabetes

Despite similar rates of FFA uptake across the leg between subjects who were lean and subjects who were obese, rates of fat oxidation across the leg during fasting conditions have been shown to be slower in obesity (Kelley et al. 1999). In both groups of subjects the fasting rates of fatty acid uptake across the leg exceeded rates of lipid oxidation, but the net storage of fatty acids was greater in obesity. Thus, this study suggested a paradigm in the setting of obesity: Skeletal muscle accrues triglyceride due to a reduced rate of lipid oxidation in obesity despite similar rates of fatty acid uptake compared with leanness.

An increased fatty acid content in skeletal muscle has been associated with insulin resistance (Goodpaster and Kelley 2002). Excess fatty acyl-CoAs can lead to increased muscle triglyceride and diacylglycerol (DAG) concentrations. DAG, in turn, activates many isoforms of PKC. PKC, a serine kinase, can phosphorylate and inhibit the tyrosine kinase activity of the insulin receptor as well as the tyrosine phosphorylation of IRS1 (Laybutt et al. 1999; Schmitz-Peiffer 2000, 2002; Schmitz-Peiffer, Craig, and Biden 1999). Other fatty acid derivatives have been implicated in altered insulin signaling. For example, ceramide, a sphingolipid derivative of palmitate, inhibits insulin stimulation of glycogen synthase kinase 3 (GSK3) and Akt in a manner similar to that produced by palmitate itself (Schmitz-Peiffer, Craig, and Biden 1999). It is very likely that these lipid metabolites within muscle are more directly associated with lipid-induced insulin resistance than are triglycerides themselves.

Decreased reliance on lipid oxidation during fasting conditions and excess lipid storage within muscle extend the phenotype of insulin resistance in skeletal muscle because they reveal metabolic defects beyond those of insulin-stimulated metabolism. However, it is impossible to determine whether these observations in humans reflect mechanistic links between impaired fatty acid metabolism and insulin resistance or are merely underlying causes of both components of metabolic flexibility. Moreover, it is not clear as to what extent these defects are modifiable. Evidence from clinical interventions involving weight loss and physical activity can perhaps shed some light on these provocative questions.

Effects of Weight Loss on Metabolic Flexibility in Obesity and Type 2 Diabetes

Modest weight loss of 5% to 10% can significantly improve hyperglycemia and reduce other cardiovascular risk factors in persons with type 2 diabetes (Kelley et al. 1993b; Sjostrom et al. 2000; Williams and Kelley 2000). Similar amounts of weight loss can also substantially improve insulin resistance in obese individuals with (Williams and Kelley 2000; Dagogo-Jack and Santiago 1997; Kelley et al. 2004) and without (Kelley et al. 1999; Colman et al. 1995; Su et al. 1995; Houmard et al. 2002) type 2 diabetes. These observations emphasize that negative energy balance benefits insulin resistance and hyperglycemia as well as hypertriglyceridemia, as becomes evident almost at the outset of negative caloric balance. The fact that negative energy balance can rapidly ameliorate insulin resistance and reduce hyperglycemia demonstrates the potential reversibility of insulin resistance. Thus, weight loss can have a profound effect on type 2 diabetes, as has been revealed by the use of bariatric surgery to reverse type 2 diabetes. Sustained weight reduction over several years as occurs following gastric bypass or gastric banding procedures reverses, or at least nearly resolves, type 2 diabetes in many individuals (Cummings et al. 2005).

The mechanisms governing the improvements in insulin sensitivity that occur with diet-induced weight loss are multifactorial and too numerous to review here. It is, however, pertinent to discuss here the effects of weight loss on the metabolic inflexibility of obesity, particularly the effects on both insulin resistance and fatty acid metabolism. Improvements in insulin sensitivity induced by weight loss are associated with enhanced insulin signaling and insulin-stimulated transport, storage, and oxidation of glucose (Goodpaster et al. 1999; Henry, Wallace, and Olefsky 1986). It is less clear whether improved insulin resistance is mechanistically linked with alterations in fatty acid metabolism accompanying weight loss.

Weight Loss and Fatty Acid Metabolism

If weight loss can improve one component of metabolic flexibility—insulin-stimulated glucose metabolism—can it also improve dysregulated fat metabolism? Clearly, weight loss can reduce fatty acid supply, as is evidenced by lower concentrations of plasma FFAs (Goodpaster et al. 1999), triglycerides (Goodpaster et al. 1999), and cholesterol (Goodpaster et al. 1999). Weight loss can also decrease the amount of lipid contained within liver (Barker et al. 2006) and skeletal muscle (Goodpaster et al. 2000; Gray et al. 2003; Greco et al. 2002), both strong correlates of insulin resistance in obesity and type 2 diabetes. As discussed, insulin resistance induced by fatty acid has been well described in human studies (Boden et al. 2001), cell cultures (Schmitz-Peiffer, Craig, and Biden 1999), and animal models (Yu et al. 2002). It is less clear, however, whether alterations in fatty acid metabolism, namely that of fatty acid oxidation by muscle as a component of metabolic flexibility, are associated with enhanced insulin sensitivity.

In individuals who were overweight or obese, diet-induced weight loss improved insulin-stimulated glucose utilization but did not alter fasting rates of lipid oxidation (figure 5.2; Kelley et al. 1999). Similar results have been observed following bariatric surgery in patients who were obese, in whom insulin sensitivity, but not skeletal muscle fatty acid metabolism, is improved with dramatic weight loss (Houmard et al. 2002; Guesbeck et al. 2001). Interestingly, however, moderate weight loss achieved without a change in aerobic fitness *does* appear to improve suppression of fat oxidation during insulin-stimulated conditions (Raben, Mygind, and Astrup 1998; Simoneau et al. 1999). The inability of weight loss to influence fat oxidation by skeletal muscle indicates that impaired fatty acid oxidation may not be a function of obesity per se. It might also suggest that physical inactivity is the culprit behind impaired fat oxidation in obesity.

Effects of Exercise Training on Metabolic Flexibility in Obesity and Type 2 Diabetes

Studies examining the effects of exercise training, with or without weight loss, have provided key insight into the metabolic inflexibility of obesity. It is often difficult to tease apart the underlying effects of physical inactivity versus the effects of energy oversupply on the development of obesity and type 2 diabetes. Put simply, a positive energy balance leading to obesity can result from reduced energy expenditure, from increased energy intake, or from both. The effects of aerobic exercise on the capacity for fat oxidation in the muscle illustrate that insulin-sensitive skeletal muscle manifests a robust reliance on fat oxidation during postabsorptive metabolism, yet it also manifests a keen reliance on glucose oxidation during insulin-stimulated conditions (figure 5.2; Goodpaster, Katsiaras, and Kelley 2003). An increase in the capacity for fat oxidation can occur concurrently with an increase in the capacity for insulin-mediated glucose uptake and metabolism.

Diet-induced weight loss increases insulin sensitivity, as is reflected by an increase in insulin-stimulated glucose oxidation (higher respiratory exchange quotient, RQ). However, weight loss has little effect on rates of fatty acid oxidation during fasting conditions. Exercise training restores both components of the metabolic inflexibility of obesity: Chronic exercise increases utilization of fat calories during fasting and also increases insulin-stimulated glucose oxidation (enhances insulin sensitivity). Indeed, in obese subjects undergoing a combination of diet-induced weight loss and exercise training, increased rates of postabsorptive fatty acid oxidation was the strongest predictor of improved insulin-stimulated glucose metabolism, even after accounting for changes in total body fat, abdominal adipose tissue, and intramyocellular lipid (Goodpaster, Katsiaras, and Kelley 2003). This finding suggests that adding exercise training to weight loss promotes the ability of the muscle to alter the utilization of both fat and glucose depending upon the metabolic demands. That is, it increases the overall metabolic flexibility in obesity. However, the question remains as to whether these concomitant benefits of increased physical activity are correlated or occur independently.

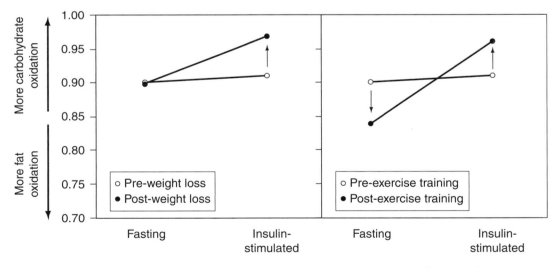

Figure 5.2 Effects of weight loss (left panel) and exercise training (right panel) on the metabolic inflexibility of obesity.

Mitochondria, Muscle Lipid Accumulation, and Insulin Resistance

It is often postulated that a diminished capacity for fatty acid utilization resulting from reduced mitochondrial function in skeletal muscle may lead to the accumulation of lipids within muscle, thereby creating insulin resistance (Lowell and Shulman 2005). We must first consider whether mitochondrial dysfunction and metabolic inflexibility are inherent or acquired defects. Cross-sectional human studies (Petersen et al. 2003, 2004) have suggested that a heritable deficit in mitochondria leads to an accumulation of muscle lipids and subsequent insulin resistance. On the other hand, studies demonstrating improved mitochondrial function, fatty acid oxidation, and insulin-stimulated glucose metabolism with exercise training support the concept of acquired defects (Goodpaster, Katsiaras, and Kelley 2003; Toledo, Watkins, and Kelley 2006).

While several different methods (histology, bioimaging, and biochemical techniques) have shown an association between insulin resistance and increased intramyocellular lipid content (Dube and Goodpaster 2006), in athletes who are lean, insulin sensitive, and endurance trained, muscle lipid content is relatively high (Bruce et al. 2003; Goodpaster et al. 2001). Thus, exercise training enhances sensitivity to insulin while at the same time increases both the storage and oxidation of triglycerides within the muscle. There are several possible explanations for why this paradox might exist.

First, one of the striking features of lipid staining patterns in skeletal muscle cells, especially human skeletal muscle cells, is the considerable heterogeneity among muscle fibers in the amount of lipid staining. This heterogeneity is related to muscle fiber type. In general, Type I, or highly oxidative, muscle fibers housing many mitochondria contain greater lipid than Type II fibers contain and are more insulin sensitive (van Loon et al. 2004; figure 5.3). This observation does not fit with the notion that a lower mitochondria content leads to a higher muscle triglyceride content and that intramuscular triglycerides themselves contribute to insulin resistance. It is, however, consistent with the observed paradox that endurance training increases the amount of intramyocellular lipid, the proportion of Type I fibers, and insulin sensitivity.

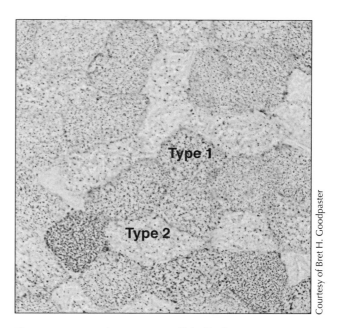

Courtesy of Bret H. Goodpaster

Figure 5.3 A paradox? Intramyocellular lipid content is greater in Type I muscle fibers. Percutaneous muscle biopsy samples are taken from skeletal muscles of interest, here in the vastus lateralis. Serial sections are stained for neutral lipid, mostly triglyceride, using Oil Red O. In serial sections for fiber type determination, the lipid content can be calculated according to fiber type. Type I, or highly oxidative, fibers contain more mitochondria and typically have a greater lipid content than Type II fibers have, as is depicted on this representative micrograph.

Muscle fiber type (Nyholm et al. 1997) and type 2 diabetes (Nyholm et al. 1996, 1997; Bogardus 1993; Lillioja and Bogardus 1988) both have a genetic component. Is the higher muscle triglyceride content observed in humans with obesity or type 2 diabetes related to the distribution of muscle fiber type? Several important studies have examined whether fiber type distribution differs in obesity and whether fiber type distribution is related to the pathogenesis of insulin resistance. Several studies have reported that a higher proportion of total body fat is associated with a low percentage of Type I muscle fibers (Helge et al. 1999; Lillioja et al. 1987; Wade, Marbut, and Round 1990). On the other hand, other studies (Simoneau and Bouchard 1995) have shown no significant relationship between the proportion of Type I fibers and obesity when adjusting for physical fitness. Utilizing a histological approach in single-fiber analyses of skeletal muscle, He and coworkers (He, Watkins, and Kelley 2001) observed that individuals who were obese and had type 2 diabetes exhibited increased lipid content regardless of fiber type composition. Skeletal muscle from subjects with obesity and from subjects with type 2 diabetes also exhibited a reduced oxidative enzyme activity in all fiber types. This finding suggests that in obesity and type 2 diabetes, lipid storage is increased out of proportion to the capacity of these myocytes for substrate oxidation. There is considerable overlap between fiber type and oxidative capacity, suggesting that reduced overall oxidative capacity rather than fiber type may be more pertinent to obesity and insulin resistance (Simoneau et al. 1999; Kelley and Simoneau 1994; Kim et al. 2000; Simoneau et al. 1995).

Response to Acute Exercise

One aspect of the training response to aerobic exercise is increased reliance on fat oxidation during bouts of physical activity. Training also enhances muscle sensitivity to insulin-mediated glucose uptake. It seems likely that in endurance-trained athletes, a high content of muscle triglyceride provides a reservoir for high rates of oxidation (Romijn et al. 1993; van Loon 2004) and that the pool of intramyocellular triglyceride undergoes exercise-induced depletion and postprandial repletion. The question therefore arises: Does the accretion of muscle triglyceride represent different kinetics—namely, slower fractional turnover of triglycerides and a relative impairment in fat oxidation—for sedentary as opposed to overweight individuals?

Obese persons with and without type 2 diabetes do not appear to have a reduced capacity to rely on fatty acids during submaximal exercise (Colberg et al. 1996; Goodpaster, Wolfe, and Kelley 2002; Horowitz and Klein 2000). This is in contrast to the metabolic inflexibility they do exhibit during postabsorptive conditions, in which they rely relatively less on fatty acids. Indeed, men and women who are sedentary and obese have higher rates of total and intramuscular fatty acid oxidation during moderate-intensity exercise when compared with their lean sedentary counterparts. These observations highlight two very important facets of the metabolic inflexibility of obesity. First, even persons who are sedentary and obese are quite capable of oxidizing fatty acids when they engage in exercise. Second, persons who are obese utilize intramuscular triglycerides during bouts of exercise. It is tempting to speculate that the periodic turnover of stored fatty acids within muscle may be an important link

among mitochondria function, fatty acid accumulation, and skeletal muscle insulin resistance of obesity and type 2 diabetes. Although the ability of lean persons to alter substrate metabolism in response to acute exercise is well known, this feature of metabolic flexibility has not been studied in adequate detail in cases of obesity, insulin resistance, or type 2 diabetes. Further research is clearly needed in this area.

Concluding Remarks

Skeletal muscle in obesity and type 2 diabetes exhibits insulin-resistant glucose metabolism. More recent studies indicate that obesity and type 2 diabetes possess a metabolic inflexibility in the handling of fat calories. This emerging concept is that biochemical characteristics of skeletal muscle in obesity and type 2 diabetes dispose to reduced capacity for fat oxidation and fat accumulation in this tissue. These metabolic defects likely have a genetic component exacerbated by lifestyle variables including physical inactivity and, perhaps, the fatty acid composition of the diet. The present challenge is to precisely define the defects within the pathways of fat metabolism and to utilize these insights to develop effective treatment strategies. An effort to modify skeletal muscle in obesity and type 2 diabetes so that its capacity for fat oxidation is improved should be considered as a potential goal of treatment.

References

Andres, R., G. Cadar, and K. Zierler. 1956. The quantitatively minor role of carbohydrate in oxidative metabolism by skeletal muscle in intact man in the basal state. *J Clin Invest* 35:671-82.

Baltzan, M.A., R. Andres, G. Cader, and K.L. Zierler. 1962. Heterogeneity of forearm metabolism with special reference to free fatty acids. *J Clin Invest* 41:116-25.

Barker, K.B., N.A. Palekar, S.P. Bowers, J.E. Goldberg, J.P. Pulcini, and S.A. Harrison. 2006. Non-alcoholic steatohepatitis: Effect of Roux-en-Y gastric bypass surgery. *Am J Gastroenterol* 101:368-73

Blaak, E.E., A.J.M. Wagenmakers, J.F.C. Glatz, B.H.R. Wolffenbuttel, G.J. Kemerink, C.J.M. Langenberg, G.A.K. Heidendal, and W.H.M. Saris. 2000. Plasma FFA utilization and fatty acid-binding protein content are diminished in type 2 diabetic muscle. *Am J Physiol* 279:146-54.

Boden, G., B. Lebed, M. Schatz, C. Homko, and S. Lemieux. 2001. Effects of acute changes of plasma free fatty acids on intramyocellular fat content and insulin resistance in healthy subjects. *Diabetes.* 50:1612-7.

Boden, G., F. Jadali, J. White, M. Mozzoli, X. Chen, E. Coleman, and C. Smith. 1991. Effects of fat on insulin-stimulated carbohydrate metabolism in normal men. *J Clin Invest* 88:960-6.

Boden, G., X. Chen, J. Ruiz, J.V. White, and L. Rossetti. 1994. Mechanisms of fatty acid-induced inhibition of glucose uptake. *J Clin Invest* 93:2438-46.

Bogardus, C. 1993. Insulin resistance in the pathogenesis of NIDDM in Pima Indians. *Diabetes Care* 16:228-31.

Brooks, G.A. 1997. Importance of the "crossover" concept in exercise metabolism. *Clin Exp Pharmacol Physiol* 24:889-95.

Bruce, C.R., M.J. Anderson, A.L. Carey, D.G. Newman, A. Bonen, A.D. Kriketos, G.J. Cooney, and J.A. Hawley. 2003. Muscle oxidative capacity is a better predictor of insulin sensitivity than lipid status. *J Clin Endocrinol Metab* 88:5444-51.

Colberg, S.R., J.A. Simoneau, F.L. Theate, and D.E. Kelley. 1995. Skeletal muscle utilization of FFA in women with visceral obesity. *J Clin Invest* 95:1846-53.

Colberg, S.R., J.M. Hagberg, S.D. McCole, J.M. Zmuda, P.D. Thompson, and D.E. Kelley. 1996. Utilization of glycogen but not plasma glucose is reduced in individuals with NIDDM during mild-intensity exercise. *J Appl Physiol* 81:2027-33.

Colman, E., L.I. Katzel, E. Rogus, P. Coon, D. Muller, and A.P. Goldberg. 1995. Weight loss reduces abdominal fat and improves insulin action in middle-aged and older men with impaired glucose tolerance. *Metabolism* 44:1502-8.

Cummings, D.E., J. Overduin, M.H. Shannon, and K.E. Foster-Schubert. 2005. Hormonal mechanisms of weight loss and diabetes resolution after bariatric surgery. *Surg Obes Relat Dis* 1:358-68.

Dagenais, G.R., R.G. Tancredi, and K.L. Zierler. 1976. Free fatty acid oxidation by forearm muscle at rest, and evidence for an intramuscular lipid pool in the human forearm. *J Clin Invest* 58:421-31.

Dagogo-Jack, S., and J.V. Santiago. 1997. Pathophysiology of type 2 diabetes and modes of action of therapeutic interventions. *Arch Intern Med* 157:1802-17.

Dube, J., and B.H. Goodpaster. 2006. Assessment of intramuscular triglycerides: Contribution to metabolic abnormalities. *Curr Opin Clin Nutr Metab Care* 9:553-9.

Friedlander, A.L., G.A. Casazza, M.A. Horning, T.F. Buddinger, and G.A. Brooks. 1998. Effects of exercise intensity and training on lipid metabolism in young women. *Am J Physiol* 275:E853-63.

Gaster, M., A.C. Rustan, V. Aas, and H. Beck-Nielson. 2004. Reduced lipid oxidation in skeletal muscle from type 2 diabetic subjects may be of genetic origin. *Diabetes* 53:542-8.

Goodpaster, B., and D. Kelley. 2002. Skeletal muscle triglyceride: Marker or mediator of obesity-induced insulin resistance in type 2 diabetes mellitus. *Curr Diab Rep* 2:216-22.

Goodpaster, B.H., A. Katsiaras, and D.E. Kelley. 2003. Enhanced fat oxidation through physical activity is associated with improvements in insulin sensitivity in obesity. *Diabetes* 52:2191-7.

Goodpaster, B.H., D.E. Kelley, R.R. Wing, A. Meier, and F.L. Thaete. 1999. Effects of weight loss on regional fat distribution and insulin sensitivity in obesity. *Diabetes* 48:839-47.

Goodpaster, B.H., J. He, S. Watkins, and D.E. Kelley. 2001. Skeletal muscle lipid content and insulin resistance: Evidence for a paradox in endurance-trained athletes. *J Clin Endocrinol Metab* 86:5755-61.

Goodpaster, B.H., R. Theriault, S.C. Watkins, and D.E. Kelley. 2000. Intramuscular lipid content is increased in obesity and decreased by weight loss. *Metabolism* 49:467-72.

Goodpaster, B.H., R.R. Wolfe, and D.E. Kelley. 2002. Effects of obesity on substrate utilization during exercise. *Obes Res* 10:575-84.

Gray, R.E., C.J. Tanner, W.J. Pories, K.G. MacDonald, and J.A. Houmard. 2003. Effect of weight loss on muscle lipid content in morbidly obese subjects. *Am J Physiol Endocrinol Metab* 284:E726-32.

Greco, A.V., G. Mingrone, A. Giancaterini, M. Manco, M. Morroni, S. Cinti, M. Granzotto, R. Vettor, S. Camastra, and E. Ferrannini. 2002. Insulin resistance in morbid obesity: Reversal with intramyocellular fat depletion. *Diabetes* 51:144-51.

Guesbeck, N.R., M.S. Hickey, K.G. MacDonald, W.J. Pories, I. Harper, E. Ravussin, G.L. Dohm, and J.A. Houmard. 2001. Substrate utilization during exercise in formerly morbidly obese women. *J Appl Physiol* 90:1007-12.

He, J., S. Watkins, and D.E. Kelley. 2001. Skeletal muscle lipid content and oxidative enzyme activity in relation to muscle fiber type in type 2 diabetes and obesity. *Diabetes* 50:817-23.

Helge, J.W., A.M. Fraser, A.D. Kriketos, A.B. Jenkins, G.D. Calvert, K.J. Ayre, and L.H. Storlien. 1999. Interrelationships between muscle fibre type, substrate oxidation and body fat. *Int J Obes Relat Metab Disord* 23:986-91.

Henriksson, J. 1995. Muscle fuel selection: Effect of exercise training. *Proc Nutr Soc* 54:125-38.

Henry, R.R., L. Abrams, and S.N. Nikoulina. 1995. Insulin action and glucose metabolism in nondiabetic control and NIDDM subjects comparison using human skeletal muscle cell cultures. *Diabetes* 44:936-46.

Henry, R.R., P. Wallace, and J.M. Olefsky. 1986. Effects of weight loss on mechanisms of hyperglycemia in obese non-insulin-dependent diabetes mellitus. *Diabetes* 35:990-8.

Henry, R.R., T.P. Ciaraldi, L. Abrams-Carter, S. Mudaliar, K.S. Park, and S.N. Nikoulina. 1996. Glycogen synthase activity is reduced in cultured skeletal muscle cells of non-insulin-dependent diabetes mellitus subjects. *J Clin Invest* 98:1231-6.

Horowitz, J.F., and S. Klein. 2000. Oxidation of nonplasma fatty acids during exercise is increased in women with abdominal obesity. *J Appl Physiol* 89:2276-82.

Houmard, J.A., C.J. Tanner, c. Yu, P.G. Cunningham, W.J. Pories, K.G. MacDonald, and G.I. Shulman. 2002. Effect of weight loss on insulin sensitivity and intramuscular long-chain fatty acyl-CoAs in morbidly obese subjects. *Diabetes* 51:2959-63.

Kelley, D., L. Kuller, T. McKolanis, P. Harper, J. Mancino, and S. Kalhan. 2004. Effects of moderate weight loss and orlistat on insulin resistance, regional adiposity and fatty acids in type 2 diabetes mellitus. *Diabetes Care* 27:33-40.

Kelley, D.E., B. Goodpaster, R.R. Wing, and J.A. Simoneau. 1999. Skeletal muscle fatty acid metabolism in association with insulin resistance, obesity, and weight loss. *Am J Physiol* 277:E1130-41.

Kelley, D.E., and J.A. Simoneau. 1994. Impaired free fatty acid utilization by skeletal muscle in non-insulin-dependent diabetes mellitus. *J Clin Invest* 94:2349-56.

Kelley, D.E., and L.J. Mandarino. 2000. Fuel selection in human skeletal muscle in insulin resistance. *Diabetes* 49:677-83.

Kelley, D.E., and L.J. Mandarino. 1990. Hyperglycemia normalizes insulin-stimulated skeletal muscle glucose oxidation and storage in noninsulin-dependent diabetes mellitus. *J Clin Invest* 861999-2007.

Kelley, D.E., M. Mokan, J. Simoneau, and L.J. Mandarino. 1993a. Interaction between glucose and free fatty acid metabolism in human skeletal muscle. *J Clin Invest* 92:91-8.

Kelley, D.E., R. Wing, C. Buonocore, J. Sturis, K. Polonsky, and M. Fitzsimmons. 1993b. Relative effects of calorie restriction and weight loss in noninsulin-dependent diabetes mellitus. *J Clin Endocrinol Metab* 77:1287-93.

Kim, J.Y., R.C. Hickner, R.L. Cortright, G.L. Dohm, and J.A. Houmard. 2000. Lipid oxidation is reduced in obese human skeletal muscle. *Am J Physiol Endocrinol Metab* 279:E1039-44.

Laybutt, D.R., C. Schmitz-Peiffer, A.K. Saha, N.B. Rudderman, T.J. Biden, and E.W. Kraegen. 1999. Muscle lipid accumulation and protein kinase C activation in the insulin-resistant chronically glucose-infused rat. *Am J Physiol Endocrinol Metab* 277:E1070-6.

Lillioja, S., A.A. Young, C.L. Culter, J.L. Ivy, W.G. Abbott, J.K. Zawadzki, H. Yki-Jarvinen, L. Christin, T.W. Secomb, and C. Bogardus. 1987. Skeletal muscle capillary density and fiber type are possible determinants of in vivo insulin resistance in man. *J Clin Invest* 80:415-24.

Lillioja, S., and C. Bogardus. 1988. Insulin resistance in Pima Indians. A combined effect of genetic predisposition and obesity-related skeletal muscle cell hypertrophy. *Acta Med Scand Suppl* 723:103-19.

Lillioja, S., C. Bogardus, D. Mott, A. Kennedy, W. Knowler, and B. Howard. 1985. Relationship between insulin-stimulated glucose disposal and lipid metabolism in man. *J Clin Invest* 75:1106-15.

Lowell, B.B., and G.I. Shulman. 2005. Mitochondrial dysfunction and type 2 diabetes. *Science* 307:384-7.

Mandarino, L.J., A. Consoli, A. Jain, and D.E. Kelley. 1996. Interaction of carbohydrate and fat fuels in human skeletal muscle: Impact of obesity and NIDDM. *Am J Physiol Endocrinol Metab* 270:E463-70.

McGarry, J.D. 1995. The mitochondrial carnitine palmitoyl transferase system: Its broadening role in fuel homeostasis and new insights into its molecular features. *Biochem Soc Trans* 23:321-4.

Mensink, M., E.E. Blaak, M.A. van Baak, A.J.M. Wagenmakers, and W.H.M. Saris. 2001. Plasma free fatty acid uptake and oxidation are already diminished in subjects at high risk for developing type 2 diabetes. *Diabetes* 50:2548-54.

Nyholm, B., Z. Qu, A. Kaal, S.B. Pedersen, C.H. Gravholt, J.L. Andersen, B. Saltin, and O. Schmitz. 1997. Evidence of an increased number of type IIb muscle fibers in insulin-resistant first-degree relatives of patients with NIDDM. *Diabetes* 46:1822-8.

Nyholm, B., A. Mengel, S. Nielsen, C. Skjaerbaek, N. Moller, K.G. Alberti, and O. Schmitz. 1996. Insulin resistance in relatives of NIDDM patients: The role of physical fitness and muscle metabolism. *Diabetologia* 39:813-22.

Petersen, K.F., D. Befroy, S. Dufour, J. Dziura, C. Ariyan, D.L. Rothman, L. DiPietro, G.W. Cline, and G.I. Shulman. 2003. Mitochondrial dysfunction in the elderly: Possible role in insulin resistance. *Science* 300:1140-2.

Petersen, K.F., S. Dufour, D. Befroy, R. Garcia, and G.I. Shulman. 2004. Impaired mitochondrial activity in the insulin-resistant offspring of patients with type 2 diabetes. *New Engl J Med* 350:664-71.

Raben, A., E. Mygind, and A. Astrup. 1998. Lower activity of oxidative key enzymes and smaller fiber areas in skeletal muscle of postobese women. *Am J Physiol* 275:E487-94.

Randle, P.J., P.B. Garland, C.N. Hales, and E.A. Newsholme. 1963. The glucose fatty acid cycle: Its role in insulin sensitivity and the metabolic disturbances of diabetes mellitus. *Lancet* 1:785-9.

Ravussin, E., and J.F. Gautier. 1999. Metabolic predictors of weight gain. *Int J Obes Relat Metab Disord* 23:37-41.

Rennie, M.J., and J.O. Holloszy. 1977. Inhibition of glucose uptake and glycogenolysis by availability of oleate in well-oxygenated perfused skeletal muscle. *Biochem J* 168:161-70.

Romijn J.A., E.F. Coyle, L.S. Sidossis, A. Gastaldelli, J.F. Horowitz, E. Endert, and R.R. Wolfe. 1993. Regulation of endogenous fat and carbohydrate metabolism in relation to exercise intensity and duration. *Am J Physiol Endocrinol Metab* 265:E380-91.

Schmitz-Peiffer, C. 2000. Signaling aspects of insulin resistance in skeletal muscle: Mechanisms induced by lipid oversupply. *Cell Signal* 12:583-94.

Schmitz-Peiffer, C. 2002. Protein kinase C and lipid-induced insulin resistance in skeletal muscle. *Ann NY Acad Sci* 967:146-57.

Schmitz-Peiffer, C., D.L. Craig, and T.J. Biden. 1999. Ceramide generation is sufficient to account for the inhibition of the insulin-stimulated PKB pathway in C2C12 skeletal muscle cells pretreated with palmitate. *J Biol Chem* 274:24202-10.

Sidossis, L.S., C.A. Stuart, G.I. Shulman, G.D. Lopaschuk, and R.R. Wolfe. 1996. Glucose plus insulin regulate fat oxidation by controlling the rate of fatty acid entry into the mitochondria. *J Clin Invest* 98:2244-50.

Simoneau, J.A., and C. Bouchard. 1995. Genetic determinism of fiber type proportion in human skeletal muscle. *FASEB J* 9:1091-5.

Simoneau, J.A., J.H. Veerkamp, L.P. Turcotte, and D.E. Kelley. 1999. Markers of capacity to utilize fatty acids in human skeletal muscle: Relation to insulin resistance and obesity and effects of weight loss. *FASEB J* 13:2051-60.

Simoneau, J.A., S.R. Colberg, F.L. Thaete, D.E. Kelley. 1995. Skeletal muscle glycolytic and oxidative enzyme capacities are determinants of insulin sensitivity and muscle composition in obese women. *FASEB J* 9:273-8.

Sjostrom, C.D., M. Peltonen, H. Wedel, and L. Sjostrom. 2000. Differentiated long-term effects of intentional weight loss on diabetes and hypertension. *Hypertension* 36:20-25.

Storlien, L., N.D. Oakes, and D.E. Kelley. 2004. Metabolic flexibility. *Proc Nutr Soc* 63:363-8.

Su, H.Y., W.H. Sheu, H.M. Chin, C.Y. Jeng, Y.D. Chen, and G.M. Reaven. 1995. Effect of weight loss on blood pressure and insulin resistance in normotensive and hypertensive obese individuals. *Am J Hypertens* 8:1067-71.

Toledo, F.G., S. Watkins, and D.E. Kelley. 2006. Changes induced by physical activity and weight loss in the morphology of intermyofibrillar mitochondria in obese men and women. *J Clin Endocrinol Metab* 91:3224-7.

Ukropcova, B., M. McNeil, O. Sereda, L. de Jonge, H. Xie, G.A. Bray, and S.R. Smith. 2005. Dynamic changes in fat oxidation in human primary myocytes mirror metabolic characteristics of the donor. *J Clin Invest* 115:1934-41.

van Loon, L.J. 2004. Use of intramuscular triacylglycerol as a substrate source during exercise in humans. *J Appl Physiol* 97:1170-87.

van Loon, L.J., R. Koopman, R. Manders, W. van der Weegen, G.P. van Kranenburg, and H.A. Keizer. 2004. Intramyocellular lipid content in type 2 diabetes patients compared with overweight sedentary men and highly trained endurance athletes. *Am J Physiol Endocrinol Metab* 287:E558-65.

Wade, A.J., M.M. Marbut, and J.M. Round. 1990. Muscle fiber type and aetiology of obesity. *Lancet* 335:805-8.

Williams, K.V., and D.E. Kelley. 2000. Metabolic consequences of weight loss on glucose metabolism and insulin action in type 2 diabetes. *Diabetes Obes Metab* 2:121-9.

Winder, W.W., J. Arogyasami, I.M. Elayan, and D. Cartmill. 1990. Time course of exercise-induced decline in malonyl-CoA in different muscle types. *Am J Physiol* 259: E266-71.

Wolfe, B.M., S. Klein, E.J. Peters, B.F. Schmidt, and R.R. Wolfe. 1988. Effect of elevated free fatty acids on glucose oxidation in normal humans. *Metabolism* 37:323-9.

Yu, C., Y. Chen, G.W. Cline, D. Zhang, H. Zong, Y. Wang, R. Bergeron, J.K. Kim, S.W. Cushman, G.J. Cooney, B. Atcheson, M.F. White, E.W. Kraegen, and G.I. Shulman. 2002. Mechanism by which fatty acids inhibit insulin activation of insulin receptor substrate-1 (IRS-1)-associated phosphatidylinositol 3-kinase activity in muscle. *J Biol Chem* 277:50230-6.

Zorzano, A., T.W. Balon, L.J. Brady, P. Rivera, L.P. Garetto, J.C. Young, M.N. Goodman, and N.B. Ruderman. 1985. Effects of starvation and exercise on concentrations of citrate, hexose phosphates and glycogen in skeletal muscle and heart. Evidence for selective operation of the glucose-fatty acid cycle. *Biochem J* 232:585-91.

Zurlo, F., S. Lillioja, A. Esposito-DelPuente, B.L. Nyomba, I. Raz, M.F. Saad, W.C. Swiunburn, W.C. Knowler, C. Bogardus, and E. Ravussin. 1990. Low ratio of fat to carbohydrate oxidation as a predictor of weight gain: A study of 24-h RQ. *Am J Physiol Endocrinol Metab* 259:E650-7.

Chapter 6

Nutrient Sensing Links Obesity With Diabetes Risk

Sarah Crunkhorn, PhD; and Mary Elizabeth Patti, MD

Obesity and diabetes are major worldwide health problems. Today, more than 300 million adults are obese, and 170 million people have insulin resistance and diabetes, 13 million of whom live in the United States. Between 1990 and 1998, the incidence of diabetes increased by up to 70% in individuals aged 30 to 39 y (Mokdad et al. 2000), and the diabetes prevalence is projected to increase even more, affecting up to 360 million people worldwide by 2030. The rapid and parallel increase in obesity and type 2 diabetes is alarming on many levels and is a major public health and scientific challenge. Abundant epidemiological data suggest a prominent role for a westernized lifestyle, including nutrient excess and long-term positive energy balance (due to suboptimal diet, inactivity, and obesity), in the development of insulin resistance and diabetes. In this chapter, we highlight normal nutrient-sensing pathways that when disrupted by long-term overnutrition may contribute to the initiation and maintenance of insulin resistance, insulin secretory dysfunction, and type 2 diabetes.

Nutrient Sensing and Control of Food Intake

In individuals who are healthy, energy intake and expenditure are tightly controlled at a whole-body level to maintain a constant body weight. Changes in nutrient availability are sensed at a cellular level in multiple tissues. For example, activation of nutrient-sensing pathways in adipocytes stimulates the synthesis and secretion of the effector hormone leptin, inducing counterregulatory responses that suppress feeding, increase metabolic rate, and limit weight gain. While such data implicate a central role for the hypothalamus in integrating nutrient signals, other tissues also contribute to nutrient signaling and responses. These are summarized in figure 6.1.

Pancreas

Nutrient homeostasis is regulated in part by pancreatic islets. Within the pancreas, blood glucose levels are sensed by glucokinase, leading to regulated insulin release from beta cells and glucagon release from alpha cells (Schuit et al. 2001). Increased blood glucose stimulates insulin secretion, which inhibits hepatic glycogenolysis and gluconeogenesis and enhances peripheral glucose uptake. Conversely, a fall in blood glucose inhibits insulin secretion and stimulates glucagon production. Insulin secretion is also regulated by inputs from neural signals and nutrient-stimulated incretin signals from the intestine, including glucagon-like peptide-1 (GLP-1) and glucose-dependent insulin-releasing polypeptide (GIP; Gautier et al. 2005). Ultimately, insulin secretion plays the key role in regulating glucose homeostasis; the development of hyperglycemia and diabetes always indicates impaired function of beta cells.

Adipose Tissue

Adipose tissue is now recognized as an active endocrine organ that plays a major role in sensing and responding to nutritional status by secreting

The authors gratefully acknowledge support from NIH DK062948 (MEP), DK060837 (Diabetes Genome Anatomy Project), and DK36836 (Diabetes and Endocrinology Research Center, Joslin Diabetes Center).

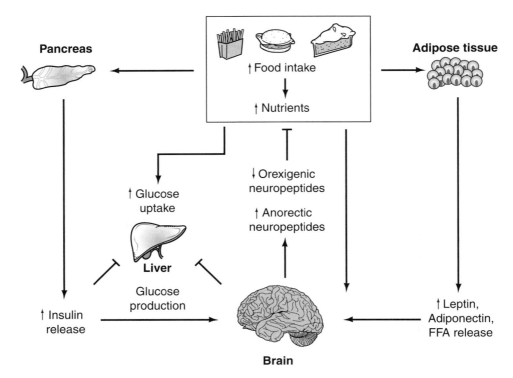

Figure 6.1   Multiple tissues within the body are responsible for sensing and controlling nutrient balance. Nutritional and endocrine signals converge to mediate hypothalamic control of food intake and energy expenditure.

FFAs, leptin, adiponectin, resistin, retinol binding protein 4 (RBP4), and proinflammatory cytokines. These adipocyte secretory products serve as both paracrine and endocrine effectors to influence systemic energy balance and glucose homeostasis. Conversely, abnormal secretion of these adipocytokines has been linked to insulin resistance and type 2 diabetes (Lazar 2005; Graham et al. 2006).

Liver

The liver is positioned centrally to sense and respond to nutrient availability within the portal circulation. Glucose sensors in the portal vein may control glucose uptake by the liver (Moore and Cherrington 1996). Furthermore, insulin may regulate glucose production in the liver both directly and indirectly by affecting gluconeogenesis and glycogenolysis (Girard 2006; Edgerton et al. 2006). Insulin acts directly by binding to hepatic insulin receptors and activating insulin signaling pathways. It acts indirectly by reducing pancreatic glucagon secretion, by inhibiting lipolysis in fat and thus reducing circulating lipid and glycerol availability

for gluconeogenesis, and by decreasing protein catabolism in muscle, which further reduces the availability of gluconeogenesic precursors. In addition, nutrient activation of hypothalamic neuronal pathways inhibits glucose production in the liver. The overall regulation of cholesterol and lipid metabolism, gluconeogenesis, and glycogenolysis also reflects an integration of nutrient signals mediating transcriptional responses.

Hypothalamus

The hypothalamus is uniquely responsive to a variety of nutrient signals, including macronutrients such as glucose and fatty acids and hormones such as leptin and insulin. A positive nutrient balance inhibits the expression of orexigenic peptides (such as neuropeptide Y and agouti-related protein), thus suppressing food intake, and also inhibits hepatic glucose production. Therefore, the appropriate integration of hormonal and nutrient excess signals suppresses food intake, stimulates energy expenditure, and decreases the output of nutrients from endogenous sources such as the liver (Gerozissis 2004; Lam, Schwartz, and Rossetti 2005).

Overnutrition, Disruption of Homeostatic Control, and Insulin Resistance

Despite multiple functionally overlapping nutrient-sensing systems, either acute nutrient excess or chronic overnutrition can overwhelm metabolic homeostatic mechanisms, induce systemic insulin resistance, and increase diabetes risk. For example, experimental increases in plasma FFA achieved through lipid infusion decrease whole-body insulin-stimulated glucose uptake and glycogen synthesis and impair insulin secretion in humans (Dresner et al. 1999; Kashyap et al. 2004). Likewise, infusion of amino acids into healthy volunteers decreases whole-body glucose oxidation and disposal (Flakoll et al. 1992; Rossetti et al. 1989). Such effects reflect a complex interplay between direct cellular responses and abnormal patterns of endocrine and neuronal signals.

Cellular Nutrient Sensing

At the cellular level, alterations in the nutrient supply exert direct, hormone-independent effects on cellular metabolism, gene expression, and signal transduction. Exposing cultured cells to glucose, fatty acids, and amino acids may induce insulin resistance, either via direct effects on gene transcription or inhibition of insulin signaling or via indirect effects mediated through nutrient-sensing pathways. Multiple pathways involved in sensing and responding to macronutrients are now associated with insulin resistance (figure 6.2).

Multiple cellular pathways link nutrient excess with both obesity and insulin resistance. Chronic activation of the pathway involving mammalian target of rapamycin (mTOR) and S6 kinase (S6K) may play a central role by reducing PGC-1 expression, causing mitochondrial dysfunction, and decreasing energy expenditure, all of which

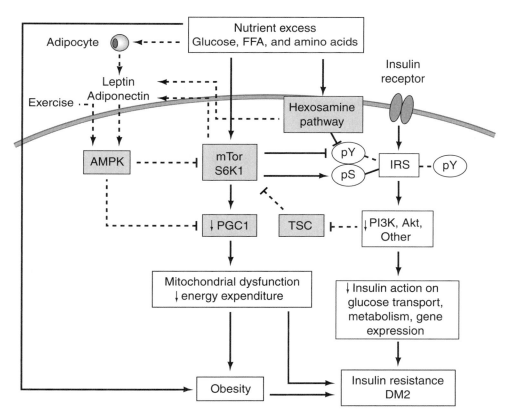

Figure 6.2   Cellular pathways linking nutrient excess with obesity and insulin resistance. Solid arrows indicate pathways linking nutrient excess to insulin resistance and obesity, while dashed arrows indicate pathways typically contributing to insulin sensitivity and metabolic homeostasis.

From M.E. Patti and B.B. Kahn, 2004, "Nutrient sensor links obesity with diabetes risk," *Nature* 10(10): 1049-1050.

enhance the risk for obesity and insulin resistance. Activation of mTOR and S6K also reduces IRS tyrosine phosphorylation (pY) and increases serine phosphorylation (pS), thus inhibiting downstream insulin effects on metabolism and transcription. Likewise, chronic activation of the hexosamine biosynthetic pathway by high concentrations of glucose or FFA also inhibits insulin signaling. Conversely, AMPK-dependent pathways activated by exercise, leptin, and adiponectin may counteract the effects of nutrient excess on mTOR, S6K, and PGC-1. Direct modulation of gene expression by nutrients may also occur (not shown in figure 6.2).

Hexosamine Biosynthetic Pathway

The hexosamine biosynthetic pathway (HBP) was first identified as a nutrient-sensing pathway responsive to glucose flux in adipocytes (Marshall, Bacote, and Traxinger 1991). Following transport into the cell, glucose is phosphorylated by hexokinase to glucose-6-phosphate, which can be stored via glycogen synthesis or used for ATP production via glycolysis and oxidative metabolism. However, between 1% and 3% of the intracellular glucose enters the HBP. Flux through this pathway is controlled by glutamine:fructose-6-phosphate aminotransferase (GFAT), which converts fructose-6-phosphate to glucosamine-6-phosphate. Subsequent steps lead to the formation of N-linked glycosylation end products, including UDP-N-acetylglucosamine (UDP-GlcNAc; Rossetti 2000). Thus, under normal conditions, the metabolism of glucose by the HBP may generate a signal of cellular satiety in a variety of tissues, including muscle, fat, liver, and pancreatic islets. However, chronic HBP overactivity induced by nutrient excess may mediate insulin resistance. In fact, experimental infusion of glucose or chronic hyperglycemia is associated with increased GlcNAc content in skeletal muscle. Glucosamine (which enters the pathway but bypasses the rate-limiting GFAT enzyme) induces insulin resistance in 3T3-L1 adipocytes, in isolated rat skeletal muscle, and in healthy rats (Nelson, Robinson, and Buse 2000; Robinson, Sens, and Buse 1993; Rossetti et al. 1995; Virkamaki et al. 1997). In addition, glucosamine induces mild defects in insulin secretion by pancreatic beta cells and insulin resistance in humans who are healthy (Monauni et al. 2000).

While these effects are often considered in the setting of glucose excess, exposure to fatty acids also activates the HBP, via increased flux of fructose-6-phosphate. Prolonged lipid infusion results in the accumulation of HBP end products, an effect which precedes the onset of insulin resistance induced by fatty acids (Hawkins et al. 1997). Furthermore, incubating primary human skeletal muscle myotubes with saturated fatty acids increases mRNA and protein expression of GFAT and intracellular concentrations of UDP-GlcNAc as well as promotes DNA binding of the transcription factor Sp1, one target of the HBP (Weigert et al. 2003).

In vivo insulin resistance is also associated with increased activity of GFAT, as demonstrated in skeletal muscle from both insulin-resistant ob/ob mice (Buse et al. 1997) and patients with type 2 diabetes (Yki-Jarvinen et al. 1996). In transgenic mouse models, overexpression of GFAT in fat alone or in both muscle and fat results in insulin resistance (Crook et al. 1993; Hebert Jr. et al. 1996). Furthermore, insulin-stimulated glucose uptake is decreased in muscle isolated from mice overexpressing GFAT in adipose tissue, a finding that indicates cross talk between adipocytes and skeletal muscle (Hazel et al. 2004). Overexpression of GFAT in liver also induces glucose intolerance, glycogen accumulation, hyperlipidemia, and obesity (Veerababu et al. 2000). Conversely, inhibition of GFAT expression and activity prevents glucose-induced impairments in glucose transport and insulin resistance (Marshall, Bacote, and Traxinger 1991).

Several mechanisms for the induction of insulin resistance via HBP flux have been proposed, including impaired insulin signaling that decreases translocation and activation of the glucose transporter GLUT4, increased proteoglycan formation, glycosylation of lipids, complex N-linked glycosylation, and increased O-linked GlcNAc (O-GlcNAc) modification of proteins (Baron et al. 1995; Slawson, Housley, and Hart 2006; Brownlee 2001). Glucosamine infusion in rats impairs insulin action, reducing insulin-stimulated IRS1 tyrosine phosphorylation, activation of PI3K, and glycogen synthase activity in muscle (Patti et al. 1999). These and other effects may be linked to alterations in O-GlcNAc modification of proteins. The end product of the HBP (UDP-GlcNAc) is a donor for O-GlcNAc modification of serine and threonine residues.

Such posttranslational modification of enzymes, transporters, or transcription factors may modulate the activity or stability of proteins or the general transcriptional responses in a nutrient-dependent fashion (Slawson, Housley, and Hart 2006). For example, elevated levels of O-GlcNAc within 3T3-L1 cells are linked to defects in insulin signaling, including a reduction in insulin-stimulated IRS1, Akt, and GSK3 phosphorylation (Vosseller et al. 2002). Furthermore, overexpression of O-GlcNAc transferase in mouse muscle and fat results in insulin resistance and hyperleptinemia (McClain et al. 2002).

Hyperglycemia-mediated HBP activation also modulates transcription. For example, O-GlcNAc glycosylation of the transcription factor Sp1 in endothelial cells decreases serine and threonine phosphorylation of Sp1, increases transcriptional activity, and increases expression of plasminogen activator inhibitor 1 (Du et al. 2000). HBP-mediated Sp1 modification may also contribute to the dysregulation of genes controlling oxidative metabolism. Glucosamine infusion or 3 d of high-fat feeding in rats decreases the expression of multiple nuclear-encoded mitochondrial genes within the tricarboxylic acid cycle and oxidative phosphorylation pathways in skeletal muscle (Obici et al. 2002). It also increases the expression of fatty acid synthase (FAS), acetyl-coa carboxylase (ACC), and glycerol-3-phosphate dehydrogenase (Rumberger et al. 2003). In parallel, whole-body oxygen consumption and energy expenditure decrease. These findings are even more striking in light of recent data suggesting that impaired mitochondrial gene expression and function are critical early events in the development of insulin resistance (Morino et al. 2005; Petersen et al. 2003, 2004; Patti et al. 2003; Mootha et al. 2003). Thus, activation of the HBP by nutrient excess may contribute to the pathophysiology of both insulin resistance and obesity.

HBP and Energy Balance

In addition to acting as a cellular nutrient-sensing pathway involved in regulating glucose metabolism and storage and peripheral insulin sensitivity, the HBP may also contribute to whole-body metabolism via actions on the central nervous system and other tissues. HBP flux can regulate leptin expression in fat and muscle in both cultured cells and animals (Obici et al. 2002; Wang et al. 1998). In addition, transgenic mice overexpressing GFAT in muscle and fat are hyperleptinemic, exhibiting increased levels of leptin mRNA in fat tissue (McClain et al. 2000). The HBP, therefore, can serve as an integrator of intracellular satiety in adipocytes, thus modulating leptin secretion and effects on nutrient balance.

5' AMP-Activated Protein Kinase

AMPK is a serine and threonine kinase that is evolutionarily conserved from yeast to mammals. It is a trimeric enzyme comprising a catalytic alpha subunit, a noncatalytic beta subunit, and a noncatalytic gamma subunit. AMPK is activated by adenosine monophosphate (AMP) or by Thr-172 phosphorylation of the alpha subunit by an upstream kinase (AMPKK). In mammalian tissues, AMPK activity is regulated by metabolic and physiological stresses that deplete cellular ATP and increase AMP, including hypoxia, ischemia, glucose deprivation, uncoupling of oxidative phosphorylation, exercise, and muscle contraction (see chapter 13 for additional information). Thus, AMPK can be considered as a metabolic stress-sensing enzyme that monitors the balance between energy supply and energy demand and then appropriately modulates metabolic pathways involved in glucose, fatty acid, and protein metabolism to ensure homeostasis. For example, low concentrations of intracellular glucose activate AMPK, whereas relatively high levels of glycogen inhibit AMPK. In exercising muscle consuming ATP, AMPK increases glucose uptake via increased GLUT4 expression and translocation and increases glucose and fatty acid oxidation, thus ensuring that fuels are available for exercise.

The potent effects of AMPK on fatty acid oxidation are mediated by AMPK phosphorylation and inactivation of ACC, which reduces malonyl-CoA levels, relieves inhibition of CPT1 and fatty acid transport into mitochondria, and stimulates beta-oxidation (Ruderman et al. 1999). In addition, AMPK regulates the expression of many genes critical for metabolic control, including fructose 2,6-bisphosphatase (glycolysis) and PGC-1α, a coactivator of PPARδ, nuclear respiratory factor 1 (NRF1), and other transcription factors.

In the liver, activation of AMPK by AICAR also has positive effects, suppressing gluconeogenesis and glucose production. Both AICAR and constitutive overexpression of the catalytic alpha isoform of AMPK have major effects on the expression of several glycolytic and lipogenic genes that are normally under the control of nutrients and insulin (Leclerc, Kahn, and Doiron 1998; Woods et al. 2000). These include genes for phosphoenolpyruvate carboxykinase, glucose-6-phosphatase, pyruvate kinase, fatty acid synthase, ACC, sterol regulatory element binding protein-1c (SREPB-1c), and hepatocyte nuclear factor 4 (HNF4α). Thus, the net effects of AMPK activation in the liver are to decrease fatty acid, triglyceride, and sterol biosythesis and to increase fatty acid oxidation and ketogenesis.

AMPK may play an important role in regulating insulin synthesis and secretion by pancreatic beta cells. Incubating INS-1 cells with low levels of glucose increases AMPK activity and decreases ACC activity, decreasing insulin release (Salt et al. 1998).

AMPK may also mediate diverse hormonal effects on glucose and lipid metabolism in peripheral tissues. For example, in muscle leptin (Minokoshi et al. 2002) and adiponectin (Yamauchi et al. 2002) stimulate AMPK activity both directly and indirectly via sympathetic activation. Conversely, leptin inhibits AMPK in the hypothalamus (Minokoshi et al. 2004) and may regulate the expression of a variety of neuropeptides, ultimately reducing food intake. The diverse effects of AMPK on specific tissues are summarized in figure 6.3.

AMPK and Insulin Resistance

Since AMPK plays such a key role in regulating carbohydrate and fat metabolism, impaired AMPK action could contribute to many of the metabolic defects associated with type 2 diabetes. Mice with a disrupted alpha 2 subunit (AMPKα–/–) are glucose intolerant and have reduced insulin sensitivity (Viollet et al. 2003). Similarly, prolonged glucose infusion in rats is associated with decreased AMPK and ACC phosphorylation and activity in muscle and liver and is temporally linked to subsequent increases in malonyl-CoA and DAG content (Kraegen et al. 2006).

AMPK is also an important regulator of mitochondrial biogenesis. Expression of key mitochondrial enzymes increases following AMPK activation by AICAR (Winder et al. 2000) or by exercise (Pilegaard, Saltin, and Neufer 2003; Norrbom et al. 2004; Russell et al. 2003). Moreover, exercise training and the resultant activation of AMPK enhance mitochondrial gene expression, oxidative capacity, and insulin sensitivity, perhaps via increased expression of PGC-1, mitochondrial transcription factor A (TFAM), and NRF1 (Pilegaard, Saltin, and Neufer 2003; Norrbom et al. 2004; Russell et al. 2003). Conversely, insulin resistance and type 2 diabetes are characterized by decreased expression of nuclear-encoded mitochondrial gene expression and by mitochondrial dysfunction (Morino et al. 2005; Petersen et al. 2003, 2004; Patti et al. 2003; Mootha et al. 2003). Inactivity, poor physical fitness, and reduced activation of AMPK may contribute to these expression patterns and to impairments in oxidative metabolism. There is a striking correlation between muscle expression of genes regulated by PGC-1 and $\dot{V}O_2$max in Caucasian sub-

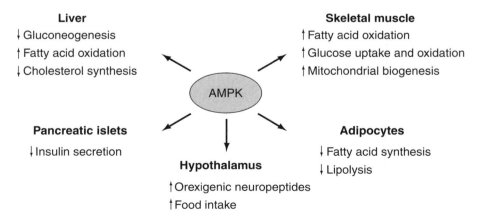

Figure 6.3 The multiple effects of AMPK on the hypothalamus, liver, adipose tissue, skeletal muscle, and pancreatic islets.

Adapted, by permission, from W.W. Winder and D.G. Hardie, 1999, "AMP-activated protein kinase, a metabolic master switch: Possible roles in type 2 diabetes," *Am J Physiol Endocrinol Metab* 277(1): E1-E10.

jects with type 2 diabetes (Mootha et al. 2003). Thus, reductions in exercise and physical fitness are likely to be important contributors to insulin resistance and diabetes, as demonstrated by the potent ability of regular exercise to diminish the incidence of diabetes in subjects at high risk (Knowler et al. 2002).

It is known from extensive animal studies that caloric restriction protects the body against age-related diseases such as cancer, hypertension, diabetes, and cardiovascular disease. Under conditions of caloric restriction, when energy supplies are lower, AMPK is activated to control gene transcription and to facilitate adaptations to the reduced availability of energy. These same effects may protect against diabetes. Moreover, drugs used to treat obesity and diabetes, including rosiglitazone and metformin, act in part by activating AMPK (Fryer, Parbu-Patel, and Carling 2002; Zhou et al. 2001).

Mammalian Target of Rapamycin

Another key cellular sensing pathway is the mammalian target of rapamycin (mTOR), which integrates signals from nutrients, energy, and growth factors, including insulin. TOR was originally identified in yeast as a spontaneous gene mutation that conferred resistance to the antifungal and immunosuppressive reagent rapamycin (Heitman, Movva, and Hall 1991). Since then, TOR has been recognized as a highly conserved, ubiquitously expressed serine and threonine kinase found in all eukaryotes (Schmelzle and Hall 2000). TOR is a major regulator of cell size and protein synthesis and also regulates many aspects of cell growth, including translation initiation and elongation, ribosomal biogenesis, autophagy, progression through the cell cycle, and transcription in response to various environmental signals. Upon activation, mTOR complexes with regulatory associated protein of mTOR (raptor) and a third protein, GβL, that stabilizes the interactions between mTOR and raptor (Kim et al. 2003). This complex is the target of rapamycin. The activated mTOR complex phosphorylates key downstream components of the pathway, including the ribosomal S6K (S6K1 and S6K2) and the eukaryotic initiation factor 4E (eIF4E)-binding protein 1 (4EBP1). These events lead to the activation of S6K and the release of 4EBP1 from the cap-dependent translation initiation factor eIF4E (Kimball et al. 1999) and thus increase ribosomal biogenesis and translation of specific mRNA populations. TOR also complexes with rapamycin-insensitive companion of mTOR (rictor) and GβL. This complex (TORC2) has no effect on S6K1 regulation; it regulates the cytoskeleton as well as proliferation and survival in cancer cell lines (Sarbassov et al. 2004, 2005).

The mTOR pathway is activated by nutrient-rich conditions, particularly high levels of amino acids and FFA. When GβL is absent or disabled, cells become insensitive to nutrient levels and display abnormal growth patterns. Amino acids (branched-chain amino acids, particularly leucine) are required for basal as well as mitogen-stimulated mTOR activation. Many studies in both cell cultures and animal models have demonstrated activation of mTOR and its downstream target p70 S6K following exposure to amino acids, typically in a rapamycin-dependent pattern (Patti et al. 1998; Kimball et al. 1999; Jefferson and Kimball 2003). While the mechanisms by which mTOR senses the availability of amino acids remain largely unknown, an intracellular or a membrane-bound sensor for mTOR activation by amino acids has been implicated (Beugnet et al. 2003). High-fat feeding also increases the activation of mTOR in both skeletal muscle and liver (Khamzina et al. 2005). Similarly, insulin and other hormones activate TOR by activating Akt, resulting in site-specific phosphorylation of TSC2 (a member of the tuberous sclerosis complex TSC1/2) and inhibition of Rheb (Hardie 2005).

The mTOR pathway may be linked reciprocally with AMPK. Conditions that activate AMPK also promote the dephosphorylation of TOR substrates S6K1 and 4EBP1. In addition, AICAR reduces phosphorylation of S6K1 but has no effect in cells with rapamycin-resistant S6K1 (Kimura et al. 2003). Whereas AMPK activation strongly inhibits protein synthesis, mTOR activation triggers protein synthesis. Depletion of ATP or mitochondrial dysfunction inhibits mTOR-mediated phosphorylation of S6K1 kinase via pathways dependent on AMPK and TSC2 (Tokunaga, Yoshino, and Yonezawa 2004; Hardie 2005).

mTOR and Insulin Resistance

Amino acids, particularly branched-chain amino acids, stimulate insulin secretion, increase hepatic glucose production, and induce insulin resistance in both hepatic tissue and skeletal muscle. High-protein diets are also associated with glucose intolerance (Rossetti et al. 1989). Amino acids modulate insulin signaling through mTOR-dependent effects on IRS1, reducing IRS tyrosine phosphorylation, increasing serine phosphorylation, and decreasing activation of PI3K (Patti et al. 1998; Tremblay and Marette 2001); these effects can be blocked with rapamycin (Tremblay et al. 2005; Tzatsos and Kandror 2006). Therefore, through mTOR regulation of IRS1-dependent signaling cascades, circulating amino acids may contribute to the complex regulation of peripheral insulin sensitivity.

The important role for the TOR pathway in insulin action is also demonstrated in TSC1–/– and TSC2–/– mouse embryonic fibroblasts (MEFs), which develop insulin resistance and insulin-like growth factor 1 (IGF-1) resistance due to decreased IRS1 and IRS2 protein expression as a consequence of sustained and unregulated mTOR signaling. Similarly, long-term treatment of cells with rapamycin or small interfering RNA (siRNA) against S6K1–/– or S6K2–/– completely restores IRS1 protein levels and the responsiveness of the PI3K and Akt pathway to insulin and IGF-1 (Shah, Wang, and Hunter 2004; Harrington et al. 2004). Therefore, constitutive mTOR activity strongly downregulates insulin and IGF-1 signaling via its effects on IRS expression, phosphorylation, and protein stability.

The rapamycin-insensitive mTOR complex with rictor may also mediate mTOR effects on insulin signaling. In drosophila and human cancer cell lines, this complex is necessary for direct Akt phosphorylation on Ser-473 and facilitates Thr-308 phosphorylation by PDK1.

Chronic activation of the mTOR pathway, as is observed in liver and muscle of obese rats (Khamzina et al. 2005), may play an important role in systemic regulation of energy homeostasis. For example, leucine activation of mTOR in isolated rat adipocytes stimulates leptin secretion in a rapamycin-sensitive fashion (Roh et al. 2003). Furthermore, mice with ablation of S6K1 are protected from obesity and insulin resistance when fed a high-fat diet (Um et al. 2004). These mice exhibit increased lipolysis, increased mitochondrial density in both muscle and adipose tissues, and increased systemic energy expenditure. At a molecular level, these effects are accompanied by increased expression of genes critical for mitochondrial function (including PGC-1), thermogenesis, and fatty acid oxidation and by enhanced insulin signaling at both receptor and postreceptor levels.

Despite these important observations, the primary tissue site underlying this metabolic phenotype is unknown; altered signaling in other tissues, such as the hypothalamus and liver, and reduced insulin secretion may also contribute to this phenotype. Moreover, protection from insulin resistance may result from the lean phenotype itself.

Taken together, these data demonstrate that TOR-dependent pathways can mediate maladaptive metabolic and transcriptional responses to chronic nutrient excess. More importantly, these data further support the integration of two key factors in the development of insulin resistance in humans: mitochondrial dysfunction and impaired insulin signaling.

SIRT1

Another potential nutrient sensor is SIRT1, the mammalian homolog of the yeast protein Sir2. Sir2 and SIRT1 are protein deacetylases that depend on nicotinamide adenine dinucleotide (NAD^+). In yeast, caloric restriction alters glucose metabolism and extends life span, effects that have been shown to require Sir2 activity (Lin, Defossez, and Guarente 2000). In mammalian cells, acute nutrient withdrawal increases the expression of SIRT1 via FOXO3a-dependent pathways that require interaction with p53 (Nemoto, Fergusson, and Finkel 2004). Caloric restriction or fasting increases fatty acid oxidation and gluconeogenesis in the liver via alterations in PGC-1α expression; SIRT1 is required for this response in mice. Once SIRT1 is induced, it deacetylates PGC-1α at specific lysine residues in an NAD^+-dependent manner to induce genes for gluconeogenesis and hepatic glucose output. Thus, SIRT1 induction during nutrient deprivation modulates PGC-1α function (Rodgers et al. 2005) and may contribute to the whole-body effects of caloric restriction.

PGC-1 as a Key Effector Responsive to Nutrition in Muscle

Muscle expression and PGC-1α activity and mitochondrial dysfunction may be regulated by many of the nutrient-sensing pathways within the cell, including positive regulation by AMPK or SIRT1 activation, and negative regulation by the HBP and mTOR. These observations implicate PGC-1α as a key effector molecule involved in mediating cellular nutrient responses in muscle. Indeed, expression of PGC-1α and the related coactivator PGC-1β is decreased in skeletal muscle from subjects who are insulin resistant, obese, or type 2 diabetic (Patti et al. 2003; Mootha et al. 2003). Such decreases in PGC-1 may contribute to the mitochondrial dysfunction characteristic of insulin resistance and diabetes (Morino et al. 2005; Petersen et al. 2003, 2004) and may play a primary role in the development of obesity, insulin resistance, and type 2 diabetes. In turn, it is likely that diet and overnutrition play a major role in regulating PGC-1α expression and the resultant metabolic effects. Interestingly, in skeletal muscle from subjects who are obese, both severe caloric restriction and exercise increase PGC-1α expression in parallel with improved insulin sensitivity (Larrouy et al. 1999). In contrast, lipid infusion into humans who are healthy decreases the expression of skeletal muscle PGC-1α and nuclear-encoded mitochondrial genes (Richardson et al. 2004). Similarly, expression of PGC-1α and mitochondrial genes in skeletal muscle falls after just 3 d of high-fat feeding in humans who are healthy (Sparks et al. 2005). At a cellular level, fatty acids reduce transcription of PGC-1α via p38-dependent pathways (Crunkhorn et al. 2007).

These data all suggest that PGC-1a is a key effector of cellular nutrient responses. Therefore, PGC-1a may be a potential therapeutic target in the prevention of insulin resistance and type 2 diabetes.

Concluding Remarks

Nutrient excess and enhanced signaling through cellular nutrient-sensing pathways may play key roles in the development of obesity, insulin resistance, and type 2 diabetes. Elucidating the molecular interactions within these pathways and identifying the key effector molecules will be critical for understanding the pathogenesis of insulin resistance and discovering new targets for the prevention and treatment of diabetes.

References

Baron, A.D., J.S. Zhu, J.H. Zhu, H. Weldon, L. Maianu, and W.T. Garvey. 1995. Glucosamine induces insulin resistance in vivo by affecting GLUT 4 translocation in skeletal muscle: Implications for glucose toxicity. *J Clin Invest* 96:2792-2801.

Beugnet, A., A.R. Tee, P.M. Taylor, and C.G. Proud. 2003. Regulation of targets of mTOR (mammalian target of rapamycin) signalling by intracellular amino acid availability. *Biochem J* 372:555-66.

Brownlee, M. 2001. Biochemistry and molecular cell biology of diabetic complications. *Nature.* 414:813-20.

Buse, M.G., K.A. Robinson, T.W. Gettys, E.G. McMahon, and E.A. Gulve. 1997. Increased activity of the hexosamine synthesis pathway in muscles of insulin-resistant ob/ob mice. *Am J Physiol* 272:E1080-8.

Crook, E.D., M.C. Daniels, T.M. Smith, and D.A. McClain. 1993. Regulation of insulin-stimulated glycogen synthase activity by overexpression of glutamine: Fructose-6-phosphate amidotransferase in rat-1 fibroblasts. *Diabetes* 42:1289-96.

Crunkhorn et al. 2007. Peroxisome proliferator activator receptor gamma coactivator-1 expression is reduced in obesity: potential pathogenic role of saturated fatty acids and p38 mitogen-activated protein kinase activation. *J Biol Chem* 282(21):15439-50.

Dann, S.G., and G. Thomas. 2006. The amino acid sensitive TOR pathway from yeast to mammals. *FEBS Lett* 580:2821-9.

Dresner, A., D. Laurent, M. Marcucci, M.E. Griffin, S. Dufour, G.W. Cline, L.A. Slezak, D.K. Andersen, R.S. Hundal, D.L. Rothman, K.F. Petersen, and G.I. Shulman. 1999. Effects of free fatty acids on glucose transport and IRS-1-associated phosphatidylinositol 3-kinase activity. *J Clin Invest* 103:253-9.

Du, X.L, D. Edelstein, L. Rossetti, I.G. Fantus, H. Goldberg, F. Ziyadeh, J. Wu, and M. Brownlee. 2000. Hyperglycemia-induced mitochondrial superoxide overproduction activates the hexosamine pathway and induces plasminogen activator inhibitor-1 expression by increasing Sp1 glycosylation. *Proc Natl Acad Sci USA* 97:12222-6.

Edgerton, D.S., M. Lautz, M. Scott, C.A. Everett, K.M. Stettler, D.W. Neal, C.A. Chu, and A.D. Cherrington. 2006. Insulin's direct effects on the liver dominate the control of hepatic glucose production. *J Clin Invest* 116:521-7.

Flakoll, P.J., L.S. Wentzel, D.E. Rice, J.O. Hill, and N.N. Abumrad. 1992. Short-term regulation of insulin-mediated glucose utilization in four-day fasted human volunteers: Role of amino acid availability. *Diabetologia* 35:357-66.

Fryer, L.G., A. Parbu-Patel, and D. Carling. 2002. The antidiabetic drugs rosiglitazone and metformin stimulate AMP-activated protein kinase through distinct signaling pathways. *J Biol Chem* 277:25226-32.

Gautier, J.F., S. Fetita, E. Sobngwi, and C. Salaun-Martin. 2005. Biological actions of the incretins GIP and GLP-1 and therapeutic perspectives in patients with type 2 diabetes. *Diabetes Metab* 31:233-42.

Gerozissis, K. 2004. Brain insulin and feeding: A bi-directional communication. *Eur J Pharmacol* 490:59-70.

Girard, J. 2006. Insulin's effect on the liver: "Direct or indirect?" continues to be the question. *J Clin Invest* 116:302-4.

Graham, T.E., Q. Yang, M. Bluher, A. Hammarstedt, T.P. Ciaraldi, R.R. Henry, C.J. Wason, A. Oberbach, P.A. Jansson, U. Smith, and B.B. Kahn. 2006. Retinol-binding protein 4 and insulin resistance in lean, obese, and diabetic subjects. *New Engl J Med* 354:2552-63.

Hardie, D.G. 2005. New roles for the LKB1-->AMPK pathway. *Curr Opin Cell Biol* 17:167-73.

Harrington, L.S., G.M. Findlay, A. Gray, T. Tolkacheva, S. Wigfield, H. Rebholz, J. Barnett, N.R. Leslie, S. Cheng, P.R. Shepherd, I. Gout, C.P. Downes, and R.F. Lamb. 2004. The TSC1-2 tumor suppressor controls insulin-PI3K signaling via regulation of IRS proteins. *J Cell Biol* 166:213-23.

Hawkins, M., N. Barzilai, R. Liu, M. Hu, W. Chen, and L. Rossetti. 1997. Role of the glucosamine pathway in fat-induced insulin resistance. *J Clin Invest* 99:2173-82.

Hazel, M., R.C. Cooksey, D. Jones, G. Parker, J.L. Neidigh, B. Witherbee, E.A. Gulve, and D.A. McClain. 2004. Activation of the hexosamine signaling pathway in adipose tissue results in decreased serum adiponectin and skeletal muscle insulin resistance. *Endocrinology* 145:2118-2128.

Hebert Jr., L.F., M.C. Daniels, J. Zhou, E.D. Crook, R.K. Turner, S.T. Simmons, J.L. Neidigh, J.S. Zhu, A.D. Baron, and D.A. McClain. 1996. Overexpression of glutamine: Fructose-6-phosphate amidotransferase in transgenic mice leads to insulin resistance. *J Clin Invest* 98:930-6.

Heitman, J., N.R. Movva, and M.N. Hall. 1991. Targets for cell cycle arrest by the immunosuppressant rapamycin in yeast. *Science* 253:905-9.

Jefferson, L.S., and S.R. Kimball. 2003. Amino acids as regulators of gene expression at the level of mRNA translation. *J Nutr* 133:2046S-51S.

Kashyap, S., D. Richardson, C.P. Jenkinson, J. Finlayson, J. Bynum, S. Suraamornkuul, R.A. DeFronzo, M.E. Patti, and L.J. Mandarino. 2003. Increased plasma FFA alter gene expression in human muscle in vivo. Diabetes 522:A244.

Kashyap S.R. et al. *Am J Physiol Endocrinol Metab.* 2004 Sep: 287(3):E537-46

Khamzina, L., A. Veilleux, S. Bergeron, and A. Marette. 2005. Increased activation of the mammalian target of rapamycin pathway in liver and skeletal muscle of obese rats: Possible involvement in obesity-linked insulin resistance. *Endocrinology* 146:1473-81.

Kim, D.H., D.D. Sarbassov, S.M. Ali, R.R. Latek, K.V. Guntur, H. Erdjument-Bromage, P. Tempst, and D.M. Sabatini. 2003. GbetaL, a positive regulator of the rapamycin-sensitive pathway required for the nutrient-sensitive interaction between raptor and mTOR. *Mol Cell* 11:895-904.

Kimball, S.R., L.M. Shantz, R.L. Horetsky, and L.S. Jefferson. 1999. Leucine regulates translation of specific mRNAs in L6 myoblasts through mTOR-mediated changes in availability of eIF4E and phosphorylation of ribosomal protein S6. *J Biol Chem* 274:11647-52.

Kimura, N., C. Tokunaga, S. Dalal, C. Richardson, K. Yoshino, K. Hara, B.E. Kemp, L.A. Witters, O. Mimura, and K. Yonezawa. 2003. A possible linkage between AMP-activated protein kinase (AMPK) and mammalian target of rapamycin (mTOR) signalling pathway. *Genes Cells* 8:65-79.

Knowler, W.C., E. Barrett-Connor, S.E. Fowler, R.F. Hamman, J.M. Lachin, E.A. Walker, and D.M. Nathan. 2002. Reduction in the incidence of type 2 diabetes with lifestyle intervention or metformin. *New Engl J Med* 346:393-403.

Kraegen, E.W., A.K. Saha, E. Preston, D. Wilks, A.J. Hoy, G.J. Cooney, and N.B. Ruderman. 2006. Increased malonyl-CoA and diacylglycerol content and reduced AMPK activity accompany insulin resistance induced by glucose infusion in muscle and liver of rats. *Am J Physiol Endocrinol Metab* 290:E471-9.

Lam, T.K., G.J. Schwartz, and L. Rossetti. 2005. Hypothalamic sensing of fatty acids. *Nat Neurosci* 8:579-84.

Larrouy, D., H. Vidal, F. Andreelli, M. Laville, and D. Langin. 1999.

Cloning and mRNA tissue distribution of human PPARgamma coactivator-1. *Int J Obes Relat Metab Disord* 23:1327-32.

Lazar, M.A. 2005. How obesity causes diabetes: Not a tall tale. *Science* 307:373-5.

Leclerc, I., A. Kahn, and B. Doiron. 1998. The 5'-AMP-activated protein kinase inhibits the transcriptional stimulation by glucose in liver cells, acting through the glucose response complex. *FEBS Lett* 431:180-4.

Lin, S.J., P.A. Defossez, and L. Guarente. 2000. Requirement of NAD and SIR2 for life-span extension by calorie restriction in Saccharomyces cerevisiae. *Science* 289:2126-8.

Marshall, S., V. Bacote, and R.R. Traxinger. 1991. Discovery of a metabolic pathway mediating glucose-induced desensitization of the glucose transport system. Role of hexosamine biosynthesis in the induction of insulin resistance. *J Biol Chem* 266:4706-12.

McClain, D.A., T. Alexander, R.C. Cooksey, and R.V. Considine. 2000. Hexosamines stimulate leptin production in transgenic mice. *Endocrinology* 141: 1999-2002.

McClain, D.A., W.A. Lubas, R.C. Cooksey, M. Hazel, G.J. Parker, D.C. Love, and J.A. Hanover. 2002. Altered glycan-dependent signaling induces insulin resistance and hyperleptinemia. *Proc Natl Acad Sci USA* 99:10695-9.

Minokoshi, Y., T. Alquier, N. Furukawa, Y.B. Kim, A. Lee, B. Xue, J. Mu, F. Foufelle, P. Ferre, M.J. Birnbaum, B.J. Stuck, and B.B. Kahn. 2004. AMP-kinase regulates food intake by responding to hormonal and nutrient signals in the hypothalamus. *Nature* 428:569-74.

Minokoshi, Y., Y.B. Kim, O.D. Peroni, L.G. Fryer, C. Muller, D. Carling, and B.B. Kahn. 2002. Leptin stimulates fatty-acid oxidation by activating AMP-activated protein kinase. *Nature* 415:339-43.

Mokdad, A.H., E.S. Ford, B.A. Bowman, D.E. Nelson, M.M. Engelgau, F. Vinicor, and J.S. Marks. 2000. Diabetes trends in the U.S.: 1990-1998. *Diabetes Care* 23:1278-83.

Monauni, T., M.G. Zenti, A. Cretti, M.C. Daniels, G. Targher, B. Caruso, M. Caputo, D. McClain, S. Del Prato, A. Giaccari, M. Muggeo, E. Bonora, and R.C. Bonadonna. 2000. Effects of glucosamine infusion on insulin secretion and insulin action in humans. *Diabetes* 49:926-35.

Moore, M.C., and A.D. Cherrington. 1996. Regulation of net hepatic glucose uptake: Interaction of neural and pancreatic mechanisms. *Reprod Nutr Dev* 36:399-406.

Mootha, V.K., C.M. Lindgren, K.F. Eriksson, A. Subramanian, S. Sihag, J. Lehar, P. Puigserver, E. Carlsson, M. Ridderstrale, E. Laurila, N. Houstis, M.J. Daly, N. Patterson, J.P. Mesirov, T.R. Golub, P. Tamayo, B. Spiegelman, E.S. Lander, J.N. Hirschhorn, D. Altshuler, and L.C. Groop. 2003. PGC-1alpha-responsive genes involved in oxidative phosphorylation are coordinately downregulated in human diabetes. *Nat Genet* 34:267-73.

Morino, K., K.F. Petersen, S. Dufour, D. Befroy, J. Frattini, N. Shatzkes, S. Neschen, M.F. White, S. Bilz, S. Sono, M. Pypaert, and G.I. Shulman. 2005. Reduced mitochondrial density and increased IRS-1 serine phosphorylation in muscle of insulin-resistant offspring of type 2 diabetic parents. *J Clin Invest* 115:3587-93.

Nelson, B.A., K.A. Robinson, and M.G. Buse. 2000. High glucose and glucosamine induce insulin resistance via different mechanisms in 3T3-L1 adipocytes. *Diabetes* 49:981-91.

Nemoto, S., M.M. Fergusson, and T. Finkel. 2004. Nutrient availability regulates SIRT1 through a forkhead-dependent pathway. *Science* 306:2105-8.

Norrbom, J., C.J. Sundberg, H. Ameln, W.E. Kraus, E. Jansson, and T. Gustafsson. 2004. PGC-1alpha mRNA expression is influenced by metabolic perturbation in exercising human skeletal muscle. *J Appl Physiol* 96:189-94.

Obici, S., J. Wang, R. Chowdury, Z. Feng, U. Siddhanta, K. Morgan, and L. Rossetti. 2002. Identification of a biochemical link between energy intake and energy expenditure. *J Clin Invest* 109:1599-605.

Patti, M.E., E. Brambilla, L. Luzi, E.J. Landaker, and C.R. Kahn. 1998. Bidirectional modulation of insulin action by amino acids. *J Clin Invest* 101:1519-29.

Patti, M.E., A.J. Butte, S. Crunkhorn, K. Cusi, R. Berria, S. Kashyap, Y. Miyazaki, I. Kohane, M. Costello, R. Saccone, E.J. Landaker, A.B. Goldfine, E. Mun, R. DeFronzo, J. Finlayson, C.R. Kahn, and L.J. Mandarino. 2003. Coordinated reduction of genes of oxidative metabolism in humans with insulin resistance and diabetes: Potential role of PGC1 and NRF1. *Proc Natl Acad Sci USA* 100:8466-71.

Patti, M.E., A. Virkamaki, E.J. Landaker, C.R. Kahn, and H. Yki-Jarvinen. 1999. Activation of the hexosamine pathway by glucosamine in vivo induces insulin resistance of early postreceptor insulin signaling events in skeletal muscle. *Diabetes* 48:1562-71.

Patti, M.E., and B.B. Kahn. 2004. Nutrient sensor links obesity with diabetes risk. *Nat Med* 10:1049-50.

Petersen, K.F., D. Befroy, S. Dufour, J. Dziura, C. Ariyan, D.L. Rothman, L. DiPietro, G.W. Cline, and G.I. Shulman. 2003. Mitochondrial dysfunction in the elderly: Possible role in insulin resistance. *Science* 300:1140-2.

Petersen, K.F., S. Dufour, D. Befroy, R. Garcia, and G.I. Shulman. 2004. Impaired mitochondrial activity in the insulin-resistant offspring of patients with type 2 diabetes. *New Engl J Med* 350:664-71.

Pilegaard, H., B. Saltin, and P.D. Neufer. 2003. Exercise induces transient transcriptional activation of the PGC-1alpha gene in human skeletal muscle. *J Physiol* 546:851-8.

Potter, C.J., L.G. Pedraza, and T. Xu. 2002. Akt regulates growth by directly phosphorylating Tsc2. *Nat Cell Biol* 4:658-65.

Richardson, D.K., S. Kashyap, M. Bajaj, K. Cusi, S.J. Mandarino, J. Finlayson, R.A. DeFronzo, C.P. Jenkinson, and L.J. Mandarino. 2004. Lipid infusion decreases the expression of nuclear encoded mitochondrial genes and increases expression of extracellular matrix genes in human skeletal muscle. *J Biol Chem* 280:10290.

Robinson, K.A., D.A. Sens, and M.G. Buse. 1993. Pre-exposure to glucosamine induces insulin resistance of glucose transport and glycogen synthesis in isolated rat skeletal muscles. Study of mechanisms in muscle and in rat-1 fibroblasts overexpressing the human insulin receptor. *Diabetes* 42:1333-46.

Rodgers, J.T., C. Lerin, W. Haas, S.P. Gygi, B.M. Spiegelman, and P. Puigserver. 2005. Nutrient control of glucose homeostasis through a complex of PGC-1alpha and SIRT1. *Nature* 434:113-8.

Roh, C., J. Han, A. Tzatsos, and K.V. Kandror. 2003. Nutrient-sensing mTOR-mediated pathway regulates leptin production in isolated rat adipocytes. *Am J Physiol Endocrinol Metab* 284:E322-30.

Rossetti, L. 2000. Perspective: Hexosamines and nutrient sensing. *Endocrinology* 141:1922-5.

Rossetti, L., M. Hawkins, W. Chen, J. Gindi, and N. Barzilai. 1995. In vivo glucosamine infusion induces insulin resistance in normoglycemic but not in hyperglycemic conscious rats. *J Clin Invest* 96:132-40.

Rossetti, L., D.L. Rothman, R.A. DeFronzo, and G.I. Shulman. 1989. Effect of dietary protein on in vivo insulin action and liver glycogen repletion. *Am J Physiol* 257:E212-9.

Ruderman, N.B., A.K. Saha, D. Vavvas, and L.A. Witters. 1999. Malonyl-CoA, fuel sensing, and insulin resistance. *Am J Physiol* 276:E1-18.

Rumberger, J.M., T. Wu, M.A. Hering, and S. Marshall. 2003. Role of hexosamine biosynthesis in glucose-mediated up-regulation of lipogenic enzyme mRNA levels: Effects of glucose, glutamine, and glucosamine on glycerophosphate dehydrogenase, fatty acid synthase, and acetyl-CoA carboxylase mRNA levels. *J Biol Chem* 278:28547-52.

Russell, A.P., J. Feilchenfeldt, S. Schreiber, M. Praz, A. Crettenand, C. Gobelet, C.A. Meier, D.R. Bell, A. Kralli, J.P. Giacobino, and O. Deriaz. 2003. Endurance training in humans leads to fiber type-specific increases in levels of peroxisome proliferator-activated receptor-gamma coactivator-1 and peroxisome proliferator-activated receptor-alpha in skeletal muscle1. *Diabetes* 52:2874-81.

Salt, I.P., G. Johnson, S.J. Ashcroft, and D.G. Hardie. 1998. AMP-activated protein kinase is activated by low glucose in cell lines derived from pancreatic beta cells, and may regulate insulin release. *Biochem J* 335:533-9.

Sarbassov, D.D., S.M. Ali, D.H. Kim, D.A. Guertin, H. Erdjument-Bromage, P. Tempst, and D.M. Sabatini. 2004. Rictor, a novel binding partner of mTOR, defines a rapamycin-insensitive and raptor-independent pathway that regulates the cytoskeleton. *Curr Biol* 14:1296-302.

Sarbassov, D.D., D.A. Guertin, S.M. Ali, and D.M. Sabatini. 2005. Phosphorylation and regulation of Akt/PKB by the rictor-mTOR complex. *Science* 307:1098-101.

Schmelzle, T., and M.N. Hall. 2000. TOR, a central controller of cell growth. *Cell* 103:253-62.

Schuit, F.C., P. Huypens, H. Heimberg, and D.G. Pipeleers. 2001. Glucose sensing in pancreatic beta-cells: A model for the study of other glucose-regulated cells in gut, pancreas, and hypothalamus. *Diabetes* 50:1-11.

Shah, O.J., Z. Wang, and T. Hunter. 2004. Inappropriate activation of the TSC/Rheb/mTOR/S6K cassette induces IRS1/2 depletion, insulin resistance, and cell survival deficiencies. *Curr Biol* 14:1650-6.

Slawson, C., M.P. Housley, and G.W. Hart. 2006. O-GlcNAc cycling: How a single sugar post-translational modification is changing the way we think about signaling networks. *J Cell Biochem* 97:71-83.

Sparks, L.M., H. Xie, R.A. Koza, R. Mynatt, M.W. Hulver, G.A. Bray, and S.R. Smith. 2005. A high-fat diet coordinately downregulates genes required for mitochondrial oxidative phosphorylation in skeletal muscle. *Diabetes* 54:1926-33.

Tokunaga, C., K. Yoshino, and K. Yonezawa. 2004. mTOR integrates amino acid- and energy-sensing pathways. *Biochem Biophys Res Commun* 313:443-6.

Tremblay, F., A. Gagnon, A. Veilleux, A. Sorisky, and A. Marette. 2005. Activation of the mammalian target of rapamycin pathway acutely inhibits insulin signaling to Akt and glucose transport in 3T3-L1 and human adipocytes. *Endocrinology* 146:1328-37.

Tremblay, F., and A. Marette. 2001. Amino acid and insulin signaling via the mTOR/p70 S6 kinase pathway. A negative feedback mechanism leading to insulin resistance in skeletal muscle cells. *J Biol Chem* 276:38052-60.

Tzatsos, A., and K.V. Kandror. 2006. Nutrients suppress phosphatidylinositol 3-kinase/Akt signaling via raptor-dependent mTOR-mediated insulin receptor substrate 1 phosphorylation. *Mol Cell Biol* 26:63-76.

Um, S.H., F. Frigerio, M. Watanabe, F. Picard, M. Joaquin, M. Sticker, S. Fumagalli, P.R. Allegrini, S.C. Kozma, J. Auwerx, and G. Thomas. 2004. Absence of S6K1 protects against age- and diet-induced obesity while enhancing insulin sensitivity. *Nature* 431:200-5.

Veerababu, G., J. Tang, R.T. Hoffman, M.C. Daniels, L.F. Hebert Jr., E.D. Crook, R.C. Cooksey, and D.A. McClain. 2000. Overexpression of glutamine: Fructose-6-phosphate amidotransferase in the liver of transgenic mice results in enhanced glycogen storage, hyperlipidemia, obesity, and impaired glucose tolerance. *Diabetes* 49:2070-8.

Viollet, B., F. Andreelli, S.B. Jorgensen, C. Perrin, A. Geloen, D. Flamez, J. Mu, C. Lenzner, O. Baud, M. Bennoun, E. Gomas, G. Nicolas, J.F. Wojtaszewski, A. Kahn, D. Carling, F.C. Schuit, M.J. Birnbaum, E.A. Richter, R. Burcelin, and S. Vaulont. 2003. The AMP-activated protein kinase alpha2 catalytic subunit controls whole-body insulin sensitivity. *J Clin Invest* 111:91-8.

Virkamaki, A., M.D. Daniels, S. Hamalainen, T. Utriainen, D. McClain, and H. Yki-Jarvinen. 1997. Activation of the hexosamine pathway by glucosamine in vivo induces insulin resistance in multiple insulin sensitive tissues. *Endocrinology* 138:2501-7.

Vosseller, K., L. Wells, M.D. Lane, and G.W. Hart. 2002. Elevated nucleocytoplasmic glycosylation by O-GlcNAc results in insulin resistance associated with defects in Akt activation in 3T3-L1 adipocytes. *Proc Natl Acad Sci USA* 99:5313-8.

Wang, J., R. Liu, M. Hawkins, N. Barzilai, and L. Rossetti. 1998. A nutrient-sensing pathway regulates leptin gene expression in muscle and fat. *Nature* 393:684-8.

Weigert, C., K. Klopfer, C. Kausch, K. Brodbeck, M. Stumvoll, H.U. Haring, and E.D. Schleicher. 2003. Palmitate-induced activation of the hexosamine pathway in human myotubes: Increased expression of glutamine:fructose-6-phosphate aminotransferase. *Diabetes* 52:650-6.

Winder, W.W., and D.G. Hardie. 1999. AMP-activated protein kinase, a metabolic master switch: Possible roles in type 2 diabetes. *Am J Physiol* 277:E1-10.

Winder, W.W., B.F. Holmes, D.S. Rubink, E.B. Jensen, M. Chen, and J.O. Holloszy. 2000. Activation of AMP-activated protein kinase increases mitochondrial enzymes in skeletal muscle. *J Appl Physiol* 88:2219-26.

Woods, A., D. Azzout-Marniche, M. Foretz, S.C. Stein, P. Lemarchand, P. Ferre, F. Foufelle, and D. Carling. 2000. Characterization of the role of AMP-activated protein kinase in the regulation of glucose-activated gene expression using constitutively active and dominant negative forms of the kinase. *Mol Cell Biol* 20:6704-11.

Yamauchi, T., J. Kamon, Y. Minokoshi, Y. Ito, H. Waki, S. Uchida, S. Yamashita, M. Noda, S. Kita, K. Ueki, K. Eto, Y. Akanuma, P. Froguel, F. Foufelle, P. Ferre, D. Carling, S. Kimura, R. Nagai, B.B. Kahn, and T. Kadowaki. 2002. Adiponectin stimulates glucose utilization and fatty-acid oxidation by activating AMP-activated protein kinase. *Nat Med* 8:1288-95.

Yki-Jarvinen, H., M.C. Daniels, A. Virkamäki, S. Makimattila, R.A. DeFronzo, and D.A. McClain. 1996. Increased glutamine:fructose-6-phosphate amidotransferase activity in skeletal muscle of patients with NIDDM. *Diabetes* 45:302-7.

Zhou, G., R. Myers, Y. Li, Y. Chen, X. Shen, J. Fenyk-Melody, M. Wu, J. Ventre, T. Doebber, N. Fujii, N. Musi, M.F. Hirshman, L.J. Goodyear, and D.E. Moller. 2001. Role of AMP-activated protein kinase in mechanism of metformin action. *J Clin Invest* 108:1167-74.

Chapter 7

Inflammation-Induced Insulin Resistance in Obesity
When Immunity Affects Metabolic Control

Phillip James White, MS; and André Marette, PhD

Obesity: A Chronic Low-Grade Inflammatory State

It is well established that obesity is a chronic inflammatory disorder (Wellen and Hotamisligil 2005). Obesity-linked type 2 diabetes is associated with a cytokine-mediated acute-phase response or stress response (Pickup et al. 1997; Visser et al. 1999; Yudkin et al. 1999; Cook et al. 2000). Interestingly, levels of C-reactive protein (CRP), an acute-phase response protein and a sensitive marker of low-grade inflammation, are associated with higher adiposity in children (Cook et al. 2000) and young adults who are healthy (Yudkin et al. 1999) and are independently related to insulin sensitivity in subjects who are nondiabetic (Festa et al. 2000). These findings confirm that an early onset, low-level systemic inflammation exists in persons who are overweight or obese. This chronic inflammatory state is in line with observations of elevated plasma levels of interleukin (IL)-6 and tumor necrosis factor (TNF)α (Yudkin et al. 1999) and of overexpression of TNFα, IL-1β, IL-6, and interferon (IFN)-γ in adipose tissues of humans and animals exhibiting obesity (Hotamisligil, Shargill, and Spiegelman 1993; Perreault and Marette 2001; see table 7.1).

The underlying cause of inflammation in obesity remains poorly understood but is suspected to lie within the origin of fat cells. Indeed, metabolic and immune pathways have evolved to be closely linked and interdependent. An ever-increasing number of molecules that are best known for their roles in immune and inflammatory cells are now considered as key modulators of energy metabolism in insulin target cells. Moreover, a growing number of adipose-specific molecules termed *adipokines,* including leptin, resistin, and adiponectin, have been shown to modulate both metabolism and inflammation through a complex interplay of signal transduction mechanisms that we are just beginning to fully appreciate. The recent findings that obesity is characterized by macrophage accumulation in adipose tissue and that macrophages and fat cells share the expression of multiple genes have added another dimension to our understanding of the development of adipose tissue inflammation in obesity (Weisberg et al. 2003; Xu et al. 2003). Macrophages in adipose tissue are now believed to help produce inflammatory mediators either alone or in concert with adipocytes. This view suggests that macrophages play a critical role in promoting obesity-linked insulin resistance. However, direct evidence to establish this connection is still warranted. In this chapter, we review the evidence linking obesity and inflammation and discuss the underlying events in this interaction that eventually lead to type 2 diabetes and cardiovascular disease.

The work described in this chapter was supported by grants from the Canadian Institutes of Health Research (CIHR), the Canadian Diabetes Association, and the Fonds de la Recherche en Santé du Québec (FRSQ) to A.M. A.M. is also the recipient of a CIHR investigator award and a FRSQ National Scientist.

Table 7.1 Immune Factors Influenced by Obesity

Molecule	Class	Physiological function	Obesity effect
Adiponectin	Adipokine	Suppresses macrophage function, NK cell cytotoxicity, and myelomonocytic proliferation; induces anti-inflammatory cytokine production and insulin sensitization (Yokota et al. 2000; Kim et al. 2006; Wolf et al. 2004; Yamauchi et al. 2001)	↓ plasma [ ] (Arita et al. 1999)
Adipsin	Adipokine	Analogue of human complement factor D (Esterbauer et al. 1999)	↑ blood [ ] (Napolitano et al. 1994)
CRP	APP	Induces cytokine production, complement activation, phagocytosis, and antigenic and apoptotic cell clearance (Marnell, Mold, and Du Clos 2005)	↑ serum [ ] (Visser et al. 1999)
Ghrelin	Peptide hormone	Negative regulator of proinflammatory cytokine production; orexigenic; stimulates growth hormone release and neurogenesis (Dixit and Taub 2005)	↓ plasma [ ] (Tschop et al. 2001)
Haptoglobin	APP	Prevents iron loss and superoxide production; stimulates angiogenesis (Dobryszycka 1997)	↑ serum [ ] (Chiellini et al. 2004)
IFN-γ	Cytokine	Proinflammatory; promotes activation of the innate immune system; enhances antigen presentation and iNOS expression; regulates balance between TH1 and TH2; controls cellular proliferation and apoptosis (Tau and Rothman 1999)	↑ adipose tissue [ ] (Perreault and Marette 2001)
IL-1β	Cytokine	Proinflammatory; induces COX-2, PLA2, and iNOS (Dinarello 2002)	↑ serum [ ] (Aygun et al. 2005)
IL-1RA	Cytokine	Anti-inflammatory; competitively inhibits IL-1 (Dinarello 2002)	↑ serum [ ] (Meier et al. 2002)
IL-6	Cytokine	Activates acute-phase response; stimulates lymphocytes; enhances hematopoietic colony formation (Kishimoto et al. 1995)	↑ serum [ ] (Roytblat et al. 2000)
IL-8	Chemokine	Recruits and activates neutrophils (Remick 2005)	↑ plasma [ ] (Straczkowski et al. 2002)
IL-10	Cytokine	Anti-inflammatory; prevents IFN-γ, IL-1β, TNFα, IL-8, IL-12, and NO production by macrophages, monocytes, and TH1 cells (Moore et al. 2001)	↑ serum [ ] in obesity, ↓ serum [ ] with MS (Esposito et al. 2003)
Leptin	Adipokine	Centrally regulates energy balance; regulates puberty and reproduction; functions in hematopoiesis and chemotaxis; modulates adaptive immune response; induces synthesis of proinflammatory cytokines, NO, and eicosanoids (Otero et al. 2005)	↑ serum [ ] (Considine et al. 1996)
MCP-1	Chemokine	Induces monocyte, basophil, NK cell, and T lymphocyte chemotaxis and IL-4, IL-5, and IFN-γ production (Lu et al. 1998; Le et al. 2004)	↑ plasma [ ] (Takahashi et al. 2003)
M-CSF	Cytokine	Primary regulator of mononuclear phagocyte survival, proliferation, differentiation, and function; mediates adipose tissue growth (Chitu and Stanley 2006)	↑ expression in adipose tissue (Levine et al. 1998)
MIF	Cytokine	Proinflammatory; enhances phagocytosis; induces TNFα, IL-8, and IL-12; inhibits apoptosis; antagonizes the anti-inflammatory effects of glucocorticoids (Morand 2005)	↑ plasma [ ] (Dandona et al. 2004)
RANTES	Chemokine	Induces leukocyte chemotaxis (Appay and Rowland-Jones 2001)	↑ serum [ ] (Herder et al. 2006)
Resistin	Adipokine	Proinflammatory; induces TNFα, IL-1β, IL-6, and IL-12; diabetogenic (McTernan, Kusminski, and Kumar 2006)	↑ serum [ ] (Steppan et al. 2001)
SAA	APP	Promotes monocyte and T lymphocyte chemotaxis and adhesion; induces expression of extracellular matrix degrading metalloproteinases (O'Brien and Chait 2006)	↑ plasma [ ] (O'Brien et al. 2005)
TGF-β	Cytokine	Inhibits lymphoid and myeloid cell proliferation; induces differentiation of myeloid dendritic cells; suppresses tumor formation (Letterio 2000)	↑ plasma [ ] (Romano et al. 2003)
TNFα	Cytokine	Proinflammatory; activates neutrophils and platelets; enhances macrophage and NK cell phagocytosis; stimulates immune system (Idriss and Naismith 2000)	↑ plasma [ ] (Hotamisligil, Shargill, and Spiegelman 1993)
Visfatin	Adipokine	Proinflammatory; B cell growth factor; inhibits neutrophil apoptosis; promotes smooth muscle cell maturation; possesses insulin mimetic activity (Stephens and Vidal-Puig 2006)	↑ plasma [ ] (Fukuhara et al. 2005)

APP, acute-phase protein; COX-2, cyclooxygenase-2; CRP, C-reactive protein; IFN, interferon; IL, interleukin; iNOS, inducible nitric oxide synthase; MCP-1, monocyte chemoattractant protein-1; M-CSF, monocyte colony stimulating factor; MIF, macrophage migration inhibitory factor; MS, metabolic syndrome; NK, natural killer; NO, nitric oxide; PLA2, phospholipase A2; RA, receptor antagonist; RANTES, regulated upon activation, normal T-cell expressed, and secreted; SAA, serum amyloid A; TGF-β, transforming growth factor beta; TH, T helper; TNFα, tumor necrosis factor alpha; [ ], concentration; ↓, decrease; ↑, increase.

Evolution of Inflammation in Obesity

Macrophage accumulation in adipose tissue is one of the earliest known events leading to the inflammatory state in obesity. This portion of the chapter discusses the information surrounding macrophage accrual in adipose tissue and the processes through which inflammation evolves alongside a growing adipose tissue mass.

Macrophage Accumulation: A Tale of Adipocyte Crossover or Monocyte Migration?

The resident macrophage population arises from circulating monocytes that originate in the bone marrow from a common myeloid progenitor. Forming part of the reticuloendothelial system and playing a major role in innate immunity, the macrophage is localized in tissues, where it is involved in inflammation, tissue repair, and host defense against infectious agents (Van Furth and Cohn 1968). The circulating monocyte population responsible for backup support and maintenance of tissue macrophage numbers displays heterogeneity in size, granularity, and nuclear morphology (for review, see Gordon and Taylor 2005). The type of chemokine receptors expressed by blood monocytes indicates their physiological function. The CCR2+ subset, for instance, is sensitive to the CC chemokine monocyte chemoattractant protein-1 (MCP-1) and rapidly recruits to sites of inflammation, where it differentiates into macrophages and antigen-presenting dendritic cells. In contrast, the CCR2- population does not respond to the MCP-1 signal and is believed to be responsible for replenishing the resident macrophage and dendritic cell populations in native tissue in the absence of inflammation (Geissman, Jung, and Littman 2003).

Evidence of a close link between adipocyte and macrophage lineages has been present since the earliest studies into the biological nature of adipocytes. The early pioneers of adipocyte biology often proposed a place for the adipocyte in the reticuloendothelial system, citing the intimate spatial relationship between adipose tissue and bone marrow, blood vessels, and connective tissue as an indicator of a functional role in this system (Bell 1909; Inglis 1927; Portis 1924). Further

works in this early era also uncovered functional physiological similarities between the adipocyte and the macrophage, the most notable of which was the phagocytic potential of preadipose cells (Latta and Rutledge 1935; McCullough 1944). These early experiments revealed that these cells, then described as *lean adipose cells* and now recognized as preadipocytes, behave similarly to macrophages, while larger, mature adipocytes display a reduced phagocytic potential that is regained upon fat depletion by means of fasting.

Although our understanding of adipocyte biology has progressed and more than 60 y have passed since these pioneering reports, the results have stood the test of time. Indeed, Cousin and colleagues (1999) recently demonstrated once again the macrophage-like activities of preadipocytes. In accordance with the findings of Latta, Rutledge, and McCullough, this work revealed that the majority of preadipocytes exhibit the same phagocytic potential displayed by peritoneal macrophages, while only a much smaller number of mature adipocytes display this capacity. Cousin and colleagues (1999) also demonstrated that preadipocytes contain microbicidal activity identical to that of the peritoneal macrophage, furthering the functional link between these two lineages. In addition, both growing and mature adipocytes express the cell surface antigen MOMA-2, a specific marker of monocyte and macrophage lineage. This observation suggests substantial genetic crossover in these two lineages and the possibility that both originally derive from the same precursor cell.

In consideration of this possibility, Charriere and colleagues (2003) examined the extent of commonality between the adipocyte and the macrophage lineages. This comparison of adipocytes, preadipocytes, and macrophages revealed an extraordinary extent of genetic crossover. Indeed, the gene expression profile of the preadipocyte resembled that of the macrophage to a greater extent than it resembled that of its own successor, the mature adipocyte. Preadipocytes also displayed great plasticity when in a macrophage-laden environment. In fact, when labeled preadipocytes were injected into the peritoneal cavity of nude mice, they rapidly attained high phagocytic activity, and 60% to 70% began expressing five macrophage-specific antigens: F4/80, Mac-1, CD80, CD86, and CD45. These expression values were much the same as those recorded for peritoneal macrophages, suggesting the induction of complete phenotypic crossover.

Subsequent in vitro analysis revealed that macrophages and adipocytes required cell-to-cell contact for such phenotypic conversion to occur.

Alongside the shared capacity of these two lineages to perform phagocytosis and the potential of adipocytes to dedifferentiate into macrophages lies an astonishing amount of physiological crossover. For instance, macrophages, like adipocytes, are capable of storing large quantities of triglycerides, and they actually become foam cells in atherosclerotic lesions (Funk et al. 1993). Conversely, adipocyte-derived hormones such as adiponectin, adipsin, leptin, and resistin have physiological roles in regulating the immune response (Yokota et al. 2000; White et al. 1992; Loffreda et al. 1998; McTernan, Kuminski, and Kumar 2006). Furthermore, macrophages express the nuclear receptor PPARγ (Braissant et al. 1996) and the fatty acid binding protein aP2 (Fu, Luo, and Lopez-Virella 2000), both of which have metabolic functions in adipose tissue differentiation. Meanwhile, adipocytes and preadipocytes express proinflammatory mediators such as TNFα, IL-6, and inducible nitric oxide synthase (iNOS; Hotamisligil, Shargill, and Speigelman 1993; Perreault and Marette 2001; Pilon, Dallaire, and Marette 2004). What is more, a recently identified circulating adipocyte progenitor cell has been shown to express the chemokine receptor CCR2 of the monocyte and macrophage lineage (Hong et al. 2005).

The extent of this crossover, which is both functional and genetic in nature, might lead to the hypothesis that the adipose tissue inflammation seen in obesity arises directly from the adipocytes and preadipocytes themselves. However, works by Weisberg and colleagues (2003) and Xu and colleagues (2003) have ruled this out. These studies have defined macrophage accumulation in the stromal vascular fraction of the expanding adipose tissue as a hallmark of obesity in mice and also in humans. In each case, the number of macrophages accumulated in the expanding adipose tissue correlated positively with BMI and adipocyte size. Furthermore, histological reports of the macrophages in the obese adipose tissue described regular formations of aggregates resembling the macrophage syncytia that are characteristic of chronic inflammatory states. Although adipocytes can produce TNFα, IL-6, and iNOS, expression analysis of macrophage and nonmacrophage cell populations isolated from adipose tissue demonstrated that almost all TNFα expression and considerable amounts of IL-6 and iNOS expression in the inflamed adipose tissue were macrophage derived. These reports suggest that the macrophage is the major proinflammatory body in the expanding adipose tissue.

To determine the origin of the proinflammatory macrophage population in the expanding adipose tissue, Weisberg and colleagues (2003) performed elegant bone marrow transplant studies. Bone marrow from C57BL/6J mice expressing the CD45.1 leukocyte marker was transplanted into 6 wk old lethally irradiated C57BL/6J mice expressing the CD45.2 leukocyte marker. Following 6 wk of a high-fat diet, approximately 85% of the macrophage cells in the adipose tissue were donor derived (i.e., CD45.1+). This finding suggests that most of the macrophages accumulated in adipose tissue with obesity are derived from bone marrow and are not the result of preadipocyte conversion to macrophages. It is still plausible that preadipocyte conversion occurs in response to the changing nature of the expanding adipose tissue following the establishment of an active inflammatory macrophage population derived from bone marrow, but this is yet to be investigated. Thus it appears at present that the story of macrophage accumulation in adipose tissue during obesity is a migratory tale.

Why Are Circulating Monocytes Attracted to Fat?

The accumulation of leukocytes at sites of inflammation is mediated by chemotactic proteins termed *chemokines* (for review, see Rossi and Zlotnik 2000). MCP-1, otherwise known as *CCL2*, is the most studied member of its CC chemokine subfamily. MCP-1 binds specifically to the G protein-coupled chemokine receptor CCR2, which it shares with MCP-2 (CCL8), MCP-3 (CCL7), MCP-4 (CCL13), and MCP-5 (CCL12; Le et al. 2004). Although functional crossover among these chemokines seems likely, evidence from KO models suggests that the role of MCP-1 in CCR2 signaling is not redundant, particularly regarding monocyte migration into sites of inflammation (Lu et al. 1998). MCP-1 also directs chemotactic signals toward memory T lymphocytes, natural killer cells, and basophils (Le et al. 2004) and influences the expression of cytokines related to T helper responses, specifically IL-4, IL-5, and IFN-γ (Lu et al. 1998). Thus MCP-1 is an important mediator of the inflammatory response.

The overexpression of MCP-1 has been reported in multiple pathogenic inflammatory conditions characterized by macrophage infiltration, including atherosclerosis (Yla-Herttuala et al. 1991), rheumatoid arthritis (Koch et al. 1992), and experimental autoimmune encephalomyelitis (Ransohoff et al. 1993). Recently obesity has been added to this list. Plasma MCP-1 is enhanced in animal models of obesity, and weight reduction dramatically reduces the circulating concentrations of MCP-1 (Takahashi et al. 2003). Furthermore, human adipose tissue expresses MCP-1, and this expression correlates positively with BMI (Christiansen, Richelsen, and Bruun 2005). Work in visceral adipose tissue explants taken from subjects who are morbidly obese has revealed that MCP-1 expression is regulated by nuclear factor NF-κβ and p38 MAPK signaling and can be induced by inflammatory factors that the adipose tissue overexpresses in cases of obesity, including TNFα and IL-1β (Fain and Madan 2005). Other factors, including IL-6, insulin, and growth hormone, also influence MCP-1 expression (Fasshauer et al. 2004; Sartipy and Loskutoff 2003a). Interestingly, MCP-1 has been shown to induce insulin resistance in adipocytes and myocytes in vitro (Sartipy and Loskutoff 2003b; Sell, Dietze-Schroeder, Kaiser et al. 2006). What is more, the A-2518G polymorphism in the regulatory element of the MCP-1 gene that results in reduced MCP-1 expression has been associated with a decreased prevalence of insulin resistance and diabetes in the affected population. Taken together, these data lend support to the role of MCP-1 in metabolic disorders related to obesity (Simeoni et al. 2004).

Functional evidence that MCP-1 and CCR2 play a role in macrophage recruitment to the expanding adipose tissue and in the development of obesity-related metabolic disorders has been established. Weisberg and colleagues (2006) investigated the role of CCR2, while Kanda and colleagues (2006) investigated the role of MCP-1. In these studies, genetic ablation of either MCP-1 or CCR2 yielded very similar phenotypes in obese mice and thus confirmed suspicion of a uniform role. These mice weighed as much as their control counterparts and yet displayed reduced inflammation and mac-

rophage content in adipose tissue. Furthermore, these changes improved metabolic function, as displayed by enhanced insulin sensitivity and the amelioration of hepatic steatosis. Similar results were obtained using antagonists of either CCR2 or MCP-1. In contrast, the model of adipose-specific MCP-1 overexpression used by Kanda and colleagues displayed inflammation, macrophage accumulation in adipose tissue, and impaired metabolic function without the presence of obesity. Thus the integral role of MCP-1 and CCR2 signaling in macrophage–adipocyte coupling in obesity is clear. Moreover, evidence of a link between macrophage infiltration into expanding adipose tissue and metabolic complications such as insulin resistance and hepatic steatosis is now well established.

Interestingly, the obesity-linked changes in the expression of more traditional adipokines appear to complement the enhanced expression of MCP-1. For instance, the expansion of adipose tissue downregulates the insulin-sensitizing adiponectin responsible for negatively regulating myelomonocytic proliferation and macrophage function (Yokota et al. 2000), whereas it greatly enhances leptin, a satiety factor known to stimulate multilineage expansion of hematopoietic progenitor cells in the bone marrow (Bennett et al. 1996) and to upregulate the expression of endothelial cell adhesion molecules and thus enhance the chemotaxis of blood monocytes (Curat et al. 2004). This complementary adipokine regulation boosts both the monocytic pool and the macrophage activity and thereby maximizes the potential effects of the MCP-1 signal coming from the expanding adipose tissue. Thus the modulation of macrophage function and migration in obesity involves not only the upregulation of MCP-1 but also the coordinate regulation of the entire adipokine milieu.

Induction of the First Chemotactic Signal by Adipose Tissue

Although the integral contribution of MCP-1 and CCR2 signaling to macrophage migration is well established, it is not yet understood how MCP-1 comes to be overexpressed by the expanding adipose tissue. One interesting explanation for this phenomenon is the endoplasmic reticulum (ER) stress theory (Wellen and Hotamisligil 2005). ER stress occurs when changes in cellular metabolism result in the accumulation of unfolded or misfolded proteins in the ER. In mammals, ER overload leads to the activation of NF-κβ and c-Jun N-terminal kinase (JNK) pathways and to the transcription of inflammatory genes such as those encoding the proinflammatory cytokines. A recent investigation by Ozcan and colleagues (2004) has linked ER stress to obesity and the development of insulin resistance. These researchers demonstrated the occurrence of ER stress in peripheral tissues during obesity and showed that the genetic induction of ER stress induces insulin resistance in the absence of obesity. In accordance with these data, it is postulated that expanding adipocytes enter into ER stress in response to the higher requirement for protein and lipid synthesis demanded by the changes in tissue architecture. Such ER overload activates both the NF-κβ and JNK pathways and leads to the translation of various proinflammatory cytokines, including TNFα, IL-6, and possibly MCP-1. Should MCP-1 not be induced directly by ER stress in the expanding adipocyte, the autocrine and paracrine actions of TNFα and IL-6 would induce the translation of MCP-1 locally in the surrounding adipose tissue.

Alternatively, Cinti and colleagues (2005) have suggested that hypertrophy-induced necrosis-like adipocyte death induces macrophage accumulation in expanding adipose tissue. In support of this theory, they cite the higher prevalence of necrosis-like adipocyte death that occurs in the adipose tissue of obese mice and humans and the observation of what appear to be macrophage syncytia and multinucleate giant cells forming around the residual lipid droplets. In this model, the proinflammatory and chemotactic factors required for macrophage infiltration could be produced by the adipocyte necrosis-like cell death sequence or by the surrounding adipocytes, preadipocytes, and resident macrophages in response to the neighboring cell death. Conversely, the observed adipocyte death may be a feature of extended ER stress or may even occur secondarily to macrophage infiltration as a result of macrophage efforts to clear toxic nonfunctional cells from the adipose tissue. These concepts are quite thought provoking; however, neither model is yet proven. Thus the origin of the first chemotactic signal in the expanding adipose tissue remains in question.

Macrophages Might Start the Fire, But How Does the Fire Spread?

As adipose tissue expands in obesity, what begins as local inflammation induced by macrophage infiltration dependent on MCP-1 spreads into other tissues like a wildfire and eventually becomes chronic low-grade systemic inflammation. Indeed, by the time insulin resistance has developed, inflammatory signaling pathways have already been activated in most tissues throughout the body, and vascular damage has likely begun (Murdolo and Smith 2006). It appears that macrophage infiltration in adipose tissue leads to such pathogenic global inflammation by means of a proinflammatory positive feedback loop. In essence, the primary invading macrophages in the adipose tissue produce proinflammatory factors, including cytokines, chemokines, and nitric oxide, that induce insulin resistance and promote proinflammatory adipokine, chemokine, and cytokine production in the surrounding adipose and preadipose cells. This in turn results in further monocyte infiltration in the adipose tissue, greater production of proinflammatory mediators, and the induction of the acute-phase response (see figure 7.1). Interestingly, inflammation induced lipolysis in adipose tissue also likely contributes to the global spread of inflammation via free fatty acid induced Toll-Like Receptor 4 activation of macrophages and adipocytes (Shi et al. 2006), further underscoring the remarkable crossover present between metabolism and immunity.

The acute-phase response induced by proinflammatory cytokines such as IL-6 and TNFα involves a series of reactions, occurring in sites distal to the inflammatory foci, that promote the neutralization of the inflammatory agents (Baumann and Gauldie 1994). Upon the induction of the acute-phase response, hepatic protein synthesis switches to produce major plasma elements required for proper immune function. These elements include complement factors, cytokines, coagulation proteins, metal-binding proteins, proteinase inhibitors, and major acute-phase reactants such as CRP and serum amyloid A (SAA; Gabay and Kushner 1999). These factors further the spread of inflammation in obesity and are involved in the pathogenesis of obesity-related metabolic complications (Pickup 2004). Indeed, while the acute-phase response is seen to be necessary for resolving acute cases of

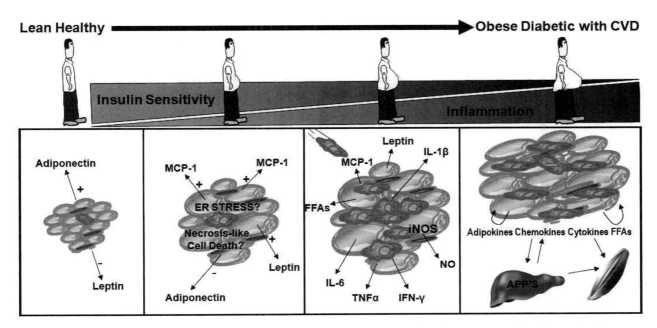

Figure 7.1 The evolution of inflammation in obesity is paralleled by the progression of insulin resistance and cardiovascular disease (CVD). It involves proinflammatory changes that commence locally in the expanding adipose tissue and spread to induce global low-grade inflammation. APPs, acute-phase proteins; NO, nitric oxide; FFAs, free fatty acids.

inflammation, such as occurs in bacterial infection, it appears to be harmful when sustained over prolonged durations, such as occurs in the chronic low-grade inflammation linked to obesity, in which circulating concentrations of acute-phase reactants such as CRP correlate positively with both markers of adiposity and metabolic disease risk. Thus the chronic feed-forward production of proinflammatory cytokines, chemokines, adipokines, acute-phase reactants, and FFAs mediates the spread of inflammation in obesity.

Mechanisms Linking Inflammation to Insulin Resistance in Obesity

Several lines of evidence support a causal link between inflammation and the development of insulin resistance. Indeed, acute inflammatory stimuli (e.g., LPS [lippolysaccharide]) as well as some proinflammatory cytokines, adipokines, and chemokines (e.g., TNFα, IL-1β, IL-6, IFN-γ, resistin, and MCP-1) overexpressed in obesity are known to promote insulin resistance (Portoles et al. 1989; Hotamisligil, Shargill, and Spiegelman 1993; Perreault and Marette 2001; Steppan et al. 2001; Sartipy and Loskutoff 2003b). This pathogenic effect of

inflammation is mediated by the activation of protein kinases, lipid mediators, and transcriptional pathways in insulin target tissues that may impair insulin signal transduction in sites both proximal and distal to the insulin receptor. Some of these pathways, particularly the kinase and lipid mediated, are also involved in insulin resistance induced by FFAs and amino acids (Gao et al. 2004; Tremblay and Marette 2001; Schmitz-Peiffer, Craig, and Biden 1999). This observation further highlights the existence of a crossover between metabolism and immunity. This portion of the chapter discusses the involvement of these pathways and the mechanisms potentially contributing to inflammation-induced insulin resistance.

Lipid Mediators

Inflammatory factors exert their inhibitory effect on insulin signaling partly by generating lipid mediators, particularly ceramides and gangliosides. The following section describes the potential mechanisms by which these lipid mediators inhibit insulin signal transduction.

Sphingomyelin Pathway

The sphingomyelin pathway is a lipid signaling pathway that is initiated when various stress signals activate sphingomyelinases (see Hannun 1996). The activated sphingomyelinases cleave the membrane phospholipid sphingomyelin, forming ceramide and phosphorylcholine. The newly formed ceramide is metabolically active and functions as a second messenger in the cell. In humans, the ceramide content of skeletal muscle is inversely related to insulin sensitivity (Straczkowski et al. 2004), and substantial evidence suggests that the proinflammatory cytokines TNFα, IL-1β, and IFN-γ, which are all overexpressed in obesity, employ the sphingomyelin pathway to effect signal transduction in target tissues (Kim et al. 1991; Kolesnick and Golde 1994). Accordingly, a growing body of research suggests that this pathway is necessary for the induction of TNFα-mediated insulin resistance. Indeed, TNFα signaling through the p55 TNFα receptor leads to sphingomyelinase activation and the formation of ceramides. When these ceramides are introduced to 32D cells in culture, they increase IRS1 serine phosphorylation and subsequently impair the insulin receptor kinase (Peraldi et al. 1996). Other studies suggest that ceramides act downstream of IRS1 (Summers et al. 1998), causing insulin resistance by blocking Akt recruitment to the plasma membrane (Stratford, DeWald, and Summers 2001), by stimulating dephosphorylation of Akt induced by protein phosphatase 2A (PP2A; Stratford et al. 2004), or by promoting the transcriptional repression of GLUT4 (Long and Pekala 1996). Thus, substantial support exists for the role of this lipid derivative in inflammation-induced insulin resistance (see figure 7.2).

The death domain of the p55 TNFα receptor, which is responsible for the TNFα inhibition of IRS1, also activates the acid sphingomyelinase (A-SMase; Csehi et al. 2005). Interestingly, the generation of ceramide induced by A-SMase can lead to the activation of other stress signaling pathways, including JNK and PKCζ (Westwick et al. 1995; Muller et al. 1995). These pathways may then act downstream of ceramide to induce insulin resistance. This is particularly true for the suppressive effect of ceramide on Akt recruitment to the plasma membrane, in which ceramide induces PKCζ-mediated phosphorylation of Akt on the PH domain necessary for Akt recruitment to the membrane (Powell et al. 2003). It is likely that a similar mechanism involving the activation of stress pathways is responsible for the ceramide-induced serine phosphorylation

of IRS1, but this remains to be elucidated. Thus, inflammation-induced insulin resistance involves the sphingomyelin pathway, which may directly or indirectly act on insulin signaling intermediates to promote insulin resistance.

Gangliosides

Gangliosides are another class of lipids proposed to mediate the insulin desensitizing effect of cytokines. These acidic glycosphingolipids are distributed alongside sphingomyelin and cholesterol within the cell membrane in functionally important microdomains known as *lipid rafts* (Kolter, Proia, and Sandhoff 2002). Gangliosides were presented as negative regulators of insulin receptor kinase by Nojiri, Stroud, and Hakomori in 1991, but it wasn't until more recently that Tagami and colleagues (2002) revealed their potential to mediate TNFα-induced insulin resistance. The latter study showed that TNFα administration elevates cellular ganglioside GM3 content by enhancing the expression of GM3 synthase and that this response is important for the effect of TNFα on insulin resistance. Heightened GM3 ganglioside content was also observed in the adipose tissue of two animal models of obesity. Other recent studies using genetic models of ganglioside overexpression (Sasaki et al. 2003) and depletion (Yamashita et al. 2003) have upheld these results. Thus, there is good support for the role of gangliosides in the mediation of inflammation-induced insulin resistance; however, much more work is required to determine the precise molecular mechanisms involved. It shall be interesting to see whether gangliosides, like ceramides, exert this function by interacting with other stress signaling pathways (see figure 7.2).

Protein Kinase Mediators

Protein kinases are an integral component of inflammatory signaling. In the following section we describe the protein kinases identified as molecular mediators of inflammation-induced insulin resistance and the mechanisms by which they exert their effects.

Mitogen-Activated Protein Kinases

It is well known that proinflammatory cytokines activate mitogen-activated protein kinases (MAPK) in several cell types. The classical MAPK pathways

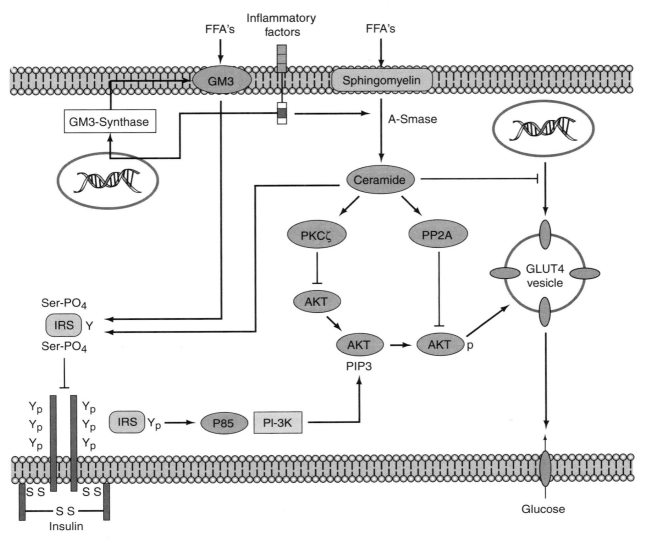

Figure 7.2 The lipid mediators of insulin resistance. Inflammatory factors such as TNFα may inhibit insulin signaling by enhancing the intracellular content of ceramide and GM3 ganglioside. The intracellular content of these lipids mediators may also be influenced by FFAs. →, leads to; ⊣, inhibits.

include p42/44 (ERK1/2), p38, and JNK. TNFα has been reported to activate p42/44 (Engelman et al. 2000), p38 (Engelman et al. 2000), and JNK (Rui et al. 2001), and the latter was reported to interact with IRS1 and increase its phosphorylation on Ser-307 (Aguirre et al. 2000). Ser-307 (Ser-312 in humans) is located near the phosphotyrosine binding (PTB) domain in IRS1 and is phosphorylated by several mechanisms, including insulin-stimulated kinases and TNFα-activated kinases like JNK1 (Rui et al. 2001; Aguirre et al. 2000). Phosphorylation of Ser-307 by JNK1 disrupts the interaction between the catalytic domain of the insulin receptor and the PTB domain of IRS1, resulting in reduced insulin

stimulation of downstream effectors such as PI3K (Aguirre et al. 2000).

In accordance with this mechanism, JNK activation is elevated in insulin target tissues in obesity, and the absence of JNK (induced by genetic ablation) confers resistance to obesity and enhances insulin signaling in insulin target tissues (Hirosumi et al. 2002). In addition, pharmacological inhibition of JNK using the cell-permeable JNK inhibitory peptide augments insulin sensitivity and glucose tolerance in diabetic mice (Kaneto et al. 2004). Somewhat interestingly, it appears that JNK interaction protein 1 (JIP1) is necessary for JNK-induced insulin resistance. Previously it was believed that

JIP1 acted as an inhibitory protein, anchoring JNK in the cytosol and thus preventing nuclear interactions; however, it is now known that JIP1 acts as a scaffolding protein to facilitate the interactions between JNK and its upstream signaling partners MAPK kinases 4 and 7 (MKK4/7; Jaeschke, Czech, and Davis 2004). Other studies investigating the role of JNK in cytokine-mediated insulin resistance have demonstrated that the lipolytic effect of TNFα in adipose tissues is JNK dependent (Ryden et al. 2004). Moreover, the hepatic expression of a dominant negative JNK decreases the expression of gluconeogenesis enzymes and reduces hepatic glucose output (Nakatani et al. 2004). Finally, adipose production of TNFα depends on JNK signaling (Nguyen et al. 2005). These findings all support a role for JNK in the development of obesity and inflammation-induced insulin resistance. It will be interesting to see whether this role is ceramide dependent.

Although less well studied, other members of the MAPK family appear to play a part in inflammation-induced insulin resistance. Indeed, TNFα-induced insulin resistance is blocked by PD98059, an inhibitor of MAPK kinase (MAP2K, or MEK), the upstream activator of p42/44. This finding suggests that the latter MAPK pathway may be involved in TNFα-linked insulin resistance (Rui et al. 2001; Engelman et al. 2000). Furthermore, the p38 MAPK is rapidly activated in adipocytes upon exposure to TNFα and IL-1, and adenovirus-mediated overexpression of p38 downregulates GLUT4 expression in these cells (Fujishiro et al. 2001). Thus the MAPK family appears to play a prominent role in inflammation-mediated insulin resistance. However, further research is necessary to elucidate how and to what extent each member is involved (see figure 7.3).

mTOR/S6K1 Pathway

First discovered as a target of the immunosuppressive drug rapamycin, mTOR is known to integrate signals arising from nutrients as well as from growth factors. mTOR and its downstream effector S6K1 both possess serine and threonine kinase activity. Activation of the mTOR/S6K1 pathway by insulin and amino acids inhibits insulin action on muscle glucose transport through increased IRS1 serine and threonine phosphorylation and accelerated deactivation of PI3K (Tremblay and Marette 2001).

More recent studies point to S6K1 as being the key mediator of both nutrient- and obesity-linked insulin resistance via phosphorylation of IRS1 on multiple serine molecules (see Um, D'Alessio, and Thomas 2006). It has also been suggested that the mTOR pathway mediates cytokine-induced insulin resistance. Indeed, both TNFα and IFN-γ can activate mTOR/S6K1 (Lekmine et al. 2004; Ozes et al. 2001) and cause inhibitory serine phosphorylation of IRS1 (Ozes et al. 2001). Moreover, endotoxin-induced phosphatidic acid accumulation leads to mTOR/S6K1 activation in macrophages, a process required for the induction of inflammatory mediators (Lim et al. 2003). Plomgaard and colleagues (2005) showed that TNFα infusion in humans who are healthy induces insulin resistance in skeletal muscle in association with increased activation of S6K1 and elevated serine phosphorylation of IRS1. Thus, there is growing evidence for cross talk between cytokine sensing pathways and nutrient sensing pathways that integrate inflammatory and nutrient excess signals to promote insulin resistance (see figure 7.3).

Janus-Activated Kinase

Janus-activated kinases (JAKs) act as tyrosine kinases for ligated receptors that lack intrinsic kinase activity. Accordingly, JAKs are involved in early signal transduction for a wide variety of polypeptides, including leptin, TNFα, most interleukins, and IFN-γ (for review, see Rane and Reddy 2000). In inflammatory signaling, MAPKs and also transcriptional mediators of inflammation-induced insulin resistance (e.g., SOCS) are often downstream of JAK, suggesting that this protein may be involved in inflammation-induced insulin resistance. Accordingly, JAK2 is upregulated in animals that are obese and insulin resistant (Gual et al. 1998; Rojas et al. 2000), and recent findings suggest that JAK2 may interact directly with insulin signaling intermediates and impair insulin signaling by depressing Akt Ser-473 phosphorylation in insulin-resistant muscle cells (Thirone et al. 2006). What is more, JAK2 partially mediates ceramide-induced defects in insulin signaling. Thus, JAKs appear to be potential mediators of inflammation-induced insulin resistance. However, the extent of their contribution to such pathogenesis in obesity remains to be fully elucidated (see figure 7.3).

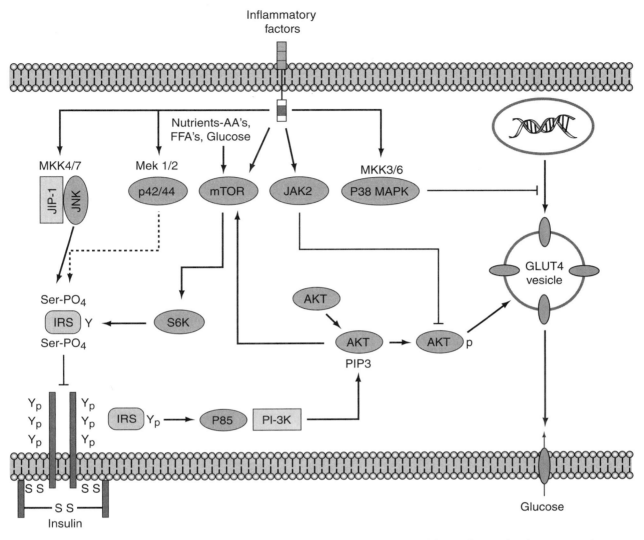

Figure 7.3   The protein kinase mediators of insulin resistance. Inflammatory factors may inhibit insulin signaling by activating the MAPK, JAK, and mTOR/S6K signaling pathways. The mTOR/S6K pathway may also be activated by nutrient excess and insulin signaling through Akt. AAs, amino acids; →, leads to; →, inhibits; possible mechanism.

Transcriptional Mediators

In addition to activating protein kinases and lipid mediators, inflammatory signaling leads to the transcription of various factors that may impair insulin signal transduction in insulin target tissues. This next section describes the key transcriptional pathways that play a role in inflammation-induced insulin resistance.

IκB Kinase and NF-κβ Pathway

NF-κβ, a collective name for inducible dimeric transcription factors of the Rel family of DNA-binding proteins, is found in essentially all cell types and is involved in activating a large number of genes in response to infection and inflammation (see Karin and Delhase 2000 for review). The subcellular location of NF-κβ is controlled by a family of inhibitory proteins termed *IκBs* that bind NF-κβ and mask its nuclear localization signal, thereby preventing its nuclear uptake. Exposure to proinflammatory adipokines, cytokines, and endotoxins leads to the rapid phosphorylation, ubiquitination, and ultimately proteolytic degradation of IκB, which frees NF-κβ to translocate to the nucleus and regulate gene transcription. The multisubunit IκB kinase (IKK) responsible for IκB phosphorylation is the point of convergence for most cytokines that activate NF-κβ.

Yuan and colleagues (2001) first suggested that IKK is a potential link between inflammation and obesity-linked insulin resistance. Using both heterozygous deletion of IKKβ (IKKβ+/−) and pharmacological inhibition of IKK (high-dose salicylate treatment), they were able to improve insulin sensitivity in rodent models of obesity. IKK likely promotes insulin resistance via its downstream transcription factor NF-κβ. Indeed, NF-κβ appears to be especially important in hepatic and myeloid cells, as it controls cytokine production and transcriptional inflammatory mediators (e.g., iNOS, SOCS) and may thereby contribute to the spread of global inflammation. Accordingly, myeloid-specific ablation of IKKβ prevents the development of insulin resistance in all insulin target tissues, further underscoring the role of

the macrophage (Arkan et al. 2005). LIKK mice, mice that contain a constitutively active form of IKKβ in their liver, display an enhanced hepatic cytokine production that is linked to the development of global insulin resistance (Cai et al. 2005). Taken together, these data point to an important role for IKK and IKK-controlled elements such as NF-κβ in the mediation and causation of inflammation-induced insulin resistance in obesity (see figure 7.4).

Inducible Nitric Oxide Synthase

The radical gas nitric oxide (NO) is synthesized from L-arginine by the enzyme nitric oxide synthase (NOS). The expression and activity of the inducible calcium-independent NO synthase (iNOS) are synergistically stimulated by bacterial endotoxins

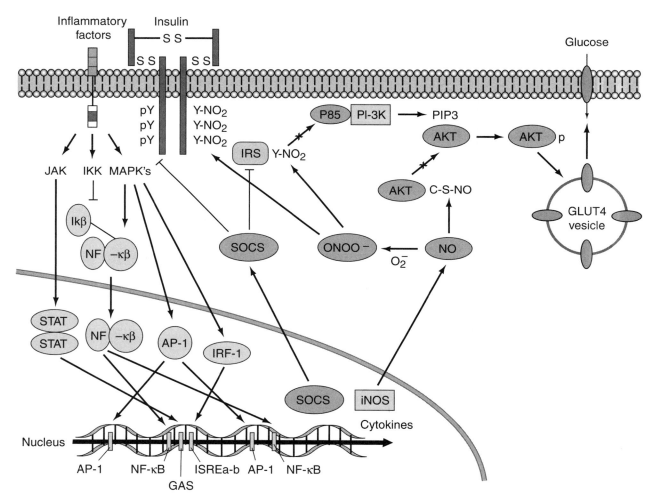

Figure 7.4 The transcriptional mediators of insulin resistance. Inflammatory signaling in metabolic tissues may impair insulin signal transduction by enhancing the expression of SOCS and iNOS. C-S-NO, S-nitrosylation of cysteine residue; Y-NO_2, nitration of tyrosine residue; →, leads to; ⊣, inhibits; X, blocks interaction. ss=disulfide bonds, GAS=IFNγ activation site, and ISREa-b=IFN stimulated response element a-b.

and also by inflammatory cytokines. The pathways regulating iNOS expression vary in different cells and species, but it is widely recognized that the IKK pathway regulates iNOS transcription through multiple NF-κβ binding sites on the iNOS promoter (see Kleinert et al. 2004). Other important transcription factors for iNOS induction include IFN regulatory factor 1 (IRF-1), signal transducer and activator of transcription 1α (STAT1α), activator protein 1 (AP-1), and CCAAT-enhancer-binding protein (C/EBP; Kleinert et al. 2004). There is also evidence that MAPK and especially JNK increase iNOS transcription in macrophages (Chan and Riches 2001). Interestingly, the JNK pathway may also regulate iNOS mRNA stability (Lahti et al. 2003), which is another major mechanism of iNOS induction by cytokines.

When induced, iNOS generates NO at a much higher rate and for longer durations than do the constitutive NOS enzymes (nNOS and eNOS; Moncada, Palmer, and Higgs 1991). Although this high-output NO pathway probably evolved to protect the host from infection, it can cause deleterious effects (e.g., hypotension, organ injury) to other normal host cells. This property confers to iNOS the protective–destructive duality inherent in every other major component of the immune response. It was first proposed a decade ago that insulin resistance is another deleterious effect of iNOS induction during systemic inflammation. Administration of the endotoxin lipopolysaccharide (LPS) in rats, a model of acute systemic inflammation, induces iNOS in muscle, liver, and adipose tissue (Kapur et al. 1997; Kapur, Marcotte, and Marette 1999). Cytokines and LPS also induce iNOS expression in cultured muscle and adipose cells, causing marked insulin resistance (Bédard, Marcotte, and Marette 1997; Kapur et al. 1997). The insulin resistance was significantly abrogated by iNOS inhibition (Bédard, Marcotte, and Marette 1997).

Perreault and Marette (2001) were the first to report that iNOS expression is induced in fat and skeletal muscle of dietary (high-fat feeding) and genetic (ob/ob mouse, Zucker diabetic rat) models of obesity. They also found that high-fat-fed obese mice lacking iNOS (iNOS KO mice) were protected from developing insulin resistance in skeletal muscle. Other groups have independently confirmed that iNOS mediates insulin resistance in obese animal models (Noronha et al. 2005; Carv-

alho-Filho et al. 2005). Importantly, iNOS is also induced in the skeletal muscle and adipose tissue of subjects with type 2 diabetes (Torres et al. 2004; Engeli et al. 2004), and its expression correlates with the occurrence of insulin resistance (Torres et al. 2004) and obesity (Engeli et al. 2004). This finding indicates that iNOS may also play a key role in the pathogenesis of human insulin resistance.

In animal models, NO produced by iNOS causes S-nitrosylation (or S-nitrosation) of IRS1 (Carvalho-Filho et al. 2005) and Akt (Carvalho-Filho et al. 2005), resulting in reduced activation of PI3K and Akt. In biological systems, NO also reacts with superoxide (O_2^-) to form the potent oxidant peroxynitrite ($ONOO^-$), which is critical for the antimicrobial and neurodestructive activities of NO (Lipton et al. 1993). When L-arginine concentrations are limited, iNOS itself can catalyze the formation of O_2^- (Xia et al. 1998). The oxidant $ONOO^-$ is known to produce nitration of tyrosine residues (Ischiropoulos et al. 1992), and tyrosine nitration is increased in several diseases, including atherosclerosis, myocardial ischemia, and sepsis (see Beckman and Koppenol 1996). Interestingly, increasing $ONOO^-$ levels in 3T3-L1 adipocytes with a NO/O_2^- donor (SIN-1) inhibited insulin-stimulated glucose transport and IRS1-associated PI3K activity in these cells (Nomiyama et al. 2004). These effects were associated with increased nitration of key tyrosine residues within IRS1 (as assessed by mass spectrometry; see figure 7.4).

iNOS may be also implicated in the deterioration of insulin secretion in type 2 diabetes. It is induced in beta cells of obese Zucker diabetic rats, and elevated NO production may cause beta-cell apoptosis or necrosis and impair insulin secretion in this model (Shimabukuro et al. 1998). Accordingly, lack of iNOS in beta cells of transgenic mice postpones cytokine-induced apoptosis and prevents necrosis (Liu et al. 2000). The expression of iNOS is also increased in cardiac muscle of Zucker diabetic rats (Zhou et al. 2000). This observation led to the proposal that cardiac dysfunction in obesity may be consequent to iNOS-mediated apoptosis. In mice fed a high-fat diet, iNOS content was elevated in the aorta, and iNOS deficiency prevented the adverse effects of the high-fat diet on vascular insulin resistance (Noronha et al. 2005). These studies strongly implicate iNOS in the occurrence of insulin resistance, cardiovascular dysfunction, and beta-cell failure in the evolution toward type 2 diabetes in obesity.

Suppressors of Cytokine Signaling: The Negative Aspect of a Negative Feedback Loop

As their name suggests, the suppressors of cytokine signaling (SOCS) play an important role in the negative feedback control of cytokine-activated pathways (Yasukawa, Sasaki, and Yoshimura 2000). SOCS expression is rapidly upregulated in response to several cytokines (TNFα, IL-1β, IL-6, and IFN-γ) and adipokines (leptin, resistin) as a means to assert feedback inhibition on inflammatory signaling. Emanuelli and colleagues (2001) first demonstrated that SOCS might be involved in inflammation-induced insulin resistance when they reported the TNFα-dependent overexpression of SOCS3 in the adipose tissue of obese mice. Since then, SOCS overexpression has been reported in the skeletal muscle of patients with diabetes (Rieusset et al. 2004). It has also been observed in experimental models of obesity and LPS-induced insulin resistance, with the expression of SOCS1 and SOCS3 being greatest in the liver, followed by muscle and adipose tissue (Ueki, Kondo, and Kahn 2004). In accordance with these data, SOCS3 appears to play a major pathogenic role in the liver as a mediator of hepatic insulin resistance induced by IL-6 (Senn et al. 2003); it also contributes to the development of hepatic steatosis (Ueki et al. 2004). The downregulation of liver IRS1 and IRS2 mediated by hepatitis C also occurs via a SOCS-dependent mechanism (Kawaguchi et al. 2004). On another interesting note, the cytokine-mediated upregulation of SOCS proteins in obesity may also affect the hypothalamic regulation of metabolism, as SOCS3 deficiency in the brain improves leptin sensitivity and confers resistance against the development of diet-induced obesity (Mori et al. 2004). Furthermore, SOCS3 is a mediator of resistin-induced insulin resistance in adipocytes (Steppan et al. 2005).

On the molecular level, SOCS is thought to act in insulin resistance by obstructing several components of the insulin signaling cascade. Indeed, SOCS proteins contain an SH2 domain that allows them to interact with phosphotyrosine molecules within the insulin receptor. This interaction leads to downstream inhibition of IRS1 tyrosine phosphorylation (Emanuelli et al. 2001; Mooney et al.

2001), to reduced association with the p85 regulatory element of PI3K (Emanuelli et al. 2001), and to impaired insulin-dependent activation of ERK1/2 and Akt (Mooney et al. 2001). In addition, evidence gathered from models of SOCS overexpression suggests that SOCS1 and SOCS3 may impair insulin signaling by targeting IRS1 and IRS2 for ubiquitin-mediated proteasomal degradation (Rui et al. 2002). However, a very recent study indicates that the physiological relevance of this mechanism is subject to debate, as the degree of SOCS3 induction does not always correlate with the degradation of IRS1 and IRS2 (He and Stephens 2006). In any case, the inflammation-mediated induction of SOCS molecules profoundly affects the liver as well as other insulin target tissues, including the brain (see figure 7.4).

In summary, inflammatory mediators such as adipokines and cytokines may employ various mechanisms such as lipid derivatives, serine and threonine kinases, and transcriptional pathways to inhibit insulin signaling. The influence of each mechanism on insulin signaling often varies among tissues, and it is most likely that global insulin resistance occurs as a result of the activation of multiple mechanisms rather than one particular mechanism. The elucidation of further mechanisms of inflammation-mediated insulin resistance and the complex cross talk occurring among these mediator pathways will provide great insight for the identification of future therapeutic targets.

AMPK: From a Gauge of Energy Status to a Novel Target for Alleviating Inflammation in Obesity

AMPK is a member of a metabolite-sensing protein kinase family that acts as a fuel gauge monitoring cellular energy levels (Carling 2004; Hardie 2004). When AMP kinase "senses" decreased energy stores, it acts to switch off ATP-consuming pathways and switch on alternative pathways for ATP regeneration. The metabolic function of AMPK has perhaps been best documented in exercising skeletal muscle, where its activation seems to contribute to increased glucose transport and fatty acid oxidation (Winder 2001; Sakamoto et al. 2005). More recent studies have also identified AMPK as the mediator of the metabolic effects of the adipose-derived hormones leptin and adiponectin in skeletal muscle and liver (Minokoshi et al. 2002; Tomas et al. 2002; Kahn et al. 2005). Since AMPK also senses cellular energy status in the brain, it is thought to play a major role in the regulation of energy balance in health and altered metabolic states (Kahn et al. 2005).

AMPK can be activated chemically with AICAR, which is taken up by cells and phosphorylated by adenosine kinase to form 5-aminoimidazole-4-carboxamide ribonucleoside (ZMP), a nucleotide that mimics AMP (Corton et al. 1995). Metformin and the thiazolidinediones (ligands of the nuclear receptor PPARγ) also activate AMPK, a mechanism believed to contribute to their insulin-sensitizing actions in subjects with diabetes (Fryer, Parbu-Patel, and Carling 2002). Since the action of AMPK in regulating metabolic processes in exercised muscle and following activation by adipokines is discussed in chapter 13, in this chapter we focus on the proposed role of AMPK as a key target for treating obesity-linked insulin resistance through its recently revealed anti-inflammatory actions.

AMPK is believed to be a promising antidiabetic drug target since its activation increases both glucose uptake and lipid oxidation (Winder and Hardie 1999). However, recent findings suggest that AMPK may also alleviate insulin resistance by blunting inflammation in obesity. AMPK activation by AICAR, metformin, and the PPARγ ligands troglitazone and 15dPGJ2 inhibits iNOS induction in macrophages, myocytes, and adipocytes (Pilon, Dallaire, and Marette 2004), three major sites of iNOS expression in obesity. These AMPK activators also reverse the iNOS-dependent inhibition of insulin-stimulated PI3K activity in cytokine-exposed muscle cells. AMPK activation by in vivo administration of metformin and AICAR also potently inhibits iNOS induction in skeletal muscle and adipose tissue of LPS-challenged rats (Pilon, Dallaire, and Marette 2004). Using siRNA to knock down AMPK in muscle cells blunted the AICAR, metformin, and troglitazone inhibition of iNOS, confirming the contribution of AMPK to the anti-inflammatory effects of these drugs. The identification of AMPK as a potent inhibitory pathway for iNOS may provide a molecular basis for the observation that exercise training, which increases AMPK expression

(Winder et al. 2000), improves exercise capacity in association with reduced iNOS expression in skeletal muscle of patients with chronic heart failure (Gielen et al. 2003).

In addition to affecting iNOS expression, AMPK may inhibit inflammation through blunting the release of inflammatory factors. Indeed, AICAR inhibits LPS-induced expression of proinflammatory cytokines (TNFα, IL-1β, and IL-6) in primary rat astrocytes, microglia, and peritoneal macrophages (Giri et al. 2004). The AMPK activator also reduces the production of IL-6, MCP-1, and macrophage inflammatory proteins MIP-1α and MIP-1β in adipocytes (Sell, Dietze-Schroeder, Eckardt et al. 2006). AICAR attenuates the LPS-induced activation of the IKK/NF-κβ (Giri et al. 2004) and PI3K/Akt pathways (Jhun et al. 2004). This effect likely explains the inhibition of cytokine production and iNOS induction. Recent studies have also shown that AMPK activation by AICAR or metformin reduces the amount of various inflammatory factors in endothelial cells, including cell adhesion molecules such as vascular cell adhesion molecule-1 (VCAM-1; Hattori et al. 2006; Prasad et al. 2006). Accordingly, AICAR was found to reduce the infiltration of leukocytes across the blood-brain barrier in an animal model of autoimmune encephalomyelitis (Prasad et al. 2006). It will be of great interest to test whether AMPK exerts anti-inflammatory effects in obesity by blunting the recruitment of macrophages into adipose tissue and other insulin target tissues.

In summary, the available data suggest that AMPK is a promising therapeutic target for inflammatory metabolic disorders such as atherosclerosis and obesity-linked diabetes. It will be crucial to elucidate the molecular mechanisms by which AMPK inhibits iNOS and other inflammatory mediators, since this understanding may help us design novel strategies for an effective and tissue-specific inhibition of inflammation in obesity-linked type 2 diabetes and cardiovascular disease.

Concluding Remarks

The complexity of the obesity-related immune response, stemming from macrophage infiltration into adipose tissue and leading to the development of inflammation-mediated metabolic complications such as insulin resistance and cardiovascular disease, is now well appreciated. It is likely, however, that many more years will pass before the true nature of this process is completely understood. Indeed, many questions regarding the origin, nature, and implications of obesity-linked inflammation are yet to be answered. The most daunting question at present is how we may intervene in obesity to prevent inflammation-linked metabolic complications while still preserving proper immune function. Whether the answer lies at the origin of inflammation; in the adipocyte, macrophage, and monocyte populations; or in the effector signaling molecules remains to be seen. An effective therapeutic avenue to follow in obesity may be to enhance the activity of anti-inflammatory mediators such as AMPK that may limit the intensity of the inflammatory response not only at the site of origin but also in the metabolic tissues.

References

Aguirre, V., T. Uchida, L. Yenush, R. Davis, and M. F. White. 2000. The c-Jun NH(2)-terminal kinase promotes insulin resistance during association with insulin receptor substrate-1 and phosphorylation of Ser(307). *J Biol Chem* 275:9047-54.

Appay, V., and S. L. Rowland-Jones. 2001. RANTES: A versatile and controversial chemokine. *Trends Immunol* 22:83-7.

Arita, Y., S. Kihara, N. Ouchi, M. Takahashi, K. Maeda, J. Miyagawa, K. Hotta, I. Shimomura, T. Nakamura, K. Miyaoka, H. Kuriyama, M. Nishida, S. Yamashita, K. Okubo, K. Matsubara, M. Muraguchi, Y. Ohmoto, T. Funahashi, and Y. Matsuzawa. 1999. Paradoxical decrease of an adipose-specific protein, adiponectin, in obesity. *Biochem Biophys Res Commun* 257:79-83.

Arkan, M.C., A.L. Hevener, F.R. Greten, S. Maeda, Z.W. Li, J.M. Long, A. Wynshaw-Boris, G. Poli, J. Olefsky, and M. Karin. 2005. IKK-beta links inflammation to obesity-induced insulin resistance. *Nat Med* 11:191-8.

Aygun, A.D., S. Gungor, B. Ustundag, M.K. Gurgoze, and Y. Sen. 2005. Proinflammatory cytokines and leptin are increased in serum of prepubertal obese children. *Mediators Inflamm* 2005:180-3.

Baumann, H., and J. Gauldie. 1994. The acute phase response. *Immunol Today* 15:74-80.

Beckman, J.S., and W.H. Koppenol. 1996. Nitric oxide, superoxide, and peroxynitrite: The good, the bad, and the ugly. *Am J Physiol* 271:C1424-37.

Bédard, S., B. Marcotte, and A. Marette. 1997. Cytokines modulate glucose transport in skeletal muscle by inducing the expression of inducible nitric oxide synthase. *Biochem J* 325:487-93.

Bell, E.T. 1909. On the histogenesis of the adipose tissue of the ox. *Am J Anat* 9:412-35.

Bennett, B.D., G.P. Solar, J.Q. Yuan, J. Mathias, G.R. Thomas, and W. Matthews. 1996. A role for leptin and its cognate receptor in hematopoiesis. *Curr Biol* 6:1170-80.

Braissant, O., F. Foufelle, C. Scotto, M. Dauca, and W. Wahli. 1996. Differential expression of peroxisome proliferator-activated receptors (PPARs): Tissue distribution of PPAR-alpha, -beta, and -gamma in the adult rat. *Endocrinology* 137:354-66.

Cai, D., M. Yuan, D.F. Frantz, P.A. Melendez, L. Hansen, J. Lee, and S.E. Shoelson. 2005. Local and systemic insulin resistance resulting from hepatic activation of IKK-beta and NF-kappaB. *Nat Med* 11:183-90.

Carling, D. 2004. The AMP-activated protein kinase cascade—a unifying system for energy control. *Trends Biochem Sci* 29:18-24.

Carvalho-Filho, M.A., M. Ueno, S.M. Hirabara, A.B. Seabra, J.B. Carvalheira, M.G. de Oliveira, L.A. Velloso, R. Curi, and M.J. Saad. 2005. S-nitrosation of the insulin receptor, insulin receptor substrate 1, and protein kinase B/Akt: A novel mechanism of insulin resistance. *Diabetes* 54:959-67.

Chan, E.D., and D.W. Riches. 2001. IFN-gamma + LPS induction of iNOS is modulated by ERK, JNK/SAPK, and p38(mapk) in a mouse macrophage cell line. *Am J Physiol Cell Physiol* 280:C441-50.

Charriere, G., B. Cousin, E. Arnaud, M. Andre, F. Bacou, L. Penicaud, and L. Casteilla. 2003. Preadipocyte conversion to macrophage. Evidence of plasticity. *J Biol Chem* 278:9850-5.

Chiellini, C., F. Santini, A. Marsili, P. Berti, A. Bertacca, C. Pelosini, G. Scartabelli, E. Pardini, J. Lopez-Soriano, R. Centoni, A.M. Ciccarone, L. Benzi, P. Vitti, S. Del Prato, A. Pinchera, and M. Maffei. 2004. Serum haptoglobin: A novel marker of adiposity in humans. *J Clin Endocrinol Metab* 89:2678-83.

Chitu, V., and E.R. Stanley. 2006. Colony-stimulating factor-1 in immunity and inflammation. *Curr Opin Immunol* 18:39-48.

Christiansen, T., B. Richelsen, and J.M. Bruun. 2005. Monocyte chemoattractant protein-1 is produced in isolated adipocytes, associated with adiposity and reduced after weight loss in morbid obese subjects. *Int J Obes (Lond)* 29:146-50.

Cinti, S., G. Mitchell, G. Barbatelli, I. Murano, E. Ceresi, E. Faloia, S. Wang, M. Fortier, A.S. Greenberg, and M.S. Obin. 2005. Adipocyte death defines macrophage localization and function in adipose tissue of obese mice and humans. *J Lipid Res* 46:2347-55.

Considine, R.V., M.K. Sinha, M.L. Heiman, A. Kriauciunas, T.W. Stephens, M.R. Nyce, J.P. Ohannesian et al. 1996. Serum immunoreactive-leptin concentrations in normal-weight and obese humans. *New Engl J Med* 334:292-5.

Cook, D.G., M.A. Mendall, P.H. Whincup, I.M. Carey, L. Ballam, J.E. Morris, G.J. Miller, and D.P. Strachan. 2000. C-reactive protein concentration in children: Relationship to adiposity and other cardiovascular risk factors. *Atherosclerosis* 149:139-50.

Corton, J.M., J.G. Gillespie, S.A. Hawley, and D.G. Hardie. 1995. 5-aminoimidazole-4-carboxamide ribonucleoside. A specific method for activating AMP-activated protein kinase in intact cells? *Eur J Biochem* 229:558-65.

Cousin, B., O. Munoz, M. Andre, A.M. Fontanilles, C. Dani, J.L. Cousin, P. Laharrague, L. Casteilla, and L. Penicaud. 1999. A role for preadipocytes as macrophage-like cells. *FASEB J* 13:305-12.

Csehi, S.B., S. Mathieu, U. Seifert, A. Lange, M. Zweyer, A. Wernig, and D. Adam. 2005. Tumor necrosis factor (TNF) interferes with insulin signaling through the p55 TNF receptor death domain. *Biochem Biophys Res Commun* 329:397-405.

Curat, C.A., A. Miranville, C. Sengenes, M. Diehl, C. Tonus, R. Busse, and A. Bouloumie. 2004. From blood monocytes to adipose tissue-resident macrophages: Induction of diapedesis by human mature adipocytes. *Diabetes* 53:1285-92.

Dandona, P., A. Aljada, H. Ghanim, P. Mohanty, C. Tripathy, D. Hofmeyer, and A. Chaudhuri. 2004. Increased plasma concentration of macrophage migration inhibitory factor (MIF) and MIF mRNA in mononuclear cells in the obese and the suppressive action of metformin. *J Clin Endocrinol Metab* 89:5043-7.

Dinarello, C.A. 2002. The IL-1 family and inflammatory diseases. *Clin Exp Rheumatol* 20:S1-13.

Dixit, V.D., and D.D. Taub. 2005. Ghrelin and immunity: A young player in an old field. *Exp Gerontol* 40:900-10.

Dobryszycka, W. 1997. Biological functions of haptoglobin—new pieces to an old puzzle. *Eur J Clin Chem Clin Biochem* 35:647-54.

Emanuelli, B., P. Peraldi, C. Filloux, C. Chavey, K. Freidinger, D.J. Hilton, G.S. Hotamisligil, and E. Van Obberghen. 2001. SOCS-3 inhibits insulin signaling and is up-regulated in response to tumor necrosis factor-alpha in the adipose tissue of obese mice. *J Biol Chem* 276:47944-9.

Engeli, S., J. Janke, K. Gorzelniak, J. Bohnke, N. Ghose, C. Lindschau, F.C. Luft, and A.M. Sharma. 2004. Regulation of the nitric oxide system in human adipose tissue. *J Lipid Res* 45:1640-8.

Engelman, J.A., A.H. Berg, R.Y. Lewis, M.P. Lisanti, and P.E. Scherer. 2000. Tumor necrosis factor alpha-mediated insulin resistance, but not dedifferentiation, is abrogated by MEK1/2 inhibitors in 3T3-L1 adipocytes. *Mol Endocrinol* 14:1557-69.

Esposito, K., A. Pontillo, F. Giugliano, G. Giugliano, R. Marfella, G. Nicoletti, and D. Giugliano. 2003. Association of low interleukin-10 levels with the metabolic syndrome in obese women. *J Clin Endocrinol Metab* 88:1055-8.

Esterbauer, H., F. Krempler, H. Oberkofler, and W. Patsch. 1999. The complement system: A pathway linking host defense and adipocyte biology. *Eur J Clin Invest* 29:653-6.

Fain, J.N., and A.K. Madan. 2005. Regulation of monocyte chemoattractant protein 1 (MCP-1) release by explants of human visceral adipose tissue. *Int J Obes (Lond)* 29:1299-307.

Fasshauer, M., J. Klein, S. Kralisch, M. Klier, U. Lossner, M. Bluher, and R. Paschke. 2004. Monocyte chemoattractant protein 1 expression is stimulated by growth hormone and interleukin-6 in 3T3-L1 adipocytes. *Biochem Biophys Res Commun* 317:598-604.

Festa, A., R. D'Agostino Jr., G. Howard, L. Mykkanen, R.P. Tracy, and S.M. Haffner. 2000. Chronic subclinical inflammation as part of the insulin resistance syndrome: The Insulin Resistance Atherosclerosis Study (IRAS). *Circulation* 102:42-7.

Fryer, L.G., A. Parbu-Patel, and D. Carling. 2002. The antidiabetic drugs rosiglitazone and metformin stimulate AMP-activated protein kinase through distinct signaling pathways. *J Biol Chem* 277:25226-32.

Fu, Y., N. Luo, and M.F. Lopes-Virella. 2000. Oxidized LDL induces the expression of ALBP/aP2 mRNA and protein in human THP-1 macrophages. *J Lipid Res* 41:2017-23.

Fujishiro, M., Y. Gotoh, H. Katagiri, H. Sakoda, T. Ogihara, M. Anai, Y. Onishi, H. Ono, M. Funaki, K. Inukai, Y. Fukushima, M. Kikuchi, Y. Oka, and T. Asano. 2001. MKK6/3 and p38 MAPK pathway activation is not necessary for insulin-induced glucose uptake but regulates glucose transporter expression. *J Biol Chem* 276 (23):19800-6.

Fukuhara, A., M. Matsuda, M. Nishizawa, K. Segawa, M. Tanaka, K. Kishimoto, Y. Matsuki, M. Murakami, T. Ichisaka, H. Murakami, E. Watanabe, T. Takagi, M. Akiyoshi, T. Ohtsubo, S. Kihara, S. Yamashita, M. Makishima, T. Funahashi, S. Yamanaka, R. Hiramatsu, Y. Matsuzawa, and I. Shimomura. 2005. Visfatin: A protein secreted by visceral fat that mimics the effects of insulin. *Science* 307:426-30.

Funk, J.L., K.R. Feingold, A.H. Moser, and C. Grunfeld. 1993. Lipopolysaccharide stimulation of RAW 264.7 macrophages induces lipid accumulation and foam cell formation. *Atherosclerosis* 98:67-82.

Gabay, C., and I. Kushner. 1999. Acute-phase proteins and other systemic responses to inflammation. *New Engl J Med* 340:448-54.

Gao, Z., X. Zhang, A. Zuberi, D. Hwang, M.J. Quon, M. Lefevre, and J. Ye. 2004. Inhibition of insulin sensitivity by free fatty acids requires activation of multiple serine kinases in 3T3-L1 adipocytes. *Mol Endocrinol* 18:2024-34.

Geissmann, F., S. Jung, and D.R. Littman. 2003. Blood monocytes consist of two principal subsets with distinct migratory properties. *Immunity* 19:71-82.

Gielen, S., V. Adams, S. Mobius-Winkler, A. Linke, S. Erbs, J. Yu, W. Kempf, A. Schubert, G. Schuler, and R. Hambrecht. 2003. Anti-inflammatory effects of exercise training in the skeletal muscle of patients with chronic heart failure. *J Am Coll Cardiol* 42:861-8.

Giri, S., N. Nath, B. Smith, B. Viollet, A.K. Singh, and I. Singh. 2004. 5-aminoimidazole-4-carboxamide-1-beta-4-ribofuranoside inhibits proinflammatory response in glial cells: A possible role of AMP-activated protein kinase. *J Neurosci* 24:479-87.

Gordon, S., and P.R. Taylor. 2005. Monocyte and macrophage heterogeneity. *Nat Rev Immunol* 5:953-64.

Gual, P., V. Baron, V. Lequoy, and E. Van Obberghen. 1998. Interaction of Janus kinases JAK-1 and JAK-2 with the insulin receptor and the insulin-like growth factor-1 receptor. *Endocrinology* 139:884-93.

Hannun, Y.A. 1996. Functions of ceramide in coordinating cellular responses to stress. *Science* 274:1855-9.

Hardie, D.G. 2004. The AMP-activated protein kinase pathway—new players upstream and downstream. *J Cell Sci* 117:5479-87.

Hattori, Y., K. Suzuki, S. Hattori, and K. Kasai. 2006. Metformin inhibits cytokine-induced nuclear factor kappaB activation via AMP-activated protein kinase activation in vascular endothelial cells. *Hypertension* 47:1183-8.

He, F., and J.M. Stephens. 2006. Induction of SOCS-3 is insufficient to confer IRS-1 protein degradation in 3T3-L1 adipocytes. *Biochem Biophys Res Commun* 344:95-8.

Herder, C., M. Peltonen, W. Koenig, I. Kraft, S. Muller-Scholze, S. Martin, T. Lakka, P. Ilanne-Parikka, J.G. Eriksson, H. Hamalainen, S. Keinanen-Kiukaanniemi, T.T. Valle, M. Uusitupa, J. Lindstrom, H. Kolb, and J. Tuomilehto. 2006. Systemic immune mediators and lifestyle changes in the prevention of type 2 diabetes: Results from the Finnish Diabetes Prevention Study. *Diabetes* 55:2340-6.

Hirosumi, J., G. Tuncman, L. Chang, C.Z. Gorgun, K.T. Uysal, K. Maeda, M. Karin, and G.S. Hotamisligil. 2002. A central role for JNK in obesity and insulin resistance. *Nature* 420:333-6.

Hong, K.M., M.D. Burdick, R.J. Phillips, D. Heber, and R.M. Strieter. 2005. Characterization of human fibrocytes as circulating adipocyte progenitors and the formation of human adipose tissue in SCID mice. *FASEB J* 19:2029-31.

Hotamisligil, G.S., N.S. Shargill, and B.M. Spiegelman. 1993. Adipose expression of tumor necrosis factor-alpha: Direct role in obesity-linked insulin resistance. *Science* 259:87-91.

Idriss, H.T., and J.H. Naismith. 2000. TNF alpha and the TNF receptor superfamily: Structure-function relationship(s). *Microsc Res Tech* 50:184-95.

Inglis, K. 1927. So-called interscapular gland and tumours arising therein. *J Anat* 61:452-3.

Ischiropoulos, H., L. Zhu, J. Chen, M. Tsai, J.C. Martin, C.D. Smith, and J.S. Beckman. 1992. Peroxynitrite-mediated tyrosine nitration catalyzed by superoxide dismutase. *Arch Biochem Biophys* 298:431-7

Jaeschke, A., M.P. Czech, and R.J. Davis. 2004. An essential role of the JIP1 scaffold protein for JNK activation in adipose tissue. *Genes Dev* 18:1976-80.

Jhun, B.S., Q. Jin, Y.T. Oh, S.S. Kim, Y. Kong, Y.H. Cho, J. Ha, H.H. Baik, and I. Kang. 2004. 5-Aminoimidazole-4-carboxamide riboside suppresses lipopolysaccharide-induced TNF-alpha production through inhibition of phosphatidylinositol 3-kinase/Akt activation in RAW 264.7 murine macrophages. *Biochem Biophys Res Commun* 318:372-80.

Kahn, B.B., T. Alquier, D. Carling, and D.G. Hardie. 2005. AMP-activated protein kinase: Ancient energy gauge provides clues to modern understanding of metabolism. *Cell Metab* 1:15-25.

Kanda, H., S. Tateya, Y. Tamori, K. Kotani, K. Hiasa, R. Kitazawa, S. Kitazawa, H. Miyachi, S. Maeda, K. Egashira, and M. Kasuga. 2006. MCP-1 contributes to macrophage infiltration into adipose tissue, insulin resistance, and hepatic steatosis in obesity. *J Clin Invest* 116:1494-505.

Kaneto, H., Y. Nakatani, T. Miyatsuka, D. Kawamori, T.A. Matsuoka, M. Matsuhisa, Y. Kajimoto, H. Ichijo, Y. Yamasaki, and M. Hori. 2004. Possible novel therapy for diabetes with cell-permeable JNK-inhibitory peptide. *Nat Med* 10:1128-32.

Kapur, S., S. Bedard, B. Marcotte, C.H. Cote, and A. Marette. 1997. Expression of nitric oxide synthase in skeletal muscle: A novel role for nitric oxide as a modulator of insulin action. *Diabetes* 46:1691-700.

Kapur, S., B. Marcotte, and A. Marette. 1999. Mechanism of adipose tissue iNOS induction in endotoxemia. *Am J Physiol* 276:E635-41.

Karin, M., and M. Delhase. 2000. The I kappa B kinase (IKK) and NF-kappa B: Key elements of proinflammatory signalling. *Semin Immunol* 12:85-98.

Kawaguchi, T., T. Yoshida, M. Harada, T. Hisamoto, Y. Nagao, T. Ide, E. Taniguchi, H. Kumemura, S. Hanada, M. Maeyama, S. Baba, H. Koga, R. Kumashiro, T. Ueno, H. Ogata, A. Yoshimura, and M. Sata. 2004. Hepatitis C virus down-regulates insulin receptor substrates 1 and 2 through up-regulation of suppressor of cytokine signaling 3. *Am J Pathol* 165:1499-508.

Kim, K.Y., J.K. Kim, S.H. Han, J.S. Lim, K.I. Kim, D.H. Cho, M.S. Lee, J.H. Lee, D.Y. Yoon, S.R. Yoon, J.W. Chung, I. Choi, E. Kim, and Y. Yang. 2006. Adiponectin is a negative regulator of NK cell cytotoxicity. *J Immunol* 176:5958-64.

Kim, M.Y., C. Linardic, L. Obeid, and Y. Hannun. 1991. Identification of sphingomyelin turnover as an effector

mechanism for the action of tumor necrosis factor alpha and gamma-interferon. Specific role in cell differentiation. *J Biol Chem* 266:484-9.

Kishimoto, T., S. Akira, M. Narazaki, and T. Taga. 1995. Interleukin-6 family of cytokines and gp130. *Blood* 86:1243-54.

Kleinert, H., A. Pautz, K. Linker, and P.M. Schwarz. 2004. Regulation of the expression of inducible nitric oxide synthase. *Eur J Pharmacol* 500:255-66.

Koch, A.E., S.L. Kunkel, L.A. Harlow, B. Johnson, H.L. Evanoff, G.K. Haines, M.D. Burdick, R.M. Pope, and R.M. Strieter. 1992. Enhanced production of monocyte chemoattractant protein-1 in rheumatoid arthritis. *J Clin Invest* 90:772-9.

Kolesnick, R., and D.W. Golde. 1994. The sphingomyelin pathway in tumor necrosis factor and interleukin-1 signaling. *Cell* 77:325-8.

Kolter, T., R.L. Proia, and K. Sandhoff. 2002. Combinatorial ganglioside biosynthesis. *J Biol Chem* 277:25859-62.

Lahti, A., U. Jalonen, H. Kankaanranta, and E. Moilanen. 2003. c-Jun NH2-terminal kinase inhibitor anthra(1,9-cd)pyrazol-6(2H)-one reduces inducible nitric-oxide synthase expression by destabilizing mRNA in activated macrophages. *Mol Pharmacol* 64:308-15.

Latta, J.S., and D.I. Rutledge. 1935. The reaction of omental tissues to trypan blue injected intraperitoneally, with special reference to interrelationships between cell types. *Am J Anat* 56:481-511.

Le, Y., Y. Zhou, P. Iribarren, and J. Wang. 2004. Chemokines and chemokine receptors: Their manifold roles in homeostasis and disease. *Cell Mol Immunol* 1:95-104.

Lekmine, F., A. Sassano, S. Uddin, J. Smith, B. Majchrzak, S.M. Brachmann, N. Hay, E.N. Fish, and L.C. Platanias. 2004. Interferon-gamma engages the p70 S6 kinase to regulate phosphorylation of the 40S S6 ribosomal protein. *Exp Cell Res* 295:173-82.

Letterio, J.J. 2000. Murine models define the role of TGF-beta as a master regulator of immune cell function. *Cytokine Growth Factor Rev* 11:81-7.

Levine, J.A., M.D. Jensen, N.L. Eberhardt, and T. O'Brien. 1998. Adipocyte macrophage colony-stimulating factor is a mediator of adipose tissue growth. *J Clin Invest* 101:1557-64.

Lim, H.K., Y.A. Choi, W. Park, T. Lee, S.H. Ryu, S.Y. Kim, J.R. Kim, J.H. Kim, and S.H. Baek. 2003. Phosphatidic acid regulates systemic inflammatory responses by modulating the Akt-mammalian target of rapamycin-p70 S6 kinase 1 pathway. *J Biol Chem* 278:45117-27.

Lipton, S.A., Y.-B. Choi, Z.-H. Pan, S.Z. Lei, H.-S.V. Chen, N.J. Sucher, J. Loscalzo, D.J. Singel, and J.S. Stamler. 1993. A redox-based mechanism for the neuroprotective and neurodestructive effects of nitric oxide and related nitro-compounds. *Nature* 364:626-32.

Liu, D., D. Pavlovic, M.C. Chen, M. Flodstrom, S. Sandler, and D.L. Eizirik. 2000. Cytokines induce apoptosis in beta-cells isolated from mice lacking the inducible isoform of nitric oxide synthase (iNOS-/-). *Diabetes* 49:1116-22.

Loffreda, S., S.Q. Yang, H.Z. Lin, C.L. Karp, M.L. Brengman, D.J. Wang, A.S. Klein, G.B. Bulkley, C. Bao, P.W. Noble, M.D. Lane, and A.M. Diehl. 1998. Leptin regulates proinflammatory immune responses. *FASEB J* 12:57-65.

Long, S.D., and P.H. Pekala. 1996. Lipid mediators of insulin resistance: Ceramide signalling down-regulates GLUT4 gene transcription in 3T3-L1 adipocytes. *Biochem J* 319:179-84.

Lu, B., B.J. Rutledge, L. Gu, J. Fiorillo, N.W. Lukacs, S.L. Kunkel, R. North, C. Gerard, and B. J. Rollins. 1998. Abnormalities in monocyte recruitment and cytokine expression in monocyte chemoattractant protein 1-deficient mice. *J Exp Med* 187:601-8.

Marnell, L., C. Mold, and T.W. Du Clos. 2005. C-reactive protein: Ligands, receptors and role in inflammation. *Clin Immunol* 117:104-11.

McCullough, A.W. 1944. Evidence of the macrophagal origin of adipose cells in the white rat as shown by studies on starved animals. *J Morphol* 75:193-201.

McTernan, P.G., C.M. Kusminski, and S. Kumar. 2006. Resistin. *Curr Opin Lipidol* 17:170-5.

Meier, C.A., E. Bobbioni, C. Gabay, F. Assimacopoulos-Jeannet, A. Golay, and J.M. Dayer. 2002. IL-1 receptor antagonist serum levels are increased in human obesity: A possible link to the resistance to leptin? *J Clin Endocrinol Metab* 87:1184-8.

Minokoshi, Y., Y.B. Kim, O.D. Peroni, L.G. Fryer, C. Muller, D. Carling, and B.B. Kahn. 2002. Leptin stimulates fatty-acid oxidation by activating AMP-activated protein kinase. *Nature* 415 (6869):339-43

Moncada, S., R.M.J. Palmer, and E.A. Higgs. 1991. Nitric oxide: Physiology, pathophysiology and pharmacology. *Pharmacol Rev* 43:109-42.

Mooney, R.A., J. Senn, S. Cameron, N. Inamdar, L.M. Boivin, Y. Shang, and R.W. Furlanetto. 2001. Suppressors of cytokine signaling-1 and -6 associate with and inhibit the insulin receptor. A potential mechanism for cytokine-mediated insulin resistance. *J Biol Chem* 276:25889-93.

Moore, K.W., R. de Waal Malefyt, R.L. Coffman, and A. O'Garra. 2001. Interleukin-10 and the interleukin-10 receptor. *Annu Rev Immunol* 19:683-765.

Morand, E.F. 2005. New therapeutic target in inflammatory disease: Macrophage migration inhibitory factor. *Intern Med J* 35:419-26.

Mori, H., R. Hanada, T. Hanada, D. Aki, R. Mashima, H. Nishinakamura, T. Torisu, K.R. Chien, H. Yasukawa, and A. Yoshimura. 2004. Socs3 deficiency in the brain elevates leptin sensitivity and confers resistance to diet-induced obesity. *Nat Med* 10:739-43.

Muller, G., M. Ayoub, P. Storz, J. Rennecke, D. Fabbro, and K. Pfizenmaier. 1995. PKC zeta is a molecular switch in signal transduction of TNF-alpha, bifunctionally regulated by ceramide and arachidonic acid. *EMBO J* 14:1961-9.

Murdolo, G., and U. Smith. 2006. The dysregulated adipose tissue: A connecting link between insulin resistance, type 2 diabetes mellitus and atherosclerosis. *Nutr Metab Cardiovasc Dis* 16:S35-8.

Nakatani, Y., H. Kaneto, D. Kawamori, M. Hatazaki, T. Miyatsuka, T.A. Matsuoka, Y. Kajimoto, M. Matsuhisa, Y. Yamasaki, and M. Hori. 2004. Modulation of the JNK pathway in liver affects insulin resistance status. *J Biol Chem* 279:45803-9.

Napolitano, A., B.B. Lowell, D. Damm, R.L. Leibel, E. Ravussin, D.C. Jimerson, M.D. Lesem et al. 1994. Concentrations of adipsin in blood and rates of adipsin secretion by adipose tissue in humans with normal, elevated and diminished adipose tissue mass. *Int J Obes Relat Metab Disord* 18:213-8.

Nguyen, M.T., H. Satoh, S. Favelyukis, J.L. Babendure, T. Imamura, J.I. Sbodio, J. Zalevsky, B. I. Dahiyat, N.W. Chi, and J.M. Olefsky. 2005. JNK and tumor necrosis factor-alpha mediate free fatty acid-induced insulin resistance in 3T3-L1 adipocytes. *J Biol Chem* 280:35361-71.

Nojiri, H., M. Stroud, and S. Hakomori. 1991. A specific type of ganglioside as a modulator of insulin-dependent cell growth and insulin receptor tyrosine kinase activity. Possible association of ganglioside-induced inhibition of insulin receptor function and monocytic differentiation induction in HL-60 cells. *J Biol Chem* 266:4531-7.

Nomiyama, T., Y. Igarashi, H. Taka, R. Mineki, T. Uchida, T. Ogihara, J.B. Choi, H. Uchino, Y. Tanaka, H. Maegawa, A. Kashiwagi, K. Murayama, R. Kawamori, and H. Watada. 2004. Reduction of insulin-stimulated glucose uptake by peroxynitrite is concurrent with tyrosine nitration of insulin receptor substrate-1. *Biochem Biophys Res Commun* 320:639-47.

Noronha, B.T., J.M. Li, S.B. Wheatcroft, A.M. Shah, and M.T. Kearney. 2005. Inducible nitric oxide synthase has divergent effects on vascular and metabolic function in obesity. *Diabetes* 54:1082-9.

O'Brien, K.D., B.J. Brehm, R.J. Seeley, J. Bean, M.H. Wener, S. Daniels, and D.A. D'Alessio. 2005. Diet-induced weight loss is associated with decreases in plasma serum amyloid a and C-reactive protein independent of dietary macronutrient composition in obese subjects. *J Clin Endocrinol Metab* 90:2244-9.

O'Brien, K.D., and A. Chait. 2006. Serum amyloid A: The "other" inflammatory protein. *Curr Atheroscler Rep* 8:62-8.

Otero, M., R. Lago, F. Lago, F.F. Casanueva, C. Dieguez, J.J. Gomez-Reino, and O. Gualillo. 2005. Leptin, from fat to inflammation: Old questions and new insights. *FEBS Lett* 579:295-301.

Ozcan, U., Q. Cao, E. Yilmaz, A.H. Lee, N.N. Iwakoshi, E. Ozdelen, G. Tuncman, C. Gorgun, L. H. Glimcher, and G.S. Hotamisligil. 2004. Endoplasmic reticulum stress links obesity, insulin action, and type 2 diabetes. *Science* 306:457-61.

Ozes, O.N., H. Akca, L.D. Mayo, J.A. Gustin, T. Maehama, J.E. Dixon, and D.B. Donner. 2001. A phosphatidylinositol 3-kinase/Akt/mTOR pathway mediates and PTEN antagonizes tumor necrosis factor inhibition of insulin signaling through insulin receptor substrate-1. *Proc Natl Acad Sci USA* 98:4640-5.

Peraldi, P., G.S. Hotamisligil, W.A. Buurman, M.F. White, and B.M. Spiegelman. 1996. Tumor necrosis factor (TNF)-alpha inhibits insulin signaling through stimulation of the p55 TNF receptor and activation of sphingomyelinase. *J Biol Chem* 271:13018-22.

Perreault, M., and A. Marette. 2001. Targeted disruption of inducible nitric oxide synthase protects against obesity-linked insulin resistance in muscle. *Nat Med* 7:1138-43.

Pickup, J.C. 2004. Inflammation and activated innate immunity in the pathogenesis of type 2 diabetes. *Diabetes Care* 27:813-23.

Pickup, J.C., M.B. Mattock, G.D. Chusney, and D. Burt. 1997. NIDDM as a disease of the innate immune system: Association of acute-phase reactants and interleukin-6 with metabolic syndrome X. *Diabetologia* 40:1286-92.

Pilon, G., P. Dallaire, and A. Marette. 2004. Inhibition of inducible nitric-oxide synthase by activators of AMP-activated protein kinase: A new mechanism of action of insulin-sensitizing drugs. *J Biol Chem* 279:20767-74.

Plomgaard, P., K. Bouzakri, R. Krogh-Madsen, B. Mittendorfer, J.R. Zierath, and B.K. Pedersen. 2005. Tumor necrosis factor-alpha induces skeletal muscle insulin resistance in healthy human subjects via inhibition of Akt substrate 160 phosphorylation. *Diabetes* 54:2939-45.

Portis, B. 1924. Role of omentum of rabbits, dogs and guinea pigs in antibody production. *J Infect Dis* 34:159-85.

Portoles, M.T., R. Pagani, M.J. Ainaga, I. Diaz-Laviada, and A.M. Municio. 1989. Lipopolysaccharide-induced insulin resistance in monolayers of cultured hepatocytes. *Br J Exp Pathol* 70:199-205.

Powell, D.J., E. Hajduch, G. Kular, and H.S. Hundal. 2003. Ceramide disables 3-phosphoinositide binding to the pleckstrin homology domain of protein kinase B (PKB)/Akt by a PKCzeta-dependent mechanism. *Mol Cell Biol* 23:7794-808.

Prasad, R., S. Giri, N. Nath, I. Singh, and A.K. Singh. 2006. 5-aminoimidazole-4-carboxamide-1-beta-4-ribofuranoside attenuates experimental autoimmune encephalomyelitis via modulation of endothelial-monocyte interaction. *J Neurosci Res* 15;84(3):614-25.

Rane, S.G., and E.P. Reddy. 2000. Janus kinases: Components of multiple signaling pathways. *Oncogene* 19:5662-79.

Ransohoff, R.M., T.A. Hamilton, M. Tani, M.H. Stoler, H.E. Shick, J.A. Major, M.L. Estes, D.M. Thomas, and V.K. Tuohy. 1993. Astrocyte expression of mRNA encoding cytokines IP-10 and JE/MCP-1 in experimental autoimmune encephalomyelitis. *FASEB J* 7:592-600.

Remick, D.G. 2005. Interleukin-8. *Crit Care Med* 33:S466-7.

Rieusset, J., K. Bouzakri, E. Chevillotte, N. Ricard, D. Jacquet, J.P. Bastard, M. Laville, and H. Vidal. 2004. Suppressor of cytokine signaling 3 expression and insulin resistance in skeletal muscle of obese and type 2 diabetic patients. *Diabetes* 53:2232-41.

Rojas, F.A., C.R. Carvalho, V. Paez-Espinosa, and M.J. Saad. 2000. Regulation of cardiac Jak-2 in animal models of insulin resistance. *IUBMB Life* 49:501-9.

Romano, M., M.T. Guagnano, G. Pacini, S. Vigneri, A. Falco, M. Marinopiccoli, M.R. Manigrasso, S. Basili, and G. Davi. 2003. Association of inflammation markers with impaired insulin sensitivity and coagulative activation in obese healthy women. *J Clin Endocrinol Metab* 88:5321-6.

Rossi, D., and A. Zlotnik. 2000. The biology of chemokines and their receptors. *Annu Rev Immunol* 18:217-42.

Roytblat, L., M. Rachinsky, A. Fisher, L. Greemberg, Y. Shapira, A. Douvdevani, and S. Gelman. 2000. Raised interleukin-6 levels in obese patients. *Obes Res* 8:673-5.

Rui, L., V. Aguirre, J.K. Kim, G.I. Shulman, A. Lee, A. Corbould, A. Dunaif, and M.F. White. 2001. Insulin/IGF-1 and TNF-alpha stimulate phosphorylation of IRS-1 at inhibitory Ser307 via distinct pathways. *J Clin Invest* 107:181-9.

Rui, L., M. Yuan, D. Frantz, S. Shoelson, and M.F. White. 2002. SOCS-1 and SOCS-3 block insulin signaling by ubiquitin-mediated degradation of IRS1 and IRS2. *J Biol Chem* 277:42394-8.

Ryden, M., E. Arvidsson, L. Blomqvist, L. Perbeck, A. Dicker, and P. Arner. 2004. Targets for TNF-alpha-induced lipolysis in human adipocytes. *Biochem Biophys Res Commun* 318:168-75.

Sakamoto, K., A. McCarthy, D. Smith, K.A. Green, D. Grahame, A. Ashworth, and D.R. Alessi. 2005. Deficiency of LKB1 in skeletal muscle prevents AMPK activation and glucose uptake during contraction. *EMBO J* 24:1810-20.

Sartipy, P., and D.J. Loskutoff. 2003a. Expression profiling identifies genes that continue to respond to insulin in adipocytes made insulin-resistant by treatment with tumor necrosis factor-alpha. *J Biol Chem* 278:52298-306.

Sartipy, P., and D.J. Loskutoff. 2003b. Monocyte chemoattractant protein 1 in obesity and insulin resistance. *Proc Natl Acad Sci USA* 100:7265-70.

Sasaki, A., K. Hata, S. Suzuki, M. Sawada, T. Wada, K. Yamaguchi, M. Obinata, H. Tateno, H. Suzuki, and T. Miyagi. 2003. Overexpression of plasma membrane-associated sialidase attenuates insulin signaling in transgenic mice. *J Biol Chem* 278:27896-902.

Schmitz-Peiffer, C., D.L. Craig, and T.J. Biden. 1999. Ceramide generation is sufficient to account for the inhibition of the insulin-stimulated PKB pathway in C2C12 skeletal muscle cells pretreated with palmitate. *J Biol Chem* 274:24202-10.

Sell, H., D. Dietze-Schroeder, K. Eckardt, and J. Eckel. 2006. Cytokine secretion by human adipocytes is differentially regulated by adiponectin, AICAR, and troglitazone. *Biochem Biophys Res Commun* 343:700-6.

Sell, H., D. Dietze-Schroeder, U. Kaiser, and J. Eckel. 2006. Monocyte chemotactic protein-1 is a potential player in the negative cross-talk between adipose tissue and skeletal muscle. *Endocrinology* 147:2458-67.

Senn, J.J., P.J. Klover, I.A. Nowak, T.A. Zimmers, L.G. Koniaris, R.W. Furlanetto, and R.A. Mooney. 2003. Suppressor of cytokine signaling-3 (SOCS-3), a potential mediator of interleukin-6-dependent insulin resistance in hepatocytes. *J Biol Chem* 278:13740-6.

Shi, H., M.V. Kokoeva, K. Inouye, I. Tzameli, H. Yin, and J.S. Flier. 2006. TLR4 links innate immunity and fatty acid-induced insulin resistance. *J Clin Invest* 116(11):3015-25.

Shimabukuro, M., Y.T. Zhou, M. Levi, and R.H. Unger. 1998. Fatty acid-induced beta cell apoptosis: A link between obesity and diabetes. *Proc Natl Acad Sci USA* 95:2498-502.

Simeoni, E., M.M. Hoffmann, B.R. Winkelmann, J. Ruiz, S. Fleury, B.O. Boehm, W. Marz, and G. Vassalli. 2004. Association between the A-2518G polymorphism in the monocyte chemoattractant protein-1 gene and insulin resistance and type 2 diabetes mellitus. *Diabetologia* 47:1574-80.

Stephens, J.M., and A.J. Vidal-Puig. 2006. An update on visfatin/pre-B cell colony-enhancing factor, an ubiquitously expressed, illusive cytokine that is regulated in obesity. *Curr Opin Lipidol* 17:128-31.

Steppan, C.M., S.T. Bailey, S. Bhat, E.J. Brown, R.R. Banerjee, C.M. Wright, H.R. Patel, R.S. Ahima, and M.A. Lazar. 2001. The hormone resistin links obesity to diabetes. *Nature* 409:307-12.

Steppan, C.M., J. Wang, E.L. Whiteman, M.J. Birnbaum, and M.A. Lazar. 2005. Activation of SOCS-3 by resistin. *Mol Cell Biol* 25:1569-75.

Straczkowski, M., S. Dzienis-Straczkowska, A. Stepien, I. Kowalska, M. Szelachowska, and I. Kinalska. 2002. Plasma interleukin-8 concentrations are increased in obese subjects and related to fat mass and tumor necrosis factor-alpha system. *J Clin Endocrinol Metab* 87:4602-6.

Straczkowski, M., I. Kowalska, A. Nikolajuk, S. Dzienis-Straczkowska, I. Kinalska, M. Baranowski, M. Zendzian-Piotrowska, Z. Brzezinska, and J. Gorski. 2004. Relationship between insulin sensitivity and sphingomyelin signaling pathway in human skeletal muscle. *Diabetes* 53:1215-21.

Stratford, S., D.B. DeWald, and S.A. Summers. 2001. Ceramide dissociates 3'-phosphoinositide production from pleckstrin homology domain translocation. *Biochem J* 354:359-68.

Stratford, S., K.L. Hoehn, F. Liu, and S.A. Summers. 2004. Regulation of insulin action by ceramide: Dual mechanisms linking ceramide accumulation to the inhibition of Akt/protein kinase B. *J Biol Chem* 279:36608-15.

Summers, S.A., L.A. Garza, H. Zhou, and M.J. Birnbaum. 1998. Regulation of insulin-stimulated glucose transporter GLUT4 translocation and Akt kinase activity by ceramide. *Mol Cell Biol* 18:5457-64.

Tagami, S., J. Inokuchi Ji, K. Kabayama, H. Yoshimura, F. Kitamura, S. Uemura, C. Ogawa, A. Ishii, M. Saito, Y. Ohtsuka, S. Sakaue, and Y. Igarashi. 2002. Ganglioside GM3 participates in the pathological conditions of insulin resistance. *J Biol Chem* 277:3085-92.

Takahashi, K., S. Mizuarai, H. Araki, S. Mashiko, A. Ishihara, A. Kanatani, H. Itadani, and H. Kotani. 2003. Adiposity elevates plasma MCP-1 levels leading to the increased CD11b-positive monocytes in mice. *J Biol Chem* 278:46654-60.

Tau, G., and P. Rothman. 1999. Biologic functions of the IFN-gamma receptors. *Allergy* 54:1233-51.

Thirone, A.C., L. JeBailey, P.J. Bilan, and A. Klip. 2006. Opposite effect of JAK2 on insulin-dependent activation of mitogen-activated protein kinases and Akt in muscle cells: Possible target to ameliorate insulin resistance. *Diabetes* 55:942-51.

Tomas, E., T.S. Tsao, A.K. Saha, H.E. Murrey, C. Zhang Cc, S.I. Itani, H.F. Lodish, and N.B. Ruderman. 2002. Enhanced muscle fat oxidation and glucose transport by ACRP30 globular domain: acetyl-CoA carboxylase inhibition and AMP-activated protein kinase activation. *Proc Natl Acad Sci USA* 99 (25):16309-13.

Torres, S.H., J.B. De Sanctis, L.M. de Briceno, N. Hernandez, and H.J. Finol. 2004. Inflammation and nitric oxide production in skeletal muscle of type 2 diabetic patients. *J Endocrinol* 181:419-27.

Tremblay, F., and A. Marette. 2001. Amino acid and insulin signaling via the mTOR/p70 S6 kinase pathway. A negative feedback mechanism leading to insulin resistance in skeletal muscle cells. *J Biol Chem* 276:38052-60.

Tschop, M., C. Weyer, P.A. Tataranni, V. Devanarayan, E. Ravussin, and M.L. Heiman. 2001. Circulating ghrelin levels are decreased in human obesity. *Diabetes* 50:707-9.

Ueki, K., T. Kondo, and C.R. Kahn. 2004. Suppressor of cytokine signaling 1 (SOCS-1) and SOCS-3 cause insulin resistance through inhibition of tyrosine phosphorylation of insulin receptor substrate proteins by discrete mechanisms. *Mol Cell Biol* 24:5434-46.

Ueki, K., T. Kondo, Y.H. Tseng, and C.R. Kahn. 2004. Central role of suppressors of cytokine signaling proteins in hepatic steatosis, insulin resistance, and the metabolic syndrome in the mouse. *Proc Natl Acad Sci USA* 101:10422-7.

Um, S.H., D. D'Alessio, and G. Thomas. 2006. Nutrient overload, insulin resistance, and ribosomal protein S6 kinase 1, S6K1. *Cell Metab* 3:393-402.

van Furth, R., and Z.A. Cohn. 1968. The origin and kinetics of mononuclear phagocytes. *J Exp Med* 128:415-35.

Visser, M., L.M. Bouter, G.M. McQuillan, M.H. Wener, and T.B. Harris. 1999. Elevated C-reactive protein levels in overweight and obese adults. *JAMA* 282:2131-5.

Weisberg, S.P., D. Hunter, R. Huber, J. Lemieux, S. Slaymaker, K. Vaddi, I. Charo, R.L. Leibel, and A.W. Ferrante Jr. 2006. CCR2 modulates inflammatory and metabolic effects of high-fat feeding. *J Clin Invest* 116:115-24.

Weisberg, S.P., D. McCann, M. Desai, M. Rosenbaum, R.L. Leibel, and A.W. Ferrante Jr. 2003. Obesity is associated with macrophage accumulation in adipose tissue. *J Clin Invest* 112:1796-808.

Wellen, K.E., and G.S. Hotamisligil. 2005. Inflammation, stress, and diabetes. *J Clin Invest* 115:1111-9.

Westwick, J.K., A.E. Bielawska, G. Dbaibo, Y.A. Hannun, and D.A. Brenner. 1995. Ceramide activates the stress-activated protein kinases. *J Biol Chem* 270:22689-92.

White, R.T., D. Damm, N. Hancock, B.S. Rosen, B.B. Lowell, P. Usher, J.S. Flier, and B.M. Spiegelman. 1992. Human adipsin is identical to complement factor D and is expressed at high levels in adipose tissue. *J Biol Chem* 267:9210-3.

Winder, W. W. 2001. Energy-sensing and signaling by AMP-activated protein kinase in skeletal muscle. *J Appl Physiol* 91: 1017-28.

Winder, W. W., and D. G. Hardie. 1999. AMP-activated protein kinase, a metabolic master switch: possible roles in type 2 diabetes. *Am J Physiol* 277: E1-10.

Winder, W.W., B.F. Holmes, D.S. Rubink, E.B. Jensen, M. Chen, and J.O. Holloszy. 2000. Activation of AMP-activated protein kinase increases mitochondrial enzymes in skeletal muscle. *J Appl Physiol* 88:2219-26.

Wolf, A.M., D. Wolf, H. Rumpold, B. Enrich, and H. Tilg. 2004. Adiponectin induces the anti-inflammatory cytokines IL-10 and IL-1RA in human leukocytes. *Biochem Biophys Res Commun* 323:630-5.

Xia, Y., L.J. Roman, B.S. Masters, and J.L. Zweier. 1998. Inducible nitric-oxide synthase generates superoxide from the reductase domain. *J Biol Chem* 273:22635-9.

Xu, H., G.T. Barnes, Q. Yang, G. Tan, D. Yang, C.J. Chou, J. Sole, A. Nichols, J.S. Ross, L.A. Tartaglia, and H. Chen. 2003. Chronic inflammation in fat plays a crucial role in the development of obesity-related insulin resistance. *J Clin Invest* 112:1821-30.

Yamashita, T., A. Hashiramoto, M. Haluzik, H. Mizukami, S. Beck, A. Norton, M. Kono, S. Tsuji, J.L. Daniotti, N. Werth, R. Sandhoff, K. Sandhoff, and R.L. Proia. 2003. Enhanced insulin sensitivity in mice lacking ganglioside GM3. *Proc Natl Acad Sci USA* 100:3445-9.

Yamauchi, T., J. Kamon, Y. Minokoshi, Y. Ito, H. Waki, S. Uchida, S. Yamashita, M. Noda, S. Kita, K. Ueki, K. Eto, Y. Akanuma, P. Froguel, F. Foufelle, P. Ferre, D. Carling, S. Kimura, R. Nagai, B.B. Kahn, and T. Kadowaki. 2002. Adiponectin stimulates glucose utilization and fatty-acid oxidation by activating AMP-activated protein kinase. *Nat Med* 8:1288-95.

Yamauchi, T., J. Kamon, H. Waki, Y. Terauchi, N. Kubota, K. Hara, Y. Mori, T. Ide, K. Murakami, N. Tsuboyama-Kasaoka, O. Ezaki, Y. Akanuma, O. Gavrilova, C. Vinson, M.L. Reitman, H. Kagechika, K. Shudo, M. Yoda, Y. Nakano, K. Tobe, R. Nagai, S. Kimura, M. Tomita, P. Froguel, and T. Kadowaki. 2001. The fat-derived hormone adiponectin reverses insulin resistance associated with both lipoatrophy and obesity. *Nat Med* 7:941-6.

Yasukawa, H., A. Sasaki, and A. Yoshimura. 2000. Negative regulation of cytokine signaling pathways. *Annu Rev Immunol* 18:143-64.

Yla-Herttuala, S., B.A. Lipton, M.E. Rosenfeld, T. Sarkioja, T. Yoshimura, E.J. Leonard, J.L. Witztum, and D. Steinberg. 1991. Expression of monocyte chemoattractant protein 1 in macrophage-rich areas of human and rabbit atherosclerotic lesions. *Proc Natl Acad Sci USA* 88:5252-6.

Yokota, T., K. Oritani, I. Takahashi, J. Ishikawa, A. Matsuyama, N. Ouchi, S. Kihara, T. Funahashi, A.J. Tenner, Y. Tomiyama, and Y. Matsuzawa. 2000. Adiponectin, a new member of the family of soluble defense collagens, negatively regulates the growth of myelomonocytic progenitors and the functions of macrophages. *Blood* 96:1723-32.

Yuan, M., N. Konstantopoulos, J. Lee, L. Hansen, Z.W. Li, M. Karin, and S.E. Shoelson. 2001. Reversal of obesity- and diet-induced insulin resistance with salicylates or targeted disruption of Ikkbeta. *Science* 293:1673-7.

Yudkin, J.S., C.D. Stehouwer, J.J. Emeis, and S.W. Coppack. 1999. C-reactive protein in healthy subjects: Associations with obesity, insulin resistance, and endothelial dysfunction: A potential role for cytokines originating from adipose tissue? *Arterioscler Thromb Vasc Biol* 19:972-8.

Zhou, Y.T., P. Grayburn, A. Karim, M. Shimabukuro, M. Higa, D. Baetens, L. Orci, and R.H. Unger. 2000. Lipotoxic heart disease in obese rats: Implications for human obesity. *Proc Natl Acad Sci USA* 97:1784-9.

Part III

Prevention of Type 2 Diabetes Through Exercise Training

Chapter 8

Transcription Factors Regulating Exercise Adaptation

David Kitz Krämer, PhD; and Anna Krook, PhD

Skeletal muscle is an extremely flexible organ and adapts immediately to changes in use. The extent of adaptation depends on both the nature and the quantity of the demand placed on the muscle. While single bouts of exercise transiently alter gene expression, repeated bouts of exercise lead to a range of longer lasting adjustments. Endurance exercise builds the aerobic capacity, increasing the time over which the muscle can produce energy by oxidizing carbohydrate and lipid. On the other hand, resistance exercise increases the ability of the muscle to utilize glycolytic energy, thereby increasing the capability to produce power over a short length of time (Holloszy and Booth 1976). In response to exercise and contraction, skeletal muscle also becomes more sensitive to insulin. These changes are reversible, and thus the opposite effects occur in response to inactivity. Key questions include how these changes are initiated and coordinated at the level of gene transcription and whether there is a master signal activated by exercise. In this chapter we focus our attention on the transcription factors that regulate these processes.

With the advent of gene array technology, numerous studies were undertaken in an attempt to identify changes in mRNA following exercise training, and several candidate genes have been identified that appear to be involved in one or more exercise-induced adaptations (Mahoney and Tarnopolsky 2005; Teran-Garcia et al. 2005). Although gene array technology has been instrumental in mapping exercise-responsive genes, discovering the key transcription factors remains more challenging. Transcription factors can be regulated by at least three separate mechanisms. The first mechanism is increased expression. The second, which involves the physical location of the transcription factor within the cell, is usually controlled by changes in phosphorylation and can be affected by exercise without accompanying changes in mRNA expression. Thus phosphorylation and dephosphorylation can translocate the transcription factor to the nucleus, where it can bind target DNA sequences and direct transcription. Finally, the role of coactivators and corepressors is becoming increasingly appreciated. Interaction of transcription factors with coactivators may increase translocation to the nucleus or increase binding to the target promoter. Thus, in order to fully understand how transcription factors are regulated by exercise, we need to establish full proteomic activation profiles.

Activation of MAPK Signaling

Several MAPK proteins are activated in direct response to muscle contraction and exercise training (Long, Widegren, and Zierath 2004; Zierath 2002). Although not transcription factors themselves, the different MAPK signaling cascades integrate signals from diverse extracellular stimuli, including hormones and growth factors as well as cellular stress, to regulate gene transcription and protein synthesis in various cell culture systems (Pearson et al. 2001). In skeletal muscle, at least three parallel MAPK signaling cascades are

Support from the Swedish Research Council, the Novo Nordisk Foundation, the Hedlund Foundation, and the European Union Framework 6 Integrated Project EXGENESIS LSHM-CT-2004-005272 and Network of Excellence EUGENE2 no. LSHM-CT-2004-512013 is acknowledged.

activated in direct response to exercise. These include ERK1/2 (p42/p44 MAPK), p38 MAPK, and JNK (Long, Widegren, and Zierath 2004; see also figure 8.1).

Exercise results in several adaptations in the muscle cell and leads to the activation of different signaling components, as summarized in figure 8.1. Cellular stress and stretch or injury induce MAPK signaling cascades, leading to the activation of several transcription factors such as nuclear factor of activated T cells (NFAT) and myocyte enhancer factor (MEF). The p38 MAPK also appears to mediate coactivation or alteration of a number of other transcription factors. MAPK activation is thought to control diverse responses ranging from muscle growth to expression of glucose transport to regulation of mitochondrial biogenesis. The neuronal input initiating contraction releases Ca++ ions that alter the activity of calcium-sensitive kinases targeting NFAT, MEF2, and NRFs, leading to changes in fiber growth and differentiation and increases in oxidative capacity via mitochondria biogenesis.

The demands of exercise change the energy status of the muscle fiber, leading to the activation of energy-sensing kinases like AMPK. AMPK is thought to mediate signaling that regulates the expression of PGC-1, an important coactivator involved in MEF2, NFAT, ERRα, and PPAR activation. These transcription factors play important roles in regulating mitochondrial density and activ-

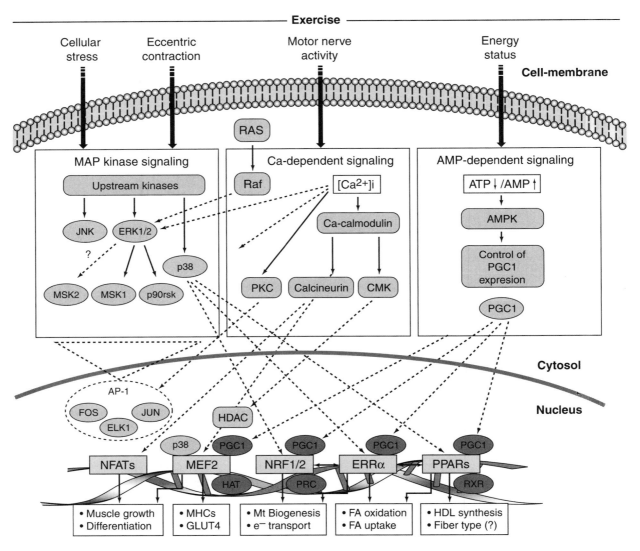

Figure 8.1 Summary of transcription factors regulated by exercise. MHC = myosin heavy chain; FA = fatty acid; HDL = high density lipoprotein; MT = mitochondria.

ity and thus also in regulating various aspects of lipid metabolism.

The ERK1/2 pathway is both rapidly and profoundly activated following acute cycling exercise (Widegren et al. 1998; Yu et al. 2003). By investigating muscle biopsies obtained from subjects performing one-legged cycling (the other leg being kept at rest), we have demonstrated that the activation of ERK1/2 MAPK signaling is specific to the exercising muscle. ERK1/2 is rapidly activated in the exercising muscle, and activity returns to basal levels within minutes of exercise cessation (Krook et al. 2000; Widegren et al. 1998). Furthermore, in vitro contraction of isolated rat skeletal muscle is sufficient to elicit ERK1/2 phosphorylation (Ryder et al. 2000; Wretman et al. 2001). Thus, local contraction-dependent effects rather than systemic exercise effects activate ERK1/2.

In human skeletal muscle, p38 MAPK was activated after acute cycling exercise (Widegren et al. 1998) and following marathon running (Boppart et al. 2000; Yu et al. 2001). In marked contrast to the profound but highly transient ERK1/2 activation, p38 MAPK activation by one-legged cycling is smaller but more persistent. Furthermore, p38 MAPK phosphorylation is increased in the resting leg, indicating the potential influence of a systemic factor (Widegren et al. 1998). It is possible that the mode of exercise influences the activation of different MAPK signaling pathways. In isolated rat skeletal muscle, concentric contractions increase ERK1/2 phosphorylation but do not affect p38 MAPK, whereas eccentric contractions increase phosphorylation along both kinase cascades (Wretman et al. 2001). Training status can also influence the exercise effect on MAPK signaling. Exercise-induced signaling responses for p38 MAPK are more profound in untrained men as compared to highly trained individuals (even at the same relative cycling exercise intensity; Yu et al. 2003). Activation of p38 MAPK may play an important role for the subsequent activation of the MEF2 transcription factor as well as the expression of the coactivator PGC-1α; this effect will be discussed in more detail subsequently.

Activation of the JNK pathway appears to relate somewhat to the degree of injury the muscle sustains with exercise, and JNK is affected more by eccentric as opposed to concentric exercise (Boppart et al. 1999). However, ERK kinase 1, which is an upstream activator of JNK, is activated during one-legged cycle ergometry, an exercise that is not associated with significant muscle injury (Widegren et al. 1998).

Downstream substrates of the various MAPK pathways include p90 ribosomal S6 kinase (p90rsk) and mitogen- and stress-activated kinases (MSK) 1 and 2. Activated p90rsk phosphorylates several transcription factors, including Elk, the cAMP response element-binding protein (CREB), and the AP-1 family (transcription factors consisting of homodimers and heterodimers of c-Jun and c-Fos). Exercise activates some of these downstream targets, including p90rsk and MSK1 and MSK2 (Krook et al. 2000; Yu et al. 2001). Activation of both MSK and p90rsk is rapid and limited to the exercising muscle, an observation suggesting that it occurs primarily via ERK1/2-dependent pathways. Histone H3 is a target of MSK1 (Thomson et al. 1999), and we have shown that exercise increases phosphorylation of histone H3 in human skeletal muscle (Yu et al. 2003). This finding may link contraction-activated signaling to gene expression, since the timing of histone H3 phosphorylation closely corresponds to the transient expression of activated immediate early genes (Thomson et al. 1999). Another target of MSK1 is the transcription factor CREB (Deak et al. 1998). However, CREB phosphorylation has not been reported to increase in skeletal muscle in response to exercise (Widegren et al. 2000). These combined observations underscore the difficulty in translating the results between different cell systems and the importance of studying in vivo responses to exercise in human skeletal muscle.

There is good evidence that exercise induces both MAPK signaling and changes in transcriptional activity, as MAPK has been directly implicated in the phosphorylation of transcription factors (Pearson et al. 2001). The nature of the transcription factors responsive to exercise, however, is not fully understood. Thus, although the transcription factors AP-1 and Elk-1 are targets of MAPK signaling, their role in exercise-mediated gene transcription is not known. Applying mechanical stress directly to skeletal muscle fibers does increase the DNA binding of AP-1 (Kumar et al. 2002). A number of other transcription factors known to be MAPK targets are also good candidates for directing exercise-mediated gene transcription. Some of these, including MEF2, PPAR, NFAT, and PGC-1, will be considered in more detail later in this chapter.

Muscle Hypoxia

Muscle contraction is thought to result in local hypoxia within the working muscle. The exercise effect on insulin sensitivity in muscle can be mimicked by hypoxia (Holloszy 2005). Hypoxia inducible factor 1 (HIF-1) is a transcription factor that acts as a master regulator for the expression of hypoxia inducible genes (Hoppeler et al. 2003). HIF-1 is a heterodimer composed of HIF-1α and HIF-1β, the latter of which is an aryl hydrocarbon receptor nuclear translocator (ARNT). Hypoxia stabilizes the HIF-1α protein, leading to nuclear translocation and activation of gene transcription. Acute exercise in humans enhances downstream HIF-1 function (Ameln et al. 2005). The precise effects of this activation are not fully understood, although it has been suggested to be important for HIF-1 for exercise-induced capillarization (Ameln et al. 2005). The effects of short-term exercise-induced hypoxia in muscle may differ from those of continuous reduced oxygen tension. For example, people who live at higher altitudes have a lower mitochondria density in the skeletal muscle as compared to people living at sea level (Hoppeler et al. 2003).

Calcium-Activated Signaling

Skeletal muscle contraction leads to an increase in intracellular Ca^{++} concentrations. Several signaling pathways are activated as a result of this increase in Ca^{++}, including calcineurin and its downstream targets NFAT, Ca^{++}/calmodulin-dependent protein kinases (CaMK), and Ca^{++}-dependent PKC.

Nuclear Factor of Activated T Cells

Initially the nuclear factor of activated T cells (NFAT) was identified as the transcription factor controlling the induction of the IL-2 gene during T cell activation (Shaw et al. 1988). To date, five NFAT family members have been identified, denominated NFAT1 through NFAT5. A rise in intracellular Ca^{++} activates the phosphatase calcineurin, which dephosphorylates and activates NFAT1 through NFAT4. NFAT5 is regulated by changes in osmotic tension.

In immune cells, NFAT acts in concert with other transcription factors, particularly the AP-1 proteins (Macian, Lopez-Rodriguez, and Rao 2001).

In cultured human skeletal muscle cells, NFATc1, NFATc2, and NFATc3 are expressed during distinct stages of differentiation (Abbott et al. 1998). Mice with targeted deletions of NFAT are characterized by skeletal muscle defects; nfatc3–/– animals have reduced muscle mass due to a lower number of both slow and fast myofibers (Kegley et al. 2001), while nfatc2–/– mice exhibit reduced muscle size due to a defect in skeletal muscle growth (Horsley et al. 2001). Hence, NFAT is thought to control skeletal muscle hypertrophy and muscle development and to be crucial in the establishment of fiber types. While NFAT appears to act in concert with other transcription factors such as the MEFs (Schulz and Yutzey 2004), it does not appear to regulate exercise effects on MEF2 in human muscle (McGee and Hargreaves 2004).

Although in principle exercise should augment NFAT activation via Ca^{++} and calcineurin, data on exercise-mediated NFAT activation in skeletal muscle are scarce to date. Moderate acute exercise increases mRNA expression of NFAT1, NFAT2, and NFAT3 in human skeletal muscle in the majority of subjects, and c-Fos mRNA (a component of the AP-1 complex) is significantly induced in skeletal muscle of all subjects (Hitomi et al. 2003). Electrical stimulation of mouse skeletal muscle leads to nuclear translocation of the NFATc1 isoform (Tothova et al. 2006). However, recent evidence suggests that NFAT shuttling into the nucleus is controlled not only by contraction, Ca^{++}, and calcineurin but also by other pathways (Shen et al. 2006). Thus further studies are needed to elucidate NFAT nuclear translocation and activation directly in exercising human skeletal muscle. Furthermore, the effects that appear to be regulated by calcineurin cannot be fully explained by NFAT transcription factors (Parsons et al. 2003), and it is likely that the coordinated regulation of other transcription factors such as MEF2 is necessary (Chin et al. 1998).

Regulation of GLUT4 Expression

Human skeletal muscle expresses the glucose transporters GLUT1 and GLUT4 (Zorzano et al. 1996). The GLUT4 isoform accounts for approximately 90% of the glucose transporter proteins in skeletal muscle; in fact, evidence suggests that GLUT4 may be the only glucose transporter in human skeletal muscle, with GLUT1 being expressed primarily in endothelial cells from intermuscular capillaries (Ploug et al. 1998). GLUT4 abundance depends on the developmental stage of the skeletal muscle and on the fiber type composition of oxidative versus glycolytic muscle fibers (Kern et al. 1990; Santalucia et al. 1992).

An acute bout of exercise elicits an insulin-independent translocation of GLUT4 to the cell surface and an increase in glucose transport (Douen et al. 1990; Lund et al. 1995) as well as an increase in expression of GLUT4. Expression of GLUT4 was increased approximately twofold 16 h after one prolonged (6 h) swim bout (Ren et al. 1994). GLUT4 expression can be linked directly to muscle innervation and contractile activity, and it can be elevated experimentally by in vivo and in vitro low-frequency stimulation of skeletal muscle contraction in animal models (Etgen et al. 1993; Hofmann and Pette 1994) and by exercise in humans (Dela et al. 1993; Houmard et al. 1991). The increase in GLUT4 is thought to be one of the key factors mediating enhanced insulin sensitivity in exercised skeletal muscle. Transgenic mice that overexpress GLUT4 in adipose and skeletal muscle exhibit improved oral glucose tolerance and insulin-stimulated glucose disposal (Deems et al. 1994; Liu et al. 1993; Ren et al. 1995).

Myocyte Enhancer Factor 2 and GLUT4 Enhancer Factor

Both the myocyte enhancer factor 2 (MEF2) family of transcription factors and the GLUT4 enhancer factor (GEF) have been implicated in the exercise-mediated effects on GLUT4 (McGee et al. 2006). Human skeletal muscle expresses MEF2A, MEF2C, and MEF2D. MEF2A in particular is thought to contribute to the exercise-mediated increase in GLUT4 expression (Mora and Pessin 2000). MyoD and thyroid hormone receptor (TRα1) have promoter binding locations that neighbor the MEF2 site, and, together with MEF and GEF, they appear to be required for full GLUT4 expression (Santalucia et al. 1992). While not fully elucidated, the regulation of MEF2 activity has been shown to be controlled by a variety of factors, including the MEF2 inhibitor class II histone deacetylase (HDAC) coactivators like histone acetyltransferase (HAT; McKinsey, Zhang, and Olson 2001), possibly PGC-1α (McGee et al. 2006), and p38 MAPK (Zhao et al. 1999).

DNA binding by both MEF2 and GEF increases in response to acute exercise (McGee et al. 2006; Yu et al. 2003). How exercise increases the DNA binding is not fully understood, although activation of AMPK is a possible mechanism (Al-Khalili et al. 2004; Holmes et al. 2005). The activity of p38 MAPK is elevated in direct response to exercise, and a docking domain for p38 MAPK has been found on MEF2A (Chang et al. 2002). The exercise-mediated increase in Ca^{++} has also been suggested to be a key signal mediating exercise-induced GLUT4 expression in skeletal muscle via regulation of MEF2, possibly through calcineurin. Expression of activated calcineurin in mouse skeletal muscle results in increased expression of GLUT4 (Ryder et al. 2003). The mechanisms leading to Ca^{++}-mediated activation of MEF2 are under debate; however, phosphorylation of HDAC leads to the dissociation of the MEF2-HDAC complex, allowing for at least a partial transcriptional activity of MEF2 (Lu et al. 2000). Additionally, calcineurin activation of NFAT may recruit coactivators to MEF2 (McKinsey, Zhang, and Olson 2002). However, despite a possible role for calcineurin in both developmental and long-term adjustments in GLUT4 expression following exercise training, a role for calcineurin in the short-term effects on GLUT4 expression has been challenged, and CaMK has been added to the list of possible mediators of the Ca^{++} effect on the MEF2A transcription factor (Garcia-Roves et al. 2005; Ojuka et al. 2002).

Thus there is good evidence that both GEF and MEF2 are important for the expression of GLUT4 in human skeletal muscle following exercise. The MEF2 family of transcription factors is known to be important for the expression of a large number of genes (Black and Olson 1998), and thus it is likely that exercise-mediated activation of MEF2 DNA binding has implications for a number of exercise adaptations and changes in gene expression. However, to date the bulk of our understanding has centered on MEF2 and GLUT4.

Mitochondria Biogenesis and Increased Lipid Oxidation

Endurance exercise has been shown to be of greater benefit than strength exercise in treating type 2 diabetes (Cauza et al. 2005). This finding is thought to reflect the fact that endurance exercise improves oxidative capacity by increasing mitochondrial density while strength exercise does not appear to have this effect (Davies, Packer, and Brooks 1981; Holloszy and Booth 1976). Furthermore, it has been proposed that a reduced oxidative capacity for fatty acids, possibly due to a decrease in mitochondrial density or function, contributes to the onset of type 2 diabetes in elderly individuals (Petersen and Shulman 2006). Similarly (but less frequently), hereditary mutations that impair mitochondria function may contribute to the onset of type 2 diabetes (Barazzoni 2004). The immediate targets of endurance exercise thought to signal to transcription factors leading to mitochondria biogenesis are Ca^{++} (Freyssenet, Di Carlo, and Hood 1999; Ojuka et al. 2002), AMPK (Atherton et al. 2005; Bergeron et al. 2001), and CaMK (Wu et al. 2002). Important end points of this signaling cascade are nuclear respiratory factor (NRF)-1 (Bergeron et al. 2001); NRF2 (Baar 2004); PGC-1 (Atherton et al. 2005); PPARα, PPARβ/δ; PRC; and TFAM (mTFA).

Nuclear Respiratory Factors 1 and 2

Mitochondria encode and express 13 subunits of the respiratory complexes but need some 100 proteins to function. Consequently these proteins have to be expressed in the nucleus and imported into the mitochondria. Although NRF1 regulates genes that are unrelated to mitochondrial function, it is also considered a key regulator of mitochondrial proliferation and differentiation (Xia et al. 1997). NRF1 is activated in response to an acute bout of exercise in rat skeletal muscle, suggesting that repeated bouts of exercise induce mitochondrial adaptation (Murakami et al. 1998). Similarly, NRF2 acts on a number of mitochondrial respiratory promoters and regulates the expression of several respiratory genes (Virbasius, Virbasius, and Scarpulla 1993). In humans, NRF2 mRNA has been shown to be upregulated 48 h after exercise (Cartoni et al. 2005). NRF2 is also involved in regulating TFAM (also known as *mTFA*), which is necessary for mitochondrial maintenance and biogenesis (Larsson et al. 1998).

Estrogen-Related Receptor α

Another exercise-responsive transcription factor involved in mitochondrial biogenesis and function is the estrogen-related receptor (ERR)-α. ERRs consist of three isoforms (alpha, beta, and gamma) and are orphan receptors with close homology to estrogen

receptors (Giguere et al. 1988). ERRα mRNA is upregulated following exercise (Cartoni et al. 2005) and interacts with PGC-1 coactivators controlling mitochondrial replication and expression of electron transport protein complexes (Ichida, Nemoto, and Finkel 2002). ERRα directly regulates mitochondrial beta-oxidation by interacting with PGC-1α (Sladek, Bader, and Giguere 1997; Vega and Kelly 1997). In skeletal muscle, genes important for fatty acid uptake and utilization have been shown to be regulated via the direct interaction of ERRα with PPARα (Huss et al. 2004) and NRFs (Finck and Kelly 2006). The importance of ERRα in regulating skeletal muscle oxidative phosphorylation was demonstrated using synthetic inhibitors (Mootha et al. 2004). Rather surprisingly, mice with a complete deletion of the ERRα gene exhibit a lean phenotype that resists the development of obesity, seemingly by a disruption of adipocyte development, indicating tissue specificity in the role of ERRα (Luo et al. 2003).

Peroxisome Proliferator-Activated Receptors

PPARs are nuclear receptors and transcription factors that play central roles in substrate utilization and have received attention as pharmacological targets for treating metabolic disease (Berger, Akiyama, and Meinke 2005; Smith and Muscat 2005). PPARs form heterodimers with RXRs. In the absence of an agonist, these heterodimers may recruit corepressors and silence transcription by active repression (Jepsen and Rosenfeld 2002). The PPARs are activated by dietary lipids and are therefore considered to be nutritional lipid sensors and to control lipid homeostasis (Smith and Muscat 2005). PPARs have also been implicated in mechanisms that release anti-inflammatory factors or repress the inflammatory response (Lee et al. 2003; Pascual et al. 2005). All three described PPAR isoforms are expressed in skeletal muscle; PPARδ and PPARα are the most abundant, while PPARγ appears to play a secondary role.

Exercise-Mediated Regulation of PPARs

Although some studies have reported elevated PPARγ mRNA in vastus lateralis muscle from healthy young men 3 h following cycling exercise

(Mahoney et al. 2005) and in rodents after 16 wk of treadmill exercise training (Kawamura et al. 2004), other studies have failed to show exercise effects on PPARγ mRNA levels in either rodents or humans (Gorla-Bajszczak et al. 2000; Russell et al. 2003; Tunstall et al. 2002). Despite this, the PPARγ2 Pro12Ala variant in humans has been associated with an improved exercise response. Carriers of PPARγ2 Pro12Ala demonstrated significantly better exercise-mediated improvement in fasting glucose than a control group demonstrated (Adamo et al. 2005). This finding suggests that the role of PPARγ in response to exercise requires further evaluation, and future studies may need to determine the actual transcriptional activation of PPARγ in addition to changes in mRNA expression.

Both PPARα and PPARδ mRNA are increased following an acute 3 h exercise bout (Watt et al. 2004). Endurance training has also been reported to elevate PPARα mRNA (Horowitz et al. 2000; Russell et al. 2003). Interestingly, nutritional status may influence the exercise effect on PPAR isoforms (Watt et al. 2004), as fasting dramatically increases PPARδ expression in mice (Luquet et al. 2003). Given the additional complexity of hormonal and nutrient regulation of these targets, dissecting exercise from nutritional effects on PPAR expression may be a challenge.

Recently we reported that protein expression of PPARδ in skeletal muscle increased significantly after physical exercise in patients with type 2 diabetes following a 4 mo, low-intensity exercise program (Fritz et al. 2006). Interestingly, the increase in PPARδ expression was associated with improvements in several clinical parameters, and PPARδ expression did not change in subjects who did not improve their clinical profile following exercise. Thus PPARα and δ may be key factors coordinating exercise-mediated changes in metabolism.

Peroxisome Proliferator-Activated Receptor Gamma Coactivator 1

A key feature of most of the transcription factors involved in mitochondrial biogenesis is their binding to the nuclear coactivator peroxisome proliferator-activated receptor gamma coactivator 1 (PGC-1) or PGC-1-related coactivator (PRC). These coactiva-

tors do not bind DNA themselves but interact with DNA-bound transcription factors to regulate gene expression (Finck and Kelly 2006). PGC-1α was the first of three PGC-1 homologues to be described. PGC-1α and PGC-1β share high sequence homology, whereas PRC is more distantly related.

An acute bout of exercise markedly increases PGC-1α mRNA immediately following the activity. PGC-1α then returns to pre-exercise levels within 24 h (Pilegaard, Saltin, and Neufer 2003). Several bouts of exercise training lead to a sustained increase in PGC-1α (Kuhl et al. 2006; Russell et al. 2003; Short et al. 2003). Furthermore, DNA polymorphisms in PGC-1α have been linked to reduced cardiovascular fitness (Ling et al. 2004) and to greater odds of developing type 2 diabetes (Barroso et al. 2006). Whether these polymorphisms are related to how PGC-1α responds to exercise has not yet been investigated.

Interestingly, overexpression of PGC-1α in cultured myoblasts is sufficient to increase mitochondria biogenesis (Wu et al. 1999). Overexpression of PGC-1β is also associated with an enhanced number of mitochondria (Lin et al. 2003; Meirhaeghe et al. 2003). PGC-1β transgenic mice have increased energy expenditure and are protected from obesity by increasing fat oxidation (Kamei et al. 2003).

Myogenic Development and Adaptation

Skeletal muscle utilizes both lipid and glucose as energy sources but prefers one or the other depending on a number of different factors. To a certain extent, substrate preference is programmed during skeletal muscle development. The transcription factors MyoD and MEF2 are part of a large number of proteins that regulate muscle development. These proteins are collectively denominated *myogenic regulatory factors* (MRFs; Blais et al. 2005). A comprehensive review of all MRFs, their interactions, and their targets is beyond the scope of this chapter; however, some of these targets have recently received attention due to their apparent importance in metabolic disease and type 2 diabetes. Among these are the PPARs, whose expression can be regulated by MyoD (Blais et al. 2005), and PGC-1α, whose expression is partly controlled by MEF2 (Czubryt et al. 2003).

Transcription Factors in the Regulation of Skeletal Muscle Fiber Types

Historically, muscle was classified by its appearance (as red or white) in recognition of the fact that its appearance correlates with its contractile properties (slow or fast, respectively; Spangenburg and Booth 2003). These functional properties of the skeletal muscle are closely coupled to metabolic profile: Oxidative, slow-twitch fibers (Type I fibers) carry larger amounts of mitochondria than fast-twitch, glycolytic fibers (Type II fibers) carry (Schiaffino and Serrano 2002; Spangenburg and Booth 2003). The regulation of fiber types is complex and will not be discussed in detail here. Some of this complexity derives from the difficulty in defining a fiber type. With this caveat it is still safe to argue that muscle fiber type composition directly influences exercise performance, and some evidence suggests that exercise training in turn influences fiber type. Furthermore, insulin-stimulated glucose transport is greater in slow-twitch, mitochondria-rich skeletal muscle fibers than it is in fast-twitch, glycolytic fibers (Daugaard et al. 2000; Henriksen et al. 1990; Song et al. 1999). In humans, insulin sensitivity correlates positively with the proportion of slow-twitch fibers (Lillioja et al. 1987). Patients displaying type 2 diabetes and insulin resistance, subjects exhibiting morbid obesity and insulin resistance, and first-degree relatives of patients with type 2 diabetes have a lower percentage of Type I fibers and a higher percentage of Type II fibers, particularly Type IIb fibers, when compared to insulin-sensitive subjects (Marin et al. 1994; Nyholm et al. 1997).

Transgenic animals have been instrumental in highlighting important regulators of muscle fiber type. However, the regulation of fiber type in transgenic animals is a result of altered mRNA expression during muscle development and may not necessarily reflect the regulation of these processes in mature muscle. Also, homogeneity, regulation, and degree of fiber type transformation differ substantially between rodent and human skeletal muscle (Delp and Duan 1996; Holloszy and Coyle 1984), and thus care must be taken when translating results from transgenic animals to humans.

However, some of the key factors implicated in the regulation of muscle fiber type are also known to be regulated by exercise training in mature muscle. These factors include PPARδ and PGC-1α.

PGC-1α has been implicated as a master regulator of the slow-twitch, oxidative Type I muscle phenotype in rodents (Lin et al. 2002). Transgenic expression of activated PPARδ increases the proportion of Type I fibers in mice, thereby transforming the skeletal muscle to a slow-twitch, oxidative phenotype (Luquet et al. 2003; Wang et al. 2004).

Do Genetic Variations in Transcription Factor Genes Control Exercise Response?

In this chapter we have discussed how exercise and muscle contraction affect the expression or activity of different transcription factors that subsequently regulate muscle remodeling and metabolism (summarized in figure 8.1). The challenge is to understand the balance and interplay among these different transcription factors as well as the relative importance of the signaling pathways that lead to their activation. Genetic variations in several key genes are also likely to influence the muscle response to exercise, as was exemplified in the previous section for the PPARγ gene (Adamo et al. 2005). Differences in mRNA profiles in skeletal muscle have been mapped between groups of subjects who show a marked difference in the improvement of glucose tolerance following the same 20 wk of exercise training, demonstrating the existence of exercise resistance (Teran-Garcia et al. 2005). We have noted that when subjects with type 2 diabetes exercised for 4 mo, only subjects who had an increase in skeletal muscle PPARδ expression responded to exercise by improving their clinical status (Fritz et al. 2006). Whether the response to exercise is linked to variations within the PPARδ gene or to variations in other genes is currently not known.

Concluding Remarks

Skeletal muscle responds to both use and disuse by changing its gene expression. As we improve our understanding of how skeletal muscle metabolism is regulated, we will begin to unravel how adaptations to exercise and exercise training are regulated at the transcriptional level. Greater understanding of these events has implications not only for improving sports performance but also for identifying molecular targets in the treatment of metabolic disorders such as type 2 diabetes.

References

Abbott, K.L., B.B. Friday, D. Thaloor, T.J. Murphy, and G.K. Pavlath. 1998. Activation and cellular localization of the cyclosporine A-sensitive transcription factor NF-AT in skeletal muscle cells. *Mol Biol Cell* 9:2905-16.

Adamo, K.B., R.J. Sigal, K. Williams, G. Kenny, D. Prud'homme, and F. Tesson. 2005. Influence of Pro12Ala peroxisome proliferator-activated receptor gamma2 polymorphism on glucose response to exercise training in type 2 diabetes. *Diabetologia* 48:1503-9.

Al-Khalili, L., A.V. Chibalin, M. Yu, B. Sjödin, C. Nylen, J.R. Zierath, and A. Krook. 2004. MEF2 activation in differentiated primary human skeletal muscle cultures requires coordinated involvement of parallel pathways. *Am J Physiol Cell Physiol* 286:C1410-6.

Ameln, H., T. Gustafsson, C.J. Sundberg, K. Okamoto, E. Jansson, L. Poellinger, and Y. Makino. 2005. Physiological activation of hypoxia inducible factor-1 in human skeletal muscle. *FASEB J* 04-2304fje.

Atherton, P.J., J. Babraj, K. Smith, J. Singh, M.J. Rennie, and H. Wackerhage. 2005. Selective activation of AMPK-PGC-1alpha or PKB-TSC2-mTOR signaling can explain specific adaptive responses to endurance or resistance training-like electrical muscle stimulation. *FASEB J* 19:786-8.

Baar, K. 2004. Involvement of PPAR gamma co-activator-1, nuclear respiratory factors 1 and 2, and PPAR alpha in the adaptive response to endurance exercise. *Proc Nutr Soc* 63:269-73.

Barazzoni, R. 2004. Skeletal muscle mitochondrial protein metabolism and function in ageing and type 2 diabetes. *Curr Opin Clin Nutr Metab Care* 7:97-102.

Barroso, I., J. Luan, M. Sandhu, P.W. Franks, V. Crowley, and A. Schafer. 2006. Meta-analysis of the Gly482Ser variant in PPARGC1A in type 2 diabetes and related phenotypes. *Diabetologia* 49:501-5.

Berger, J.P., T.E. Akiyama, and P.T. Meinke. 2005. PPARs: Therapeutic targets for metabolic disease. *Trends Pharmacol Sci* 26:244-51.

Bergeron, R., J.M. Ren, K.S. Cadman, I.K. Moore, P. Perret, M. Pypaert, L.H. Young, C.F. Semenkovich, and G.I. Shulman. 2001. Chronic activation of AMP kinase results in NRF-1 activation and mitochondrial biogenesis. *Am J Physiol Endocrinol Metab* 281:E1340-6.

Black, B.L., and E.N. Olson. 1998. Transcriptional control of muscle development by myocyte enhancer factor-2 (MEF2) proteins. *Annu Rev Cell Dev Biol* 14:167-96.

Blais, A., M. Tsikitis, D. Acosta-Alvear, R. Sharan, Y. Kluger, and B.D. Dynlacht. 2005. An initial blueprint for myogenic differentiation. *Genes Dev* 19:553-69.

Boppart, M.D., D. Aronson, L. Gibson, R. Roubenoff, L.W. Abad, J. Bean, L.J. Goodyear, and R.A. Fielding. 1999. Eccentric exercise markedly increases c-Jun NH 2 -terminal kinase activity in human skeletal muscle. *J Appl Physiol* 87:1668-73.

Boppart, M.D., S. Asp, J.F.P. Wojtaszewski, R.A. Fielding, T. Mohr, and L.J. Goodyear. 2000. Marathon running transiently increases c-Jun NH 2 -terminal kinase and p38 activities in human skeletal muscle. *J Physiol* 526:663-9.

Cartoni, R., B. Leger, M.B. Hock, M. Praz, A. Crettenand, S. Pich, J.L. Ziltener, F. Luthi, O. Deriaz, A. Zorzano, C. Gobelet, A. Kralli, and A.P. Russell. 2005. Mitofusins 1/2

and ERRalpha expression are increased in human skeletal muscle after physical exercise. *J Physiol* 567:349-58.

Cauza, E., U. Hanusch-Enserer, B. Strasser, B. Ludvik, S. Metz-Schimmerl, G. Pacini, O. Wagner, P. Georg, R. Prager, K. Kostner, A. Dunky, and P. Haber. 2005. The relative benefits of endurance and strength training on the metabolic factors and muscle function of people with type 2 diabetes mellitus. *Arch Phys Med Rehabil* 86:1527-33.

Chang, C.I., B.E. Xu, R. Akella, M.H. Cobb, and E.J. Goldsmith. 2002. Crystal structures of MAP kinase p38 complexed to the docking sites on its nuclear substrate MEF2A and activator MKK3b. *Mol Cell* 9:1241-9.

Chin, E.R., E.N. Olson, J.A. Richardson, Q. Yang, C. Humphries, J.M. Shelton, H. Wu, W. Zhu, R. Bassel-Duby, and R.S. Williams. 1998. A calcineurin-dependent transcriptional pathway controls skeletal muscle fiber type. *Genes Dev* 12:2499-509.

Czubryt, M.P., J. McAnally, G.I. Fishman, and E.N. Olson. 2003. Regulation of peroxisome proliferator-activated receptor gamma coactivator 1 alpha (PGC-1 alpha) and mitochondrial function by MEF2 and HDAC5. *Proc Natl Acad Sci USA* 100:1711-6.

Daugaard, J.R., J.N. Nielsen, S. Kristiansen, J.L. Andersen, M. Hargreaves, and E.A. Richter. 2000. Fiber type-specific expression of GLUT4 in human skeletal muscle: Influence of exercise training. *Diabetes* 49:1092-5.

Davies, K.J., L. Packer, and G.A. Brooks. 1981. Biochemical adaptation of mitochondria, muscle, and whole-animal respiration to endurance training. *Arch Biochem Biophys* 209:539-54.

Deak, M., A.D. Clifton, L.M. Lucocq, and D.R. Alessi. 1998. Mitogen- and stress-activated protein kinase-1 (MSK1) is directly activated by MAPK and SAPK2/p38, and may mediate activation of CREB. *EMBO J* 17:4426-41.

Deems, R.O., J.L. Evans, R.W. Deacon, C.M. Honer, D.T. Chu, K. Burki, W.S. Fillers, D.K. Cohen, and D.A. Young. 1994. Expression of human GLUT4 in mice results in increased insulin action. *Diabetologia* 37:1097-104.

Dela, F., A. Handberg, K.J. Mikines, J. Vinten, and H. Galbo. 1993. GLUT4 and insulin receptor binding and kinase activity in trained human muscle. *J Physiol* 469:615-24.

Delp, M.D., and C. Duan. 1996. Composition and size of type I, IIA, IID/X, and IIB fibers and citrate synthase activity of rat muscle. *J Appl Physiol* 80:261-70.

Douen, A.G., T. Ramlal, S.A. Rastogi, P.J. Bilan, G.D. Cartee, M. Vranic, J.O. Holloszy, and A. Klip. 1990. Exercise induces recruitment of the "insulin responsive" glucose transporter. Evidence for distinct intracellular insulin- and exercise-recruitable transporter pools in skeletal muscle. *J Biol Chem* 265:13427-30.

Etgen Jr., G.J., A.R. Memon, G.A. Thompson Jr., and J.L. Ivy. 1993. Insulin- and contraction-stimulated translocation of GTP-binding proteins and GLUT4 protein in skeletal muscle. *J Biol Chem* 268:20164-9.

Finck, B.N., and D.P. Kelly. 2006. PGC-1 coactivators: Inducible regulators of energy metabolism in health and disease. *J Clin Invest* 116:615-22.

Freyssenet, D., M. Di Carlo, and D.A. Hood. 1999. Calcium-dependent regulation of cytochrome c gene expression in skeletal muscle cells. Identification of a protein kinase c-dependent pathway. *J Biol Chem* 274:9305-11.

Fritz, T., D. Kitz Krämer, H.K.R. Karlsson, D. Galuska, P. Engfeldt, J.R. Zierath, and A. Krook. 2006. Low-intensity exercise increases skeletal muscle protein expression of PPARdelta and UCP3 in type 2 diabetic patients. *Diabetes Metab Res Rev* 22:492-8.

Garcia-Roves, P.M., T.E. Jones, K. Otani, D.H. Han, and J.O. Holloszy. 2005. Calcineurin does not mediate exercise-induced increase in muscle GLUT4. *Diabetes* 54:624-8.

Giguere, V., N. Yang, P. Segui, and R.M. Evans. 1988. Identification of a new class of steroid hormone receptors. *Nature* 331:91-4.

Gorla-Bajszczak, A., C. Siegrist-Kaiser, O. Boss, A.G. Burger, and C.A. Meier. 2000. Expression of peroxisome proliferator-activated receptors in lean and obese Zucker rats. *Eur J Endocrinol* 142:71-8.

Henriksen, E.J., R.E. Bourey, K.J. Rodnick, L. Koranyi, M.A. Permutt, and J.O. Holloszy. 1990. Glucose transporter protein content and glucose transport capacity in rat skeletal muscles. *Am J Physiol* 259:E593-8.

Hitomi, Y., T. Kizaki, T. Katsumura, M. Mizuno, C.E. Itoh, K. Esaki, Y. Fujioka, T. Takemasa, S. Haga, and H. Ohno. 2003. Effect of moderate acute exercise on expression of mRNA involved in the calcineurin signaling pathway in human skeletal muscle. *IUBMB Life* 55:409-13.

Hofmann, S., and D. Pette. 1994. Low-frequency stimulation of rat fast-twitch muscle enhances the expression of hexokinase II and both the translocation and expression of glucose transporter 4 (GLUT-4). *Eur J Biochem* 219:307-15.

Holloszy, J.O. 2005. Exercise-induced increase in muscle insulin sensitivity. *J Appl Physiol* 99:338-43.

Holloszy, J.O., and F.W. Booth. 1976. Biochemical adaptations to endurance exercise in muscle. *Annu Rev Physiol* 38:273-91.

Holloszy, J.O., and E.F. Coyle. 1984. Adaptations of skeletal muscle to endurance exercise and their metabolic consequences. *J Appl Physiol* 56:831-8.

Holmes, B.F., D.P. Sparling, A.L. Olson, W.W. Winder, and G.D. Dohm. 2005. Regulation of muscle GLUT4 enhancer factor and myocyte enhancer factor 2 by AMP-activated protein kinase. *Am J Physiol Endocrinol Metab* 289: E1071-6.

Hoppeler, H., M. Vogt, E.R. Weibel, and M. Fluck. 2003. Response of skeletal muscle mitochondria to hypoxia. *Exp Physiol* 88:109-19.

Horowitz, J.F., T.C. Leone, W. Feng, D.P. Kelly, and S. Klein. 2000. Effect of endurance training on lipid metabolism in women: A potential role for PPARalpha in the metabolic response to training. *Am J Physiol Endocrinol Metab* 279: E348-55.

Horsley, V., B.B. Friday, S. Matteson, K.M. Kegley, J. Gephart, and G.K. Pavlath. 2001. Regulation of the growth of multinucleated muscle cells by an NFATC2-dependent pathway. *J Cell Biol* 153:329-38.

Houmard, J.A., P.C. Egan, P.D. Neufer, J.E. Friedman, W.S. Wheeler, R.G. Israel, and G.L. Dohm. 1991. Elevated skeletal muscle glucose transporter levels in exercise-trained middle-aged men. *Am J Physiol* 261:E437-43.

Huss, J.M., I.P. Torra, B. Staels, V. Giguere, and D.P. Kelly. 2004. Estrogen-related receptor alpha directs peroxisome proliferator-activated receptor alpha signaling in the transcriptional control of energy metabolism in cardiac and skeletal muscle. *Mol Cell Biol* 24:9079-91.

Ichida, M., S. Nemoto, and T. Finkel. 2002. Identification of a specific molecular repressor of the peroxisome proliferator-activated receptor gamma coactivator-1 alpha (PGC-1alpha). *J Biol Chem* 277:50991-5.

Jepsen, K., and M.G. Rosenfeld. 2002. Biological roles and mechanistic actions of co-repressor complexes. *J Cell Sci* 115:689-98.

Kamei, Y., H. Ohizumi, Y. Fujitani, T. Nemoto, T. Tanaka, N. Takahashi, T. Kawada, M. Miyoshi, O. Ezaki, and A. Kakizuka. 2003. PPARgamma coactivator 1beta/ERR ligand 1 is an ERR protein ligand, whose expression induces a high-energy expenditure and antagonizes obesity. *Proc Natl Acad Sci USA* 100:12378-383.

Kawamura, T., K. Yoshida, A. Sugawara, M. Nagasaka, N. Mori, K. Takeuchi, and M. Kohzuki. 2004. Regulation of skeletal muscle peroxisome proliferator-activated receptor gamma expression by exercise and angiotensin-converting enzyme inhibition in fructose-fed hypertensive rats. *Hypertens Res* 27:61-70.

Kegley, K.M., J. Gephart, G.L. Warren, and G.K. Pavlath. 2001. Altered primary myogenesis in NFATC3(-/-) mice leads to decreased muscle size in the adult. *Dev Biol* 232:115-26.

Kern, M., J.A. Wells, J.M. Stephens, C.W. Elton, J.E. Friedman, E.B. Tapscott, P.H. Pekala, and G.L. Dohm. 1990. Insulin responsiveness in skeletal muscle is determined by glucose transporter (GLUT4) protein level. *Biochem J* 270:397-400.

Krook, A., U. Widegren, X.J. Jiang, J. Henriksson, H. Wallberg-Henriksson, D. Alessi, and J.R. Zierath. 2000. Effects of exercise on mitogen- and stress-activated kinase signal transduction in human skeletal muscle. *Am J Physiol Regul Integr Comp Physiol* 279:R1716-21.

Kuhl, J.E., N.B. Ruderman, N. Musi, L.J. Goodyear, M.E. Patti, S. Crunkhorn, D. Dronamraju, A. Thorell, J. Nygren, O. Ljungkvist, M. Degerblad, A. Stahle, T.B. Brismar, A.K. Saha, S. Efendic, and P.N. Båvenholm. 2006. Exercise training decreases the concentration of malonyl CoA and increases the expression and activity of malonyl CoA decarboxylase in human muscle. *Am J Physiol Endocrinol Metab.* 290(6): E1296-E1303.

Kumar, A., I. Chaudhry, M.B. Reid, and A.M. Boriek. 2002. Distinct signaling pathways are activated in response to mechanical stress applied axially and transversely to skeletal muscle fibers. *J Biol Chem* 277:46493-503.

Larsson, N.G., J. Wang, H. Wilhelmsson, A. Oldfors, P. Rustin, M. Lewandoski, G.S. Barsh, and D.A. Clayton. 1998. Mitochondrial transcription factor A is necessary for mtDNA maintenance and embryogenesis in mice. *Nat Genet* 18:231-6.

Lee, C.H., A. Chawla, N. Urbiztondo, D. Liao, W.A. Boisvert, R.M. Evans, and L.K. Curtiss. 2003. Transcriptional repression of atherogenic inflammation: Modulation by PPARdelta. *Science* 302:453-7.

Lillioja, S., A.A. Young, C.L. Culter, J.L. Ivy, W.G. Abbott, J.K. Zawadzki, H. Yki-Jarvinen, L. Christin, T.W. Secomb, and C. Bogardus. 1987. Skeletal muscle capillary density and fiber type are possible determinants of in vivo insulin resistance in man. *J Clin Invest* 80:415-24.

Lin, J., P.T. Tarr, R. Yang, J. Rhee, P. Puigserver, C.B. Newgard, and B.M. Spiegelman. 2003. PGC-1beta in the regulation of hepatic glucose and energy metabolism. *J Biol Chem* 278:30843-8.

Lin, J., H. Wu, P.T. Tarr, C.Y. Zhang, Z. Wu, O. Boss, L.F. Michael, P. Puigserver, E. Isotani, E.N. Olson, B.B. Lowell, R. Bassel-Duby, and B.M. Spiegelman. 2002. Transcriptional co-activator PGC-1 alpha drives the formation of slow-twitch muscle fibres. *Nature* 418:797-801.

Ling, C., P. Poulsen, E. Carlsson, M. Ridderstråle, P. Almgren, J. Wojtaszewski, H. Beck-Nielsen, L. Groop, and A. Vaag. 2004. Multiple environmental and genetic factors influence skeletal muscle PGC-1alpha and PGC-1beta gene expression in twins. *J Clin Invest* 114:1518-26.

Liu, M.L., E.M. Gibbs, S.C. McCoid, A.J. Milici, H.A. Stukenbrok, R.K. McPherson, J.L. Treadway, and J.E. Pessin. 1993. Transgenic mice expressing the human GLUT4/muscle-fat facilitative glucose transporter protein exhibit efficient glycemic control. *Proc Natl Acad Sci USA* 90:11346-50.

Long, Y.C., U. Widegren, and J.R. Zierath. 2004. Exercise-induced mitogen-activated protein kinase signalling in skeletal muscle. *Proc Nutr Soc* 63:227-32.

Lu, J., T.A. McKinsey, R.L. Nicol, and E.N. Olson. 2000. Signal-dependent activation of the MEF2 transcription factor by dissociation from histone deacetylases. *Proc Natl Acad Sci USA* 97:4070-5.

Lund, S., G.D. Holman, O. Schmitz, and O. Pedersen. 1995. Contraction stimulates translocation of glucose transporter GLUT4 in skeletal muscle through a mechanism distinct from that of insulin. *Proc Natl Acad Sci USA* 92:5817-21.

Luo, J., R. Sladek, J. Carrier, J.A. Bader, D. Richard, and V. Giguere. 2003. Reduced fat mass in mice lacking orphan nuclear receptor estrogen-related receptor alpha. *Mol Cell Biol* 23:7947-56.

Luquet, S., J. Lopez-Soriano, D. Holst, A. Fredenrich, J. Melki, M. Rassoulzadegan, and P.A. Grimaldi. 2003. Peroxisome proliferator-activated receptor delta controls muscle development and oxidative capability. *FASEB J* 17:2299-301.

Macian, F., C. Lopez-Rodriguez, and A. Rao. 2001. Partners in transcription: NFAT and AP-1. *Oncogene* 20:2476-89.

Mahoney, D.J., and M.A. Tarnopolsky. 2005. Understanding skeletal muscle adaptation to exercise training in humans: Contributions from microarray studies. *Phys Med Rehabil Clin N Am* 16:859.

Mahoney, D.J., G. Parise, S. Melov, A. Safdar, and M.A. Tarnopolsky. 2005. Analysis of global mRNA expression in human skeletal muscle during recovery from endurance exercise. *FASEB J* 19:1498-1500.

Marin, P., B. Andersson, M. Krotkiewski, and P. Bjorntorp. 1994. Muscle fiber composition and capillary density in women and men with NIDDM. *Diabetes Care* 17:382-6.

McGee, S.L., and M. Hargreaves. 2004. Exercise and myocyte enhancer factor 2 regulation in human skeletal muscle. *Diabetes* 53:1208-14.

McGee, S.L., D. Sparling, A.L. Olson, and M. Hargreaves. 2006. Exercise increases MEF2- and GEF DNA-binding activity in human skeletal muscle. *FASEB J* 20:348-9.

McKinsey, T.A., C.L. Zhang, and E.N. Olson. 2001. Control of muscle development by dueling HATs and HDACs. *Curr Opin Genet Dev* 11:497-504.

McKinsey, T.A., C.L. Zhang, and E.N. Olson. 2002. MEF2: A calcium-dependent regulator of cell division, differentiation and death. *Trends Biochem Sci* 27:40-47.

Meirhaeghe, A., V. Crowley, C. Lenaghan, C. Lelliott, K. Green, A. Stewart, K. Hart, S. Schinner, J.K. Sethi, G. Yeo, M.D. Brand, R.N. Cortright, S. O'Rahilly, C. Montague, and A.J. Vidal-Puig. 2003. Characterization of the human, mouse and rat PGC1 beta (peroxisome-proliferator-activated receptor-gamma co-activator 1 beta) gene in vitro and in vivo. *Biochem J* 373:155-65.

Mootha, V.K., C. Handschin, D. Arlow, X. Xie, J. St Pierre, S. Sihag, W. Yang, D. Altshuler, P. Puigserver, N. Patterson,

P.J. Willy, I.G. Schulman, R.A. Heyman, E.S. Lander, and B.M. Spiegelman. 2004. Erralpha and Gabpa/b specify PGC-1alpha-dependent oxidative phosphorylation gene expression that is altered in diabetic muscle. *Proc Natl Acad Sci USA* 101:6570-5.

Mora, S., and J.E. Pessin. 2000. The MEF2A isoform is required for striated muscle-specific expression of the insulin-responsive GLUT4 glucose transporter. *J Biol Chem* 275:16323-8.

Murakami, T., Y. Shimomura, A. Yoshimura, M. Sokabe, and N. Fujitsuka. 1998. Induction of nuclear respiratory factor-1 expression by an acute bout of exercise in rat muscle. *Biochim Biophys Acta* 1381:113-22.

Nyholm, B., Z. Qu, A. Kaal, S.B. Pedersen, C.H. Gravholt, J.L. Andersen, B. Saltin, and O. Schmitz. 1997. Evidence of an increased number of type IIb muscle fibers in insulin-resistant first-degree relatives of patients with NIDDM. *Diabetes* 46:1822-8.

Ojuka, E.O., T.E. Jones, D.H. Han, M. Chen, B.R. Wamhoff, M. Sturek, and J.O. Holloszy. 2002. Intermittent increases in cytosolic Ca^{2+} stimulate mitochondrial biogenesis in muscle cells. *Am J Physiol Endocrinol Metab* 283:E1040-5.

Parsons, S.A., B.J. Wilkins, O.F. Bueno, and J.D. Molkentin. 2003. Altered skeletal muscle phenotypes in calcineurin Aα and Aβ gene-targeted mice. *Mol Cell Biol* 23:4331-43.

Pascual, G., A.L. Fong, S. Ogawa, A. Gamliel, A.C. Li, V. Perissi, D.W. Rose, T.M. Willson, M.G. Rosenfeld, and C.K. Glass. 2005. A SUMOylation-dependent pathway mediates transrepression of inflammatory response genes by PPAR-gamma. *Nature* 437:759-63.

Pearson, G., F. Robinson, T. Beers Gibson, B.E. Xu, M. Karandikar, K. Berman, M.H. Cobb. 2001. Mitogen-activated protein (MAP) kinase pathways: Regulation and physiological functions. *Endocr Rev* 22:153-83.

Petersen, K.F., and G.I. Shulman. 2006. Etiology of insulin resistance. *Am J Med* 119(5 Suppl. no. 1): S10-6.

Pilegaard, H., B. Saltin, and P.D. Neufer. 2003. Exercise induces transient transcriptional activation of the PGC-1alpha gene in human skeletal muscle. *J Physiol* 546:851-8.

Ploug, T., B. van Deurs, H. Ai, S.W. Cushman, and E. Ralston. 1998. Analysis of GLUT4 distribution in whole skeletal muscle fibers: Identification of distinct storage compartments that are reduced by insulin and muscle contractions. *J Cell Biol* 142:1429-46.

Ren, J.M., B.A. Marshall, M.M. Mueckler, M. McCaleb, J.M. Amatruda, and G.I. Shulman. 1995. Overexpression of Glut4 protein in muscle increases basal and insulin-stimulated whole body glucose disposal in conscious mice. *J Clin Invest* 95:429-32.

Ren, J.M., C.F. Semenkovich, E.A. Gulve, J. Gao, and J.O. Holloszy. 1994. Exercise induces rapid increases in GLUT4 expression, glucose transport capacity, and insulin-stimulated glycogen storage in muscle. *J Biol Chem* 269:14396-401.

Russell, A.P., J. Feilchenfeldt, S. Schreiber, M. Praz, A. Crettenand, C. Gobelet, C.A. Meier, D.R. Bell, A. Kralli, J.P. Giacobino, and O. Deriaz. 2003. Endurance training in humans leads to fiber type-specific increases in levels of peroxisome proliferator-activated receptor-gamma coactivator-1 and peroxisome proliferator-activated receptor-alpha in skeletal muscle. *Diabetes* 52:2874-81.

Ryder, J.W., R. Bassel-Duby, E.N. Olson, and J.R. Zierath. 2003. Skeletal muscle reprogramming by activation of calcineurin improves insulin action on metabolic pathways. *J Biol Chem* 278:44298-304.

Ryder, J.W., R. Fahlman, H. Wallberg-Henriksson, D.R. Alessi, A. Krook, and J.R. Zierath. 2000. Effect of contraction on mitogen-activated protein kinase signal transduction in skeletal muscle. Involvement of the mitogen and stress activated protein kinase 1. *J Biol Chem* 275:1457-62.

Santalucia, T., M. Camps, A. Castello, P. Munoz, A. Nuel, X. Testar, M. Palacin, and A. Zorzano. 1992. Developmental regulation of GLUT-1 (erythroid/Hep G2) and GLUT-4 (muscle/fat) glucose transporter expression in rat heart, skeletal muscle, and brown adipose tissue. *Endocrinology* 130:837-46.

Schiaffino, S., and A. Serrano. 2002. Calcineurin signaling and neural control of skeletal muscle fiber type and size. *Trends Pharmacol Sci* 23:569-75.

Schulz, R.A., and K.E. Yutzey. 2004. Calcineurin signaling and NFAT activation in cardiovascular and skeletal muscle development. *Dev Biol* 266:1-16.

Shaw, J.P., P.J. Utz, D.B. Durand, J.J. Toole, E.A. Emmel, and G.R. Crabtree. 1988. Identification of a putative regulator of early T cell activation genes. *Science* 241:202-5.

Shen, T., Y. Liu, Z. Cseresnyes, A. Hawkins, W.R. Randall, and M.F. Schneider. 2006. Activity- and calcineurin-independent nuclear shuttling of NFATc1, but not NFATc3, in adult skeletal muscle fibers. *Mol Biol Cell* 17:1570-82.

Short, K.R., J.L. Vittone, M.L. Bigelow, D.N. Proctor, R.A. Rizza, J.M. Coenen- Schimke, and K.S. Nair. 2003. Impact of aerobic exercise training on age-related changes in insulin sensitivity and muscle oxidative capacity. *Diabetes* 52:1888-96.

Sladek, R., J.A. Bader, and V. Giguere. 1997. The orphan nuclear receptor estrogen-related receptor alpha is a transcriptional regulator of the human medium-chain acyl coenzyme A dehydrogenase gene. *Mol Cell Biol* 17:5400-9.

Smith, A.G., and G.E. Muscat. 2005. Skeletal muscle and nuclear hormone receptors: Implications for cardiovascular and metabolic disease. *Int J Biochem Cell Biol* 37:2047-63.

Song, X.M., J.W. Ryder, Y. Kawano, A.V. Chibalin, A. Krook, and J.R. Zierath. 1999. Muscle fiber type specificity in insulin signal transduction. *Am J Physiol* 277:R1690-6.

Spangenburg, E.E., and F.W. Booth. 2003. Molecular regulation of individual skeletal muscle fibre types. *Acta Physiol Scand* 178:413-24.

Teran-Garcia, M., T. Rankinen, R.A. Koza, D.C. Rao, and C. Bouchard. 2005. Endurance training-induced changes in insulin sensitivity and gene expression. *Am J Physiol Endocrinol Metab* 288:E1168-78.

Thomson, S., A.L. Clayton, C.A. Hazzalin, S. Rose, M.J. Barratt, and L.C. Mahadevan. 1999. The nucleosomal response associated with immediate-early gene induction is mediated via alternative MAP kinase cascades: MSK1 as a potential histone H3/HMG-14 kinase. *EMBO J* 18:4779-93.

Tothova, J., B. Blaauw, G. Pallafacchina, R. Rudolf, C. Argentini, C. Reggiani, and S. Schiaffino. 2006. NFATc1 nucleocytoplasmic shuttling is controlled by nerve activity in skeletal muscle. *J Cell Sci* 119:1604-11.

Tunstall, R.J., K.A. Mehan, G.D. Wadley, G.R. Collier, A. Bonen, M. Hargreaves, and D. Cameron-Smith. 2002. Exercise training increases lipid metabolism gene expression in human skeletal muscle. *Am J Physiol Endocrinol Metab* 283:E66-72.

Wang, Y.X., C.L. Zhang, R.T. Yu, H.K. Cho, M.C. Nelson, C.R. Bayuga-Ocampo, J. Ham, H. Kang, and R.M. Evans. 2004. Regulation of muscle fiber type and running endurance by PPARdelta. *PLoS Biol* 2:e294.

Watt, M.J., R.J. Southgate, A.G. Holmes, and M.A. Febbraio. 2004. Suppression of plasma free fatty acids upregulates peroxisome proliferator-activated receptor (PPAR) alpha and delta and PPAR coactivator 1alpha in human skeletal muscle, but not lipid regulatory genes. *J Mol Endocrinol* 33:533-44.

Vega, R.B., and D.P. Kelly. 1997 A role for estrogen-related receptor alpha in the control of mitochondrial fatty acid beta-oxidation during brown adipocyte differentiation. *J Biol Chem* 272:31693-9.

Virbasius, J.V., C.A. Virbasius, and R.C. Scarpulla. 1993. Identity of GABP with NRF-2, a multisubunit activator of cytochrome oxidase expression, reveals a cellular role for an ETS domain activator of viral promoters. *Genes Dev* 7:380-92.

Widegren, U., X.J. Jiang, A. Krook, A.V. Chibalin, M. Björnholm, M. Tally, R.A. Roth, J. Henriksson, H. Wallberg-Henriksson, and J.R. Zierath. 1998. Divergent effects of exercise on metabolic and mitogenic signaling pathways in human skeletal muscle. *FASEB J* 12:1379-89.

Widegren, U., C. Wretman, A. Lionikas, G. Hedin, and .J Henriksson. 2000. Influence of exercise intensity on ERK/MAP kinase signalling in human skeletal muscle. *Pflugers Arch* 441:317-22.

Wretman, C., A. Lionikas, U. Widegren, J. Lännergren, H. Westerblad, and J. Henriksson. 2001. Effects of concentric and eccentric contractions on phosphorylation of MAP-Kerk1/2 and MAPKp38 in isolated rat skeletal muscle. *J Physiol* 535:155-64.

Wu, H., S.B. Kanatous, F.A. Thurmond, T. Gallardo, E. Isotani, R. Bassel-Duby, and R.S. Williams. 2002. Regulation of mitochondrial biogenesis in skeletal muscle by CaMK. *Science* 296:349-52.

Wu, Z., P. Puigserver, U. Andersson, C. Zhang, G. Adelmant, V. Mootha, A. Troy, S. Cinti, B. Lowell, R.C. Scarpulla, and B.M. Spiegelman. 1999. Mechanisms controlling mitochondrial biogenesis and respiration through the thermogenic coactivator PGC-1. *Cell* 98:115-24.

Xia, Y., L.M. Buja, R.C. Scarpulla, and J.B. McMillin. 1997. Electrical stimulation of neonatal cardiomyocytes results in the sequential activation of nuclear genes governing mitochondrial proliferation and differentiation. *Proc Natl Acad Sci USA* 94:11399-404.

Yu, M., E. Blomstrand, A.V. Chibalin, A. Krook, and J.R. Zierath. 2001. Marathon running increases ERK1/2 and p38 MAP kinase signalling to downstream targets in human skeletal muscle. *J Physiol* 536:273-82.

Yu, M., N.K. Stepto, A.V. Chibalin, L.G.D. Fryer, D. Carling, A. Krook, J.A. Hawley, and J.R. Zierath. 2003. Metabolic and mitogenic signal transduction in human skeletal muscle after intense cycling exercise. *J Physiol* 546:327-35.

Zhao, M., L. New, V.V. Kravchenko, Y. Kato, H. Gram, F. di Padova, E.N. Olson, R.J. Ulevitch, and J. Han. 1999. Regulation of the MEF2 family of transcription factors by p38. *Mol Cell Biol* 19:21-30.

Zierath, J.R. 2002. Invited review: Exercise training-induced changes in insulin signaling in skeletal muscle. *J Appl Physiol* 93:773-81.

Zorzano, A., P. Munoz, M. Camps, C. Mora, X. Testar, and M. Palacin. 1996. Insulin-induced redistribution of GLUT4 glucose carriers in the muscle fiber. In search of GLUT4 trafficking pathways. *Diabetes* 45:S70-81.

Chapter 9

Exercise and Calorie Restriction Use Different Mechanisms to Improve Insulin Sensitivity

Gregory D. Cartee, PhD

Reducing calorie intake or increasing physical activity can improve insulin sensitivity. This chapter summarizes and interprets the research that addresses the mechanisms by which exercise or calorie restriction (CR; a reduction in calorie intake below ad libitum levels) enhances insulin-stimulated glucose transport in skeletal muscle. The chapter also presents evidence indicating that these interventions act by distinct mechanisms.

Effects of Exercise and Calorie Restriction on Skeletal Muscle Energy Status

Exercise training or CR can reduce body, total, and regional fat mass. In individuals who are overweight or obese, reduced body fat, especially reduced visceral fat, likely contributes to improved insulin sensitivity. However, insulin sensitivity can be increased after a single exercise session, an effect not attributable to reduced body fat. Also, exercise training can improve insulin sensitivity in the absence of altered body weight or composition (Lamarche et al. 1992; Nassis et al. 2005; Short et al. 2003), although training without weight loss may still reduce visceral fat. Reduced body and fat mass are inevitable with long-term CR, but the magnitude of the improvement in insulin sensitivity is not closely matched with the magnitude of weight loss. Losing visceral fat likely plays a role in the improved insulin sensitivity seen with either

exercise or CR, but other mechanisms are also involved.

Under some circumstances, a negative relationship exists between muscle triglyceride concentration and insulin sensitivity. However, exercise training can improve insulin sensitivity despite an increase in muscle triglyceride levels (van Loon and Goodpaster 2006). Furthermore, during short-term CR (consuming 60% of ad libitum intake for 20 d), muscle triglyceride concentration was not reduced concomitantly with enhanced insulin-stimulated glucose transport (Gazdag, Wetter et al. 2000). Reduced muscle triglyceride concentration is not essential for improved insulin sensitivity with either intervention.

AMPK has been described as a fuel sensor in the cell (Hardie and Sakamoto 2006). AMPK activation provides evidence for negative energy balance. Whereas vigorous exercise elevates AMPK activation (Winder 2001), CR (60% of ad libitum intake for 4 mo) does not alter skeletal muscle AMPK activity in mice (Gonzalez et al. 2004). AMPK activation appears to participate in insulin-independent glucose uptake during and shortly after acute exercise. If CR activated AMPK, insulin-independent glucose transport would increase, but CR does not increase insulin-independent glucose transport in skeletal muscle from rats (Dean et al. 1998). AMPK has also been implicated in triggering the increase in GLUT4 abundance observed with exercise training (Zheng et al. 2001), but CR does not increase GLUT4 abundance (Cartee, Kietzke, and Briggs-Tung 1994;

Research by G.D. Cartee was supported by NIH AG10026.

Gazdag, Sullivan et al. 2000). Apparently CR does not cause a sufficient energy deficit in skeletal muscle to activate AMPK, and other mechanisms trigger the CR-induced improvement in insulin sensitivity.

Vigorous exercise depletes muscle glycogen, and there are conditions in which an inverse relationship exists between glycogen stores in skeletal muscle and insulin sensitivity for glucose transport. However, muscle glycogen concentration does not differ between ad libitum rats and rats on 20 d of CR (Gazdag, Wetter et al. 2000), though insulin-stimulated glucose transport is enhanced in the CR rats. Skeletal muscle glycogen is also not lower in CR rats after 8 mo of CR (Wetter et al. 1999). Reduced muscle glycogen during exercise may play a role in the subsequent increase in insulin sensitivity, but enhanced insulin-stimulated glucose transport with CR does not depend on reduced muscle glycogen.

Insulin Signaling for Glucose Transport

Glucose transport in skeletal muscle is a pivotal process because muscle accounts for about 85% of insulin-stimulated blood glucose clearance (DeFronzo et al. 1981), and glucose transport is a rate-limiting step in muscle glucose metabolism (Ziel, Venkatesan, and Davidson 1988). Insulin receptor binding (figure 9.1) stimulates autophosphorylation on tyrosine residues, activating receptor tyrosine kinase to phosphorylate IRS (White 2003). IRS1 and IRS2 are expressed in skeletal muscle (Myers Jr., Sun, and White 1994; Tamemoto et al. 1994). Tyrosine-phosphorylated IRS (pY-IRS) binds and activates PI3K, which is essential for insulin to induce GLUT4 to translocate to cell surface membranes and increase glucose uptake (Birnbaum 1992). Functional PI3K is a heterodimer in which a regulatory subunit links pY-IRS with the catalytic subunit. A key post-PI3K activator of insulin-stimulated glucose transport, Akt (Whiteman, Cho, and Birnbaum 2002), binds PIP_3, a lipid product of PI3K. Membrane-bound Akt is phosphorylated on Thr-308 via PDK1 and Ser-473 via an unidentified kinase (Shepherd, Withers, and Siddle 1998). Akt1 and Akt2 are both highly expressed by skeletal muscle, but only Akt2 is crucial for insulin to affect glucose uptake (Cho, Mu

et al. 2001; Cho, Thorvaldsen et al. 2001). Muscles from Akt2 null mice are resistant to submaximal insulin levels, but with higher insulin levels they attain glucose uptake rates equal to those of wild-type mice (Cho, Mu et al. 2001). This finding reveals that Akt2-independent processes can substitute for Akt2 in activating glucose uptake. Atypical PKC is a likely candidate for this mechanism (Farese 2002). Insulin leads to serine and threonine phosphorylation of AS160, which is phosphorylated on multiple Akt phosphomotifs (Kane et al. 2002; Sano et al. 2003). Point mutations of AS160 on Akt-consensus phosphorylation sites resulted in a marked decline of insulin-stimulated GLUT4 translocation (Sano et al. 2003), indicating that in 3T3-L1 adipocytes, AS160 links Akt to glucose transport. Bruss and colleagues (2005) demonstrated an insulin-stimulated increase in AS160 phosphorylation in skeletal muscle. Insulin signaling is rapidly reversible with insulin withdrawal because of various mechanisms, including protein tyrosine phosphatases such as protein tyrosine phosphatase 1B (PTP1B), Src homology domain tyrosine phosphatase-2 (SHP-2), and leukocyte antigen-related (LAR) phosphatase that dephosphorylate the insulin receptor and IRS proteins and the lipid phosphatases that degrade PIP_3, including SH2-containing 5'-inositol phosphatase (SHIP2) and phosphatase and tensin homologue deleted on chromosome 10 (PTEN).

Exercise- and Contraction-Stimulated Signaling Pathway for Glucose Transport

Although insulin and exercise use different signaling pathways to stimulate glucose transport, each stimulus induces GLUT4 to translocate from the cytoplasm to the cell surface (Holloszy and Hansen 1996). The two stimuli appear to act via distinct processes; that is, muscles stimulated with supramaximal insulin plus exercise demonstrate greater glucose transport than is observed with either stimulus alone (Constable et al. 1988). PI3K inhibitors block the effects of insulin on glucose transport without altering the effects of exercise in isolated muscle (Yeh et al. 1995). Contraction-stimulated glucose transport appears to depend both on elevated Ca^{++} and on elevated AMPK (Mu et al. 2001; Wright et al. 2004).

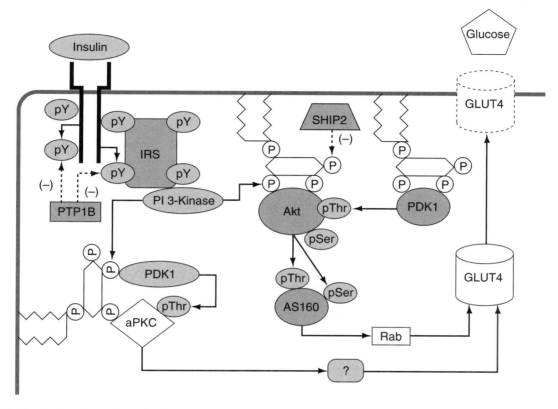

Figure 9.1 Insulin signaling pathway for glucose transport.

Insulin Signaling and Action After Acute Exercise

The insulin-independent effect on glucose transport observed during exercise and muscle contraction begins to reverse shortly after activity ends. Most of this insulin-independent effect is lost by 1 to 3 h following exercise (Cartee & Holloszy 1990; Cartee et al. 1989; Gulve et al. 1990; Wallberg-Henriksson 1987). A much more persistent increase in insulin-dependent glucose transport can last 3 to 48 h (Cartee and Holloszy 1990). Acute exercise can markedly enhance in vivo insulin-stimulated glucose disposal (Perseghin et al. 1996). This ability is likely due, in large part, to increased glucose transport.

Table 9.1 summarizes the research on insulin signaling and GLUT4 after acute exercise. At physiological insulin levels, insulin receptor signaling increased in only 2 (Cusi et al. 2000; Webster, Vigna, and Paquette 1986) of 11 studies (Bonen and Tan 1989; Bonen et al. 1985; Bonen, Tan, and Watson-Wright 1984; Cusi et al. 2000; Hansen et al. 1998; Thong et al. 2002; Treadway et al. 1989; Webster,

Vigna, and Paquette 1986; Wojtaszewski et al. 1997, 2000; Zorzano et al. 1985). Among 10 studies that assessed insulin receptor signaling at supraphysiological insulin levels (Bonen and Tan 1989; Bonen et al. 1985; Bonen, Tan, and Watson-Wright 1984; Chibalin et al. 2000; Howlett et al. 2002; Treadway et al. 1989; Ropelle et al. 2006; Webster, Vigna, and Paquette 1986; Wojtaszewski et al. 1999; Zorzano et al. 1985), only 2 studies found an exercise-induced increase (Ropelle et al. 2006; Webster, Vigna, and Paquette 1986). At physiological insulin concentrations, postreceptor signaling was elevated in 3 studies (Arias et al. 2007; Christ-Roberts et al. 2003; Thorell et al. 1999) and unchanged in 5 studies (Cusi et al. 2000; Fisher et al. 2002; Hamada et al. 2006; Thong et al. 2002; Wojtaszewski et al. 2000). IRS1-PI3K was elevated in only 1 (Christ-Roberts et al. 2003) of 4 studies (Christ-Roberts et al. 2003; Cusi et al. 2000; Fisher et al. 2002; Thong et al. 2002). The design of this study (Christ-Roberts et al. 2003) was unusual because exercise was performed during rather than before the hyperinsulinemic clamp, and the muscle biopsy was taken immediately after rather than several hours after exercise. The increased blood flow that occurred during exercise with simultaneous insulin infusion was accompanied by a plasma insulin concentration that was higher than sedentary values, perhaps secondary to reduced hepatic blood flow and insulin clearance, which likely contributed to the exercise-related increase in insulin signaling. Christ-Roberts and colleagues (2003) and Thorell and coworkers (1999) found that acute exercise enhanced the activation of Akt with physiological insulin, whereas 3 other studies did not (Fisher et al. 2002; Thong et al. 2002; Wojtaszewski et al. 2000). AS160 phosphorylation was increased in muscles dissected from rats 4 hours postexercise, and this effect was found when muscles were incubated without insulin or with a physiological insulin concentration (Arias et al. 2007). Increased postreceptor signaling (including pY-IRS2, IRS1-PI3K, IRS2-PI3K, pY-PI3K, pSerAkt, Akt activity) following exercise was found in 5 studies using supramaximal insulin (Chibalin et al. 2000; Howlett et al. 2002; Ropelle et al. 2006; Wojtaszewski et al. 1999; Zhou and Dohm 1997). The data from the supramaximal insulin studies indicate an increase in the maximal signaling capacity, but the experiments using submaximally effective insulin doses are more representative of physiological conditions.

Table 9.1 Effects of Acute Exercise on Subsequent Insulin Signaling in Skeletal Muscle

Citation	Species	Exercise mode and duration	Time postex	Insulin	Signaling and GLUT4
Arias et al. 2007	Rat	Swim 2 h	~4 h	Submax	↔ pSerAkt; ↑pTAkt; ↑pAS160
Bonen and Tan 1989	Mouse	Treadmill 2-3 h	~10 min	Range	↔ IR binding soleus; ↓ IR binding EDL
Bonen et al. 1985	Human	Cycle 1 h	<5 min and 60 min	Range	Both times: ↔ IR binding ≤60% $\dot{V}O_2$max and ↓ IR binding ≥69% $\dot{V}O_2$max
Bonen, Tan, and Watson-Wright 1984	Mouse	Treadmill 1 h	~10 min	Range	↔ IR binding soleus and EDL
Cartee, Briggs-Tung, and Kietzke 1993	Rat	Swim 2 h	~4 h	—	↔ [GLUT4]
Chibalin et al. 2000	Rat	Swim 6 h	~16 h	Max	↑ [IR]; ↔ pY-IR; ↔ [IRS1]; ↑ IRS1-PI3K; ↑ IRS2-PI3K; ↑ [IRS2]; ↔ pSerAkt; ↑ [GLUT4]
Christ-Roberts et al. 2003	Human with and without type 2 diabetes	Cycle 0.5 h	<5 min	Submax (during exercise)	↑ IRS1-PI3K and ↑ pSerAkt nondiabetics; ↔ IRS1-PI3K and pSerAkt type 2 diabetics
Cusi et al. 2000	Human, obese, with and without type 2 diabetes	Cycle 1 h	24 h	Submax	↑ pY-IR; ↔ IRS1-PI3K both groups
Fisher et al. 2002	Rat	Swim 2 h	~3.5 h	Submax	↔ IRS1-PI3K; ↔ pSerAkt
Hamada et al. 2006	Mouse	Treadmill 1 h	~85 min	Submax	↔ pSerAkt; ↔ pTAkt
Hansen et al. 1998	Rat	Swim 2 h	~3.5 h	Submax	↔ pY-IR; ↔ pY-IRS1; ↔ [GLUT4]; ↑ insulin-stimulated cell surface GLUT4
Howlett et al. 2002	Mouse	Treadmill 1 h	~5 min	Max	↔ pY-IR; ↔ pY-IRS1; ↑ pY-IRS2; ↑ IRS2-PI3K; ↑ pY-PI3K
Ropelle et al. 2006	Rats with high fat diet-induced obesity	Swim 6 h	16 h	Max	↑ pY-IR; ↑ pY-IRS1; ↑ pSerAkt; ↑ IRS1-PI3K; ↑ IRS2-PI3K

Citation	Species	Exercise mode and duration	Time postex	Insulin	Signaling and GLUT4
Thong et al. 2002	Human	One-legged knee extension 1 h	3-5 h	Submax	↔ IRTK; ↔ IRS1-PI3K; ↔ pSerAkt
Thorell et al. 1999	Human	Cycle 1 h	2 h	Submax	↑ pSerAkt
Treadway et al. 1989	Rat	Treadmill 0.75 h	~1 h	Range	↔ IR binding, autophosphorylation, and IRTK
Webster, Vigna, and Paquette 1986	Rat	Swim 1.5 h	~5 min	Range	↑ IR binding
Wojtaszewski et al. 2000	Human	One-legged knee extension 1 h	4-6 h	Submax	↔ IRTK; ↔ pSerAkt; ↔ [GLUT4]
Wojtaszewski et al. 1997	Human	One-legged knee extension 1 h	4-6 h	Submax	↔ IRTK; ↔ pY-IRS1; ↔ IRS1-PI3K
Wojtaszewski et al. 1999	Mouse	Treadmill 1 h	~5 min	Max	↔ pY-IR; ↔ IRS1-PI3K; ↑ pY-PI3K; ↑ pSerAkt; ↑ Akt activity
Zhou and Dohm 1997	Rat	Treadmill 1 h	~2 min	Max	↔ pY-IRS1; ↑ pY-PI3K
Zorzano et al. 1985	Rat	Treadmill 0.75 h	~10 min	Range	↔ IR binding

Time postex, the time after completion of exercise when muscle insulin signaling or GLUT4 measurements were made; range, a range of insulin concentrations, including submaximally (submax) and maximally (max) effective doses, was studied; ↑, a significantly greater value for exercised versus sedentary condition; ↓, a significantly lower value for exercised versus sedentary condition; ↔, no significant difference between exercised and sedentary condition; [GLUT4], [IR], [IRS1], and [IRS2], abundance of each respective protein; IR, insulin receptor; IRTK, IR tyrosine kinase; EDL, extensor digitorum longus.

Three studies (Chibalin et al. 2000; Cusi et al. 2000; Ropelle et al. 2006) assessed insulin signaling on the day after acute exercise rather than 1 to 6 h postexercise. Chibalin and coworkers (2000) found that 16 h after prolonged (6 h duration) exercise, insulin signaling (IRS1-PI3K and IRS2-PI3K) in rat muscle with supraphysiological insulin was greater compared to sedentary controls. Ropelle et al. (2006), studying diet-induced obese rats, found that prior exercise (6 h duration) resulted in increased pY-IR, pY-IRS1, IRS1-PI3K, IRS2-PI3K, and pSerAkt in muscles stimulated with a supraphysiologic insulin concentration. Cusi and colleagues (2000), studying human muscle with a physiological insulin dose, found that previous exercise (24 h earlier) led to a small but significant increase in pY-IR but not in IRS1-PI3K. More investigation is required to fully understand the role of physiological insulin signaling in improved insulin action the day after exercise.

Prior exercise may elevate insulin signaling at steps that have not yet been studied postexercise (e.g., aPKC). Incubating isolated skeletal muscle under hypoxic conditions can stimulate glucose transport by a mechanism that is distinct from the insulin-stimulated pathway (Cartee et al. 1991). Insulin-stimulated glucose transport is increased 4 h postexercise, and skeletal muscle is also more susceptible to the activation of glucose transport by in vitro hypoxia at 4 h postexercise, suggesting that exercise may act via a distal step in the machinery involved in GLUT4 translocation shared by the insulin- and hypoxia-stimulated pathways (Cartee and Holloszy 1990).

Regardless of the proximal mechanism, insulin-stimulated localization of GLUT4 at the cell surface increases after exercise (Hansen et al. 1998). In rats (Cartee, Briggs-Tung, and Kietzke 1993; Hansen et al. 1998) at 3 to 4 h postexercise (2 h duration) and in humans (Wojtaszewski et al. 2000) at 6 h postexercise (1 h duration), insulin-stimulated glucose uptake is increased with unaltered GLUT4 abundance. The day after prolonged (6 h duration) exercise, GLUT4 abundance is increased in rats (Chibalin et al. 2000), potentially contributing to the improved insulin action observed at this time.

In summary, many of the studies that used a submaximally effective insulin concentration did not find evidence that a single exercise bout enhances subsequent insulin signaling 1 to 6 h after exercise. Muscle GLUT4 abundance is not elevated at this time, but insulin-stimulated GLUT4 translocation is elevated. Acute exercise also increases susceptibility to insulin-independent activation of glucose transport, raising the possibility that exercise acts by amplifying a distal process in GLUT4 translocation that is shared by insulin-dependent and insulin-independent pathways.

Effects of Exercise Training on Insulin Sensitivity and Signaling

Exercise training consists of performing repeated bouts of acute exercise regularly over time (from a few days to many decades). Dissecting the accumulated effects of chronic exercise from the effects of the final exercise session can be challenging because even a single exercise session can improve insulin sensitivity 48 h postexercise (Cartee et al. 1989; Mikines et al. 1988; Perseghin et al. 1996), but a single session cannot improve sensitivity 5 d postexercise (Mikines et al. 1988). Thus, improved insulin action observed ≤48 h after the final training session may be due, at least in part, to the most recent exercise bout. Increased insulin-stimulated glucose uptake has been detected as long as 3 to 7 d after the final training session (Christ et al. 2002; Dolkas, Rodnick, and Mondon 1990; Nassis et al. 2005; Short et al. 2003). Other studies found increased insulin-stimulated glucose uptake at 24 to 29 h but not at 48 to 53 h after the final training bout (Etgen Jr. et al. 1993; Kump and Booth 2005).

Perseghin and coworkers (1996) evaluated acute and chronic exercise effects in initially untrained humans with normal glucose tolerance who either had no family history of type 2 diabetes or were offspring of parents with type 2 diabetes. All participants were tested for insulin-stimulated glucose disposal at baseline before beginning to exercise, at 48 h after the initial exercise session (45 min), and at 48 h after the final exercise session of a 6 wk training program (45 min/d for 4 d/wk). Acute exercise elicited an increase above baseline glucose disposal that in both groups equaled a large portion of the improved insulin action found after 6 wk of training (42% of total increase for offspring of parents with type 2 diabetes; 71% for controls).

Davis and colleagues (1986) evaluated insulin-stimulated glucose uptake in isolated muscles at 1 h and at 2 to 3 h after exercise (2 h swimming) performed by untrained and trained (4 wk) rats. Insulin-stimulated glucose uptake was elevated in both the acute and chronic exercise groups compared to sedentary controls (Davis et al. 1986). There appeared to be a greater improvement in the trained exercise group versus the acute exercise group at 2 to 3 h but not at 1 h postexercise. Together, these studies suggest that the most recent training bout can account for a significant portion of the total improvement in insulin sensitivity seen after several weeks of training.

Table 9.2 summarizes the results of published studies that have assessed insulin signaling after exercise training. Most of these studies sampled muscle 12 to 48 h after the final exercise bout, so it is possible that observed effects are partly attributable to the final training session. Among 10 studies that used submaximally effective insulin doses to assess receptor function, 6 provided evidence for greater receptor tyrosine phosphorylation, binding, or kinase activity with training (Bonen, Clune, and Tan 1986; Dohm, Sinha, and Caro 1987; Kump and Booth 2005; Santos et al. 1989; Tan and Bonen 1987; Youngren et al. 2001), and 4 did not (Bak et al. 1989; Crettaz et al. 1983; Dela et al. 1993; Grimditch et al. 1986). In 14 studies that included measurements with supraphysiological insulin levels, 7 found that receptor function was increased with training (Bonen, Clune, and Tan 1986; Chibalin et al. 2000; Dohm, Sinha, and Caro 1987; Hevener, Reichart, and Olefsky 2000; Santos et al. 1989; Tan and Bonen 1987; Youngren et al. 2001), and an equal number did not find such an increase (Bak et al. 1989; Christ et al. 2002; Crettaz et al. 1983; Dela et al. 1993; Grimditch et al. 1986; Lemieux et al. 2005; Luciano et al. 2002). Training increased postreceptor signaling in 3 studies with supraphysiological insulin (Chibalin et al. 2000; Han et al. 1998; Luciano et al. 2002) but not in 3 other investigations (Christ et al. 2002; Hevener, Reichart, and Olefsky 2000; Lemieux et al. 2005). Of the studies reporting increased maximal insulin signaling, 2 used normal rats (Chibalin et al. 2000; Luciano et al. 2002) while 1 used old rats (Han et al. 1998). Each study that did not see elevated maximal insulin signaling used insulin-resistant rats. Postreceptor signaling (pY-IRS1 or pY-IRS2, IRS1-PI3K or IRS2-PI3K, or pAkt) was increased in 2 investigations using submaximal insulin (Houmard et al. 1999; Kump and Booth 2005) but was unchanged in 4 other studies (Christ-Roberts et al. 2004; Hevener, Reichart, and Olefsky 2000; Kemppainen et al. 2003; Tanner et al. 2002). Postreceptor signaling with physiological insulin was increased by training in studies using men who were lean and healthy (aged 25 y; Houmard et al. 1999) and in studies using normal young rats (Kump and Booth 2005). Training did not improve postreceptor signal-

ing with physiological insulin in insulin-resistant groups, including older men (aged 58 y; Tanner et al. 2002), obese Zucker rats (Hevener, Reichart, and Olefsky 2000), humans who were overweight with or without type 2 diabetes (Christ-Roberts et al. 2004), and men with chronic heart failure (Kemppainen et al. 2003). Nonetheless, each of these insulin-sensitive and insulin-resistant groups was characterized by a training-induced improvement in insulin action. Processes other than the measured insulin signaling steps were apparently responsible for this effect in the insulin-resistant groups.

Table 9.2 Insulin Signaling in Skeletal Muscle After Chronic Exercise Training

Citation	Species	Training mode and duration	Time postex	Insulin	Signaling and GLUT4
Bak et al. 1989	Human with type 1 diabetes	Jog 6 wk	30 h	Range	↔ IR binding or tyrosine kinase
Bonen, Clune, and Tan 1986	Rat	Treadmill 6 wk	24 h	Range	↑ IR binding in soleus, plant, EDL; ↔ RG, WG
Chibalin et al. 2000	Rat	Swim 5 d	~16 h	Max	↑ [IR]; ↑ pY-IR; ↓ [IRS1]; ↑ pY-IRS1; ↑ IRS1-PI3K; ↔ IRS2-PI3K; ↔ [IRS2]; ↑ pSerAkt; ↑ [GLUT4]
Christ et al. 2002	Obese Zucker rat	Treadmill 7 wk	24 h, 96 h, 7 d	Max	↔ pY-IR all times; ↔ pY-IRS1 all times; ↔ IRS1-PI3K and IRS1-p85 all times; ↔ pY-PI3K all times ↔ pSerAkt; [GLUT4] ↑ 24 h ↑ 96 h ↔ 7 d
Christ-Roberts et al. 2004	Human, overweight, with and without type 2 diabetes	Cycle 8 wk	24 h	Submax	↔ IRS1-PI3K; ↔ pSerAkt; ↑ [GLUT4] in both groups
Cox et al. 1999	Human, 18-30 y and 50-70 y women and men	Cycle 7 d	15-17 h	NA	↑ [GLUT4] all four groups
Crettaz et al. 1983	Lean and obese Zucker rat	Treadmill 6-8 wk	44 h	Range	↔ IR binding both groups
Dela et al. 1993	Human	One-legged cycle 10 wk	16 h	Range	↔ IR binding or IRTK; ↑ GLUT4
Dohm, Sinha, and Caro 1987	Rat	Treadmill 4 wk	20-24 h	Range	↑ IR binding; ↓ IRTK
Etgen Jr. et al. 1997	Obese Zucker rat	Treadmill 15 d	~40 h	Max	↑ [GLUT4]; ↑ insulin-stimulated cell surface GLUT4
Grimditch et al. 1986	Rat	Treadmill 10-12 wk	24-48 h	Range	↔ IR binding

(continued)

Table 9.2 *(continued)*

Citation	Species	Training mode and duration	Time postex	Insulin	Signaling and GLUT4
Han et al. 1998	Rat, 25 mo	Wheel run 21 mo	48 h	Max	↑ IRS1-p85PI3K; ↔ [GLUT4]
Hevener, Reichart, and Olefsky 2000	Zucker fatty rat	Treadmill 3 wk	18-20 h	Max	↔ [IR]; ↑ pY-IR; ↔ pAkt; ↑ [GLUT4]
Houmard et al. 1999	Human	Cycle 7 d	15-17 h	Submax	↑ pY-PI3K
Kemppainen et al. 2003	Human with chronic heart failure	Cycle, treadmill, resistance 5 mo	≥48 h	Submax	↔ pY-IRS1; ↔ IRS1-PI3K; ↔ pSerAkt
Kump and Booth 2005	Rat	Wheel run 3 wk	5, 29, 53 h	Submax	5 and 29 h versus sed ↑ IR binding; ↑ [IR]; ↑ pY-IR; ↑ pSerAkt; ↑ [GLUT4]; all of these lost at 53 h
Lemieux et al. 2005	Rat, hypertensive	Wheel run 6 wk	>9 h	Max	↔ pY-IR; ↔ pY-IRS1; ↔ IRS1-PI3K ; ↔ pSerAkt
Luciano et al. 2002	Rat	Swim 6 wk	48 h	Max	↔ pY-IR; ↑ pY-IRS1; ↑ IRS1-PI3K;↑ pY-IRS2; ↑ IRS2-PI3K; ↑ pSerAkt1
Reynolds et al. 1997	Rat	Swim 5 d	18-22 h	Max	↑ [GLUT4]; ↑ insulin-stimulated cell surface GLUT4
Santos et al. 1989	Rat	Wheel run 6 wk	<24 h	Range	↑ IR binding, autophosphorylation, and IRTK
Tan and Bonen 1987	Mouse	Treadmill 6 wk	48 h	Range	↑ IR binding
Tanner et al. 2002	Human, 50-70 y	Cycle 7 d	15-17 h	Submax	↔ pY-PI3K and Akt activity
Willis et al. 1998	Mice, 24 mo	Wheel run 9 mo	48 h	NA	↓ [IR]; ↑ [GLUT4]
Youngren et al. 2001	Human, 20-37 y	Cycle 7 d	15-17 h	Range	↔ [IR]; ↑ IR autophosphorylation

Time postex, the time after completion of exercise when muscle insulin signaling or GLUT4 measurements were made; range, a range of insulin concentrations, including submaximally (submax) and maximally (max) effective doses, was studied; ↑, a significantly greater value for exercised versus sedentary condition; ↓, a significantly lower value for exercised versus sedentary condition; ↔, no significant difference between exercised and sedentary condition; [GLUT4], [IR], [IRS1], and [IRS2], abundance of each respective protein; IR, insulin receptor; IRTK, IR tyrosine kinase; EDL, extensor digitorum longus; RG, red gastrocnemius; WG, white gastrocnemius; sed, sedentary; NA, not applicable.

Elevated GLUT4 expression in muscle is a hallmark of exercise training and is repeatedly observed in males and females, in insulin-sensitive and insulin-resistant groups, and in humans, rats, and mice (Cox et al. 1999; Houmard et al. 1991, 1993, 1995; Rodnick et al. 1990; Slentz et al. 1992; Willis et al. 1998). Although old and young humans showed similar training-induced elevations in muscle GLUT4 (Cox et al. 1999), old rats demonstrated no training-induced increase in muscle GLUT4 levels (Han et al. 1998). In both insulin-sensitive, lean rats (Reynolds et al. 1997) and insulin-resistant, obese rats (Etgen Jr. et al. 1997), the increased GLUT4 abundance experienced with training was accompanied by greater insulin stimulation of GLUT4 recruitment to the cell surface.

In summary, approximately half of the published studies found evidence that training improved insulin receptor function. Training increased postreceptor insulin signaling with physiological insulin in healthy but not in insulin-resistant groups. Nonetheless, both individuals who are insulin sensitive and individuals who are insulin resistant can improve insulin action with training. The role of elevated insulin signaling in the training-enhanced insulin action is uncertain, but increased GLUT4 abundance probably contributes to this benefit in diverse groups.

Effects of Calorie Restriction Distinct From Weight Loss

Several studies have attempted to separate the specific effects of weight loss from the direct effects of CR in humans who are obese and have type 2 diabetes. In one study, the participants lost 11% of body weight by consuming either 400 kcal/d or 1,000 kcal/d (Wing et al. 1994). As expected, the weight loss was achieved sooner on the 400 kcal/d (9.1 wk) versus 1,000 kcal/d (12.7 wk) diet. Insulin sensitivity increased in both groups, but the magnitude of the increase was greater for the group consuming the lower calorie diet, despite the fact that both groups attained identical weight loss. All participants consumed a 1,000 kcal/d diet for about 15 additional weeks, and both groups ultimately achieved similar total weight loss (~18%) and similar insulin sensitivity. In a subsequent study, humans who were obese and had type 2 diabetes were tested with euglycemic-hyperinsulinemic clamps on multiple occasions during 3 mo of consuming 400 to 800 kcal/d (Kelley et al. 1993). After the first 7 d of CR (when consuming 800 kcal/d), the participants lost only 2.2 kg (17% of total weight loss), but they achieved 45% of the total improvement in insulin sensitivity measured after 3 mo of CR. Arciero and coworkers (1999) studied humans who were obese and exhibited impaired glucose tolerance or mild type 2 diabetes before and after 10 d of CR and found that a body weight reduction of only 3.5% was accompanied by a 35% increase in glucose disposal rate during a hyperglycemic clamp. Together, these results demonstrate that CR resulting in only modest weight loss can significantly enhance insulin sensitivity.

Effects of Calorie Restriction on Insulin Signaling in Skeletal Muscle

In contrast to the increased GLUT4 observed in muscle after exercise training, GLUT4 abundance was not affected by CR in 7 out of 8 studies (Argentino, Dominici, Al-Regaiey et al. 2005; Argentino, Munoz et al. 2005; Cartee, Kietzke, and Briggs-Tung

1994; Friedman et al. 1992; Gazdag, Sullivan et al. 2000; Wang et al. 1997; Wilkes and Nagy 1996). The only investigation that found increased GLUT4 in muscle with CR (Argentino, Dominici, Munoz et al. 2005) used mice. Two other studies using mice did not observe a CR effect on GLUT4 (Argentino, Dominici, Al-Regaiey et al. 2005; Argentino, Munoz et al. 2005). The lack of a CR effect on muscle GLUT4 abundance has been found in rats, rhesus monkeys, and humans. Although enhanced GLUT4 abundance is not required for the CR-induced increase in insulin sensitivity, Dean and coworkers (1998) found that CR increased the GLUT4 content at the cell surface in insulin-stimulated muscle of rats. The relative magnitude of increased GLUT4 translocation was very similar to the relative magnitude of CR-induced increased glucose transport.

When muscles from CR and ad libitum rats were exposed to in vitro hypoxia, there was no diet-related difference in the hypoxia-stimulated glucose transport (Dean et al. 1998). This finding, together with the CR-induced increase in insulin-stimulated glucose transport, suggests that CR may enhance glucose transport by acting specifically on the insulin signaling pathway.

Table 9.3 summarizes the studies that have investigated the effects of CR on insulin signaling in muscle. Only 1 (Wang et al. 1997) of 4 studies that evaluated insulin receptor function with physiological insulin provided evidence for elevations in CR versus ad libitum controls (Bak et al. 1992; Balage, Grizard, and Manin 1990; Cecchin et al. 1988; Wang et al. 1997). Of 10 studies that evaluated receptor function with supraphysiological insulin, 5 found a CR-related elevation in signaling (Argentino, Dominici, Munoz et al. 2005; Dean and Cartee 2000; Wang et al. 1997; Zhu et al. 2004, 2005), and 5 did not (Argentino, Dominici, Al-Regaiey et al. 2005; Argentino, Munoz et al. 2005; Bak et al. 1992; Balage, Grizard, and Manin 1990; Cecchin et al. 1988). Two studies using a supraphysiological insulin dose found significantly improved postreceptor insulin signaling with CR (Dean and Cartee 2000; Kim et al. 2003), but 4 did not (Argentino, Dominici, Al-Regaiey et al. 2005; Argentino, Dominici, Munoz et al. 2005; Argentino, Munoz et al. 2005; Dean et al. 1998). Of the studies using submaximally effective insulin, 3 demonstrated evidence for a CR-related improvement in postreceptor signaling (IRS1-p110PI3K, pSerAkt, and pThrAkt; McCurdy

Table 9.3 CR Effects on Insulin Signaling in Skeletal Muscle

Citation	Species	CR, % AL, and duration	Insulin	Signaling and GLUT4
Argentin, Dominici, Al-Regaiey et al. 2005	Mouse	70% for 14 mo	Max	↔ [IR]; ↔ pY-IR; ↑ [IRS1]; ↔ pY-IRS1; ↔ [p85PI3K]; ↔ [GLUT4]
Argentino, Dominici, Munoz et al. 2005	Mouse	70% for 14 mo	Max	↔ [IR]; ↑ pY-IR; ↔ [IRS1]; ↔ pY-IRS1; ↓ [p85PI3K]; ↑ [GLUT4]
Argentino, Munoz et al. 2005	Mouse	70% for 20 d	Max	↔ [IR]; ↔ pY-IR; ↔ [IRS1]; ↔ pY-IRS1; ↔ IRS1-p85PI3K; ↔ pSerAkt; ↔ [GLUT4]
Bak et al. 1992	Human with type 2 diabetes	59% for 6 wk	Range	↔ IR binding; ↔ IRTK
Balage, Grizard, and Manin 1990	Rat	65% for 21 d	Range	↔ IR binding
Cartee, Kietzke, and Briggs-Tung 1994	Rat	60% for 4.5 mo	NA	↔ [GLUT4]
Cecchin et al. 1988	Rat	57% for 4 wk	Range	↔ IR binding; ↔ IRTK
Davidson, Arias, and Cartee 2002	Rat	60% for 20 d	Submax	↔ IRS1-PI3K; ↔ IRS2-PI3K; ↔ pY-PI3K
Dean et al. 1998	Rat	56% for 20 d	Max	↓ [IRS1]; ↔ [IRS2]; ↔ [p85PI3K]; ↔ IRS1-PI3K; ↑ insulin-stimulated cell surface GLUT4
Dean and Cartee 2000	Rat	61% for 20 d	Max	↑ pY-IR; ↑ pY-IRS1; ↔ IRS1-PI3K
Friedman et al. 1992	Obese human before and after gastric bypass	~1 y post bypass	Max	↔ [GLUT4]
Gazdag, Sullivan et al. 2000	Rhesus monkey	70% for 6 y	NA	↔ [IRS1]; ↔ [p85PI3K]; ↔ [GLUT4]
Kim et al. 2003	Obese human	CR = 600-800 kcal/d for ~24 wk	Max	↔ [IRS1]; ↑ pY-IRS1; ↑ IRS1-PI3K; ↔ [aPKCλ/ξ]; ↑ aPKCλ/ξ activity
McCurdy and Cartee 2005	Mouse	60% for 20 d	Submax	↑ pSerAkt1; ↑ pSerAkt2; ↔ [Akt1]; ↔ [Akt2]
McCurdy, Davidson, and Cartee 2003	Rat	60% for 20 d	Submax	↓ [Akt1]; ↔ [Akt2]; ↔ pSerAkt1; ↔ pThrAkt1; ↑ pSerAkt2; ↑ pThrAkt2; ↔ [SHIP2]
McCurdy, Davidson, and Cartee 2005	Rat	60% for 8 wk	Submax	↓ [IRS1]; ↔ [p85α/βPI3K]; ↓ [p55αPI3K]; ↓ [p50αPI3K]; ↔ [p110PI3K]; ↑ IRS-p110PI3K; ↑ pSerAkt
Wang et al. 1997	Rat	60% for 7, 13, or 25 mo	Range	↑ IR binding; ↔ [GLUT4]
Wilkes and Nagy 1996	Rat	~73% for 4 wk	NA	↔ [GLUT4] in any muscle
Zhu et al. 2004	Rat	60% for 2 or 25 mo	Max	↔ pY-IR 2 mo; ↑ pY-IR 25 mo; ↔ PTP1B activity in 2 mo; ↑ PTP1B activity in 25 mo
Zhu et al. 2005	Rat	60% for 2 or 25 mo	Max	↑ [IR] 2 mo; ↔ pY-IR 2 mo; ↔ [IR] 25 mo; ↑ pY-IR 25 mo ↔ [PTP1B] for both; ↓ PTP1B activity for both

% AL, percentage of ad libitum intake that was consumed by members of the calorie restricted (CR) group; range, a range of insulin concentrations, including submaximally (submax) and maximally (max) effective doses, was studied; ↑, a significantly greater value for CR versus ad libitum condition; ↓, a significantly lower value for CR versus ad libitum condition; ↔, no significant difference between CR and ad libitum condition; [GLUT4], [IR], [IRS1], [IRS2], [PTP1B], [p85α/βPI3K], [p55αPI3K], [p50αPI3K], [p110PI3K], [aPKCλ/ξ], [SHIP2], [Akt1], [Akt2], and [p85PI3K], abundance of each respective protein; IR, insulin receptor; NA, not applicable.

and Cartee 2005; McCurdy, Davidson, and Cartee 2003, 2005), and only 1 did not (Davidson, Arias, and Cartee 2002). In 3 studies (Davidson, Arias, and Cartee 2002; Dean et al. 1998; Dean and Cartee 2000), we found a consistent but statistically insignificant trend for higher IRS1-PI3K activity in insulin-stimulated muscles from CR versus ad libitum rats. When the data from the studies using isolated epitrochlearis muscles (Davidson, Arias, and Cartee 2002; Dean et al. 1998) are pooled and expressed as a ratio of insulin-treated muscles over basal controls, the pooled values for CR (5.5 ± 0.7) are greater (P < 0.05) than the ad libitum values (3.5 ± 0.4). Consistent with a CR-induced increase in PI3K activity is the repeated demonstration that CR can enhance submaximal insulin-stimulated phosphorylation of Akt, a key post-PI3K step (McCurdy and Cartee 2005; McCurdy, Davidson, and Cartee 2003, 2005). In submaximal insulin-stimulated muscles, pSerAkt2 and pThrAkt2 were greater for CR versus ad libitum rats without any change in pSerAkt1 or pThrAkt1 (McCurdy, Davidson, and Cartee 2003). In insulin-stimulated muscles from mice, both pSerAkt1 and pSerAkt2 were greater for CR versus ad libitum values (McCurdy and Cartee 2005). Insulin-stimulated glucose uptake in the extensor digitorum longus muscle from wild-type mice was greater for CR mice than it was for their ad libitum littermates, but this CR effect was absent from Akt2 null mice. In the soleus of Akt2 null males, insulin-stimulated glucose uptake was higher for CR mice versus ad libitum controls. However, soleus glucose uptake for CR Akt2 null mice was lower than that found in CR wild-type mice. Thus, with submaximal insulin, Akt2 is essential for CR to exert its full effect on muscle glucose uptake, and, at least in the soleus of males, there appears to be an Akt2-independent mechanism for CR effects. Increased aPKC activity, as seen in humans who are obese and undergo CR-induced weight loss (Kim et al. 2003), may be this Akt2-independent mechanism.

Most studies on insulin signaling with CR have used individuals who are lean and young, but increased insulin signaling in muscle was also found after CR in humans who are morbidly obese (Kim et al. 2003) and in old rats (Wang et al. 1997; Zhu et al. 2004, 2005). CR improves insulin action in both of these groups (Cartee, Kietzke, and Briggs-Tung 1994; Kim et al. 2003). In summary, previous studies have consistently indicated that increased insulin-stimulated Akt phosphorylation is a CR effect that is likely crucial for improved insulin-stimulated GLUT4 translocation and glucose transport.

Combined Effects of Exercise and Calorie Restriction

If CR and exercise elevate insulin sensitivity by different mechanisms, their combined effects should exceed their individual effects. The result of combined CR and exercise on insulin-stimulated glucose transport and insulin signaling in skeletal muscle remains to be determined. However, several studies have assessed the effects of either exercise training (ET) or CR as well as the combined effects of both on humans who are overweight or obese (Cox et al. 2004; Dengel et al. 1996, 1998; Janssen et al. 2002; Rice et al. 1999). The dietary modifications used in these studies were not limited to CR, as both the quantity and the composition of the diet were altered. Oral glucose tolerance and plasma insulin levels during the oral glucose tolerance test (OGTT) were measured, and the area under the curve (AUC) for plasma insulin, interpreted together with plasma glucose during the OGTT, provided an approximation of insulin sensitivity. Table 9.4 summarizes the relative changes in the insulin AUC in participants assigned to CR, ET, and CR+ET interventions. Body weight was reduced by about 10% for the CR and the CR+ET groups in each study, while the ET groups showed little change in body weight. As expected, insulin AUC was reduced in the CR group from each study, consistent with improved insulin sensitivity. A similar result was found for the studies that included an ET group. In 4 of 5 studies, the combined CR+ET group demonstrated a substantially greater reduction in insulin AUC compared to the CR or ET groups. The studies that reported that combined CR+ET had greater effects on insulin sensitivity included only men as participants (Cox et al. 2004; Dengel et al. 1996, 1998; Rice et al. 1999), whereas the only study that did not find added benefit to the combined intervention evaluated women who were obese (Janssen et al. 2002). The duration of the interventions, amount of weight lost by CR and by CR+ET groups, and training protocols were similar for several of the studies using only men when compared to the study using women. Perhaps there is a gender

Table 9.4 CR and Exercise Training: Separate and Combined Effects on Insulin Area Under the Curve During an Oral Glucose Tolerance Test

Citation	Description of subjects and interventions	Decrease in insulin AUC with CR	Decrease in insulin AUC with exercise	Decrease in insulin AUC with CR and exercise
Cox et al. 2004	Overweight men ~40 y, 16 wk intervention, CR aimed to reduce intake by 1,000-1,500 kcal/d, cycle 30 min/d for 3 d/wk	28% decrease	11% decrease	67% decrease
Dengel et al. 1998	Overweight men ~60 y, 9 mo intervention, CR aimed to reduce intake by 300-500 kcal/d, cycle, walk, or jog 40 min/d for 3 d/wk	32% decrease	29% decrease	50% decrease
Dengel et al. 1996	Overweight men ~60 y, 10 mo intervention, CR aimed to reduce intake by 300-500 kcal/d, cycle, walk, or jog 40 min/d for 3 d/wk	21% decrease	19% decrease	42% decrease
Janssen et al. 2002	Obese, premenopausal women ~35 y, 16 wk intervention, CR aimed to reduce intake by 1,000 kcal/d, cycle, walk, stair-climb up to 60 min/d for 5 d/wk	17% decrease	Not determined	23% decrease
Rice et al. 1999	Obese men ~45 y, 16 wk intervention, CR aimed to reduce intake by 1,000 kcal/d, cycle, walk, stair-climb up to 60 min/d for 5 d/wk	20% decrease	Not determined	52% decrease

AUC, area under the curve; OGTT, oral glucose tolerance test. The decrease in insulin AUC refers to the relative decline compared to a given group's baseline value.

difference affecting the added benefits of combined CR+ET on insulin sensitivity, but this interpretation should be viewed with caution until additional data on women are available. Importantly, even in women, combined CR+ET did not interfere with the effects of CR alone.

Concluding Remarks

Improved insulin sensitivity following exercise or CR is well established, but the mechanisms that underlie this benefit are not completely understood. Many studies using a physiological insulin concentration have not found improved insulin signaling 1 to 6 h after acute exercise, and GLUT4 abundance is unaltered at this time even though insulin action is elevated. Chronic exercise can elevate insulin action further beyond the effects of acute exercise. Two studies have found that training leads to greater insulin signaling in individuals with normal insulin sensitivity, whereas studies of individuals who are insulin resistant have typically failed to demonstrate training effects on postreceptor insulin signaling. Nonetheless, both groups can improve insulin sensitivity after training. One training adaptation shared by groups with normal insulin sensitivity and by most groups with insulin resistance is an increase in total GLUT4 abundance in muscle. In contrast, many studies have demonstrated enhanced insulin sensitivity with CR without elevated GLUT4 content. CR, unlike acute exercise, does not increase hypoxia-stimulated glucose transport by isolated muscle, a finding that suggests that CR may act specifically on the insulin-dependent pathway. CR results in greater insulin-mediated activation of Akt in muscle, and Akt2 is essential for the full effects of CR on glucose uptake with submaximal insulin. Obesity and type 2 diabetes are characterized by diminished insulin signaling, reduced insulin-stimulated GLUT4 translocation, and decreased glucose transport, despite normal GLUT4 abundance (Bjornholm and Zierath 2005; Christ et al. 2002; Christ-Roberts et al. 2004). Results of several studies suggest that exercise training may compensate for rather than reverse these defects, whereas CR may act directly on the deficient processes. Although exercise and CR differ in their effects on insulin signaling and GLUT4 expression, each enhances insulin-stimulated translocation of GLUT4, ultimately resulting in elevated insulin-stimulated glucose transport in skeletal muscle. Additional research is needed to map out the apparently disparate paths that these interventions take to reach the same final destination.

References

Arciero, P.J., M.D. Vukovich, J.O. Holloszy, S.B. Racette, and W.M. Kohrt. 1999. Comparison of short-term diet and exercise on insulin action in individuals with abnormal glucose tolerance. *J Appl Physiol* 86:1930-5.

Argentino, D.P., F.P. Dominici, K. Al-Regaiey, M.S. Bonkowski, A. Bartke, and D. Turyn. 2005. Effects of long-term caloric restriction on early steps of the insulin-signaling system in mouse skeletal muscle. *J Gerontol A Biol Sci Med Sci* 60:28-34.

Argentino, D.P., F.P. Dominici, M.C. Munoz, K. Al-Regaiey, A. Bartke, and D. Turyn. 2005. Effects of long-term caloric restriction on glucose homeostasis and on the first steps of the insulin signaling system in skeletal muscle of normal and Ames dwarf (Prop1df/Prop1df) mice. *Exp Gerontol* 40:27-35.

Argentino, D.P., M.C. Munoz, J.S. Rocha, A. Bartke, D. Turyn, and F.P. Dominici. 2005. Short-term caloric restriction does not modify the in vivo insulin signaling pathway leading to Akt activation in skeletal muscle of Ames dwarf (Prop1(df)/Prop1(df)) mice. *Horm Metab Res* 37:672-9.

Arias, E.B., J. Kim, K. Funai, and G.D. Cartee. 2007. Prior exercise increases phosphorylation of Akt substrate of 160 kDa (AS160) in rat skeletal muscle. *Am J Physiol Endocrinol Metab* 292: E1191-200.

Bak, J.F., U.K. Jacobsen, F.S. Jorgensen, and O. Pedersen. 1989. Insulin receptor function and glycogen synthase activity in skeletal muscle biopsies from patients with insulin-dependent diabetes mellitus: Effects of physical training. *J Clin Endocrinol Metab* 69:158-64.

Bak, J.F., N. Moller, O. Schmitz, A. Saaek, and O. Pedersen. 1992. In vivo insulin action and muscle glycogen synthase activity in type 2 (non-insulin-dependent) diabetes mellitus: Effects of diet treatment. *Diabetologia* 35:777-84.

Balage, M., J. Grizard, and M. Manin. 1990. Effect of calorie restriction on skeletal muscle and liver insulin binding in growing rat. *Horm Metab Res* 22:207-14.

Birnbaum, M.J. 1992. The insulin-sensitive glucose transporter. *Int Rev Cytol* 137:239-97.

Bjornholm, M., and J.R. Zierath. 2005. Insulin signal transduction in human skeletal muscle: Identifying the defects in type II diabetes. *Biochem Soc Trans* 33:354-7.

Bonen, A., P.A. Clune, and M.H. Tan. 1986. Chronic exercise increases insulin binding in muscles but not liver. *Am J Physiol* 251:E196-203.

Bonen, A., and M.H. Tan. 1989. Dissociation between insulin binding and glucose utilization after intense exercise in mouse skeletal muscles. *Horm Metab Res* 21:172-8.

Bonen, A., M.H. Tan, P. Clune, and R.L. Kirby. 1985. Effects of exercise on insulin binding to human muscle. *Am J Physiol* 248:E403-8.

Bonen, A., M.H. Tan, and W.M. Watson-Wright. 1984. Effects of exercise on insulin binding and glucose metabolism in muscle. *Can J Physiol Pharmacol* 62:1500-4.

Bruss, M.D., E.B. Arias, G.E. Lienhard, and G.D. Cartee. 2005. Increased phosphorylation of Akt substrate of 160 kDa (AS160) in rat skeletal muscle in response to insulin or contractile activity. *Diabetes* 54:41-50.

Cartee, G.D., C. Briggs-Tung, and E.W. Kietzke. 1993. Persistent effects of exercise on skeletal muscle glucose transport across the life-span of rats. *J Appl Physiol* 75:972-8.

Cartee, G.D., A.G. Douen, T. Ramlal, A. Klip, and J.O. Holloszy. 1991. Stimulation of glucose transport in skeletal muscle by hypoxia. *J Appl Physiol* 70:1593-600.

Cartee, G.D., and J.O. Holloszy. 1990. Exercise increases susceptibility of muscle glucose transport to activation by various stimuli. *Am J Physiol* 258:E390-3.

Cartee, G.D., E.W. Kietzke, and C. Briggs-Tung. 1994. Adaptation of muscle glucose transport with caloric restriction in adult, middle-aged, and old rats. *Am J Physiol* 266:R1443-7.

Cartee, G.D., D.A. Young, M.D. Sleeper, J. Zierath, H. Wallberg-Henriksson, and J.O. Holloszy. 1989. Prolonged increase in insulin-stimulated glucose transport in muscle after exercise. *Am J Physiol* 256:E494-9.

Cecchin, F., O. Ittoop, M.K. Sinha, and J.F. Caro. 1988. Insulin resistance in uremia: Insulin receptor kinase activity in liver and muscle from chronic uremic rats. *Am J Physiol* 254:E394-401.

Chibalin, A.V., M. Yu, J.W. Ryder, X.M. Song, D. Galuska, A. Krook, H. Wallberg-Henriksson, and J.R. Zierath. 2000. Exercise-induced changes in expression and activity of proteins involved in insulin signal transduction in skeletal muscle: Differential effects on insulin-receptor substrates 1 and 2. *Proc Natl Acad Sci USA* 97:38-43.

Cho, H., J. Mu, J.K. Kim, J.L. Thorvaldsen, Q. Chu, E.B. Crenshaw, 3rd, K.H. Kaestner, M.S. Bartolomei, G.I. Shulman, and M.J. Birnbaum. 2001. Insulin resistance and a diabetes mellitus-like syndrome in mice lacking the protein kinase Akt2 (PKB beta). *Science* 292:1728-31.

Cho, H., J.L. Thorvaldsen, Q. Chu, F. Feng, and M.J. Birnbaum. 2001. Akt1/PKBalpha is required for normal growth but dispensable for maintenance of glucose homeostasis in mice. *J Biol Chem* 276:38349-52.

Christ, C.Y., D. Hunt, J. Hancock, R. Garcia-Macedo, L.J. Mandarino, and J.L. Ivy. 2002. Exercise training improves muscle insulin resistance but not insulin receptor signaling in obese Zucker rats. *J Appl Physiol* 92:736-44.

Christ-Roberts, C.Y., T. Pratipanawatr, W. Pratipanawatr, R. Berria, R. Belfort, S. Kashyap, and L.J. Mandarino. 2004. Exercise training increases glycogen synthase activity and GLUT4 expression but not insulin signaling in overweight nondiabetic and type 2 diabetic subjects. *Metabolism* 53:1233-42.

Christ-Roberts, C.Y., T. Pratipanawatr, W. Pratipanawatr, R. Berria, R. Belfort, and L.J. Mandarino. 2003. Increased insulin receptor signaling and glycogen synthase activity contribute to the synergistic effect of exercise on insulin action. *J Appl Physiol* 95:2519-29.

Constable, S.H., R.J. Favier, G.D. Cartee, D.A. Young, and J.O. Holloszy. 1988. Muscle glucose transport: Interactions of in vitro contractions, insulin, and exercise. *J Appl Physiol* 64:2329-32.

Cox, J.H., R.N. Cortright, G.L. Dohm, and J.A. Houmard. 1999. Effect of aging on response to exercise training in humans: Skeletal muscle GLUT-4 and insulin sensitivity. *J Appl Physiol* 86:2019-25.

Cox, K.L., V. Burke, A.R. Morton, L.J. Beilin, and I.B. Puddey. 2004. Independent and additive effects of energy restriction and exercise on glucose and insulin concentrations in sedentary overweight men. *Am J Clin Nutr* 80:308-16.

Crettaz, M., E.S. Horton, L.J. Wardzala, E.D. Horton, and B. Jeanrenaud. 1983. Physical training of Zucker rats: Lack of alleviation of muscle insulin resistance. *Am J Physiol* 244: E414-20.

Cusi, K., K. Maezono, A. Osman, M. Pendergrass, M.E. Patti, T. Pratipanawatr, R.A. DeFronzo, C.R. Kahn, and L.J. Mandarino. 2000. Insulin resistance differentially affects the PI 3-kinase- and MAP kinase-mediated signaling in human muscle. *J Clin Invest* 105:311-20.

Davidson, R.T., E.B. Arias, and G.D. Cartee. 2002. Calorie restriction increases muscle insulin action but not IRS-1-, IRS-2-, or phosphotyrosine-PI 3-kinase. *Am J Physiol Endocrinol Metab* 282:E270-6.

Davis, T.A., S. Klahr, E.D. Tegtmeyer, D.F. Osborne, T.L. Howard, and I.E. Karl. 1986. Glucose metabolism in epitrochlearis muscle of acutely exercised and trained rats. *Am J Physiol* 250:E137-43.

Dean, D.J., J.T. Brozinick Jr., S.W. Cushman, and G.D. Cartee. 1998. Calorie restriction increases cell surface GLUT-4 in insulin-stimulated skeletal muscle. *Am J Physiol* 275: E957-64.

Dean, D.J., and G.D. Cartee. 2000. Calorie restriction increases insulin-stimulated tyrosine phosphorylation of insulin receptor and insulin receptor substrate-1 in rat skeletal muscle. *Acta Physiol Scand* 169:133-9.

DeFronzo, R.A., E. Jacot, E. Jequier, E. Maeder, J. Wahren, and J.P. Felber. 1981. The effect of insulin on the disposal of intravenous glucose. Results from indirect calorimetry and hepatic and femoral venous catheterization. *Diabetes* 30:1000-7.

Dela, F., A. Handberg, K.J. Mikines, J. Vinten, and H. Galbo. 1993. GLUT 4 and insulin receptor binding and kinase activity in trained human muscle. *J Physiol* 469:615-24.

Dengel, D.R., A.T. Galecki, J M. Hagberg, and R.E. Pratley. 1998. The independent and combined effects of weight loss and aerobic exercise on blood pressure and oral glucose tolerance in older men. *Am J Hypertens* 11:1405-12.

Dengel, D.R., R.E. Pratley, J.M. Hagberg, E.M. Rogus, and A.P. Goldberg. 1996. Distinct effects of aerobic exercise training and weight loss on glucose homeostasis in obese sedentary men. *J Appl Physiol* 81:318-25.

Dohm, G.L., M.K. Sinha, and J.F. Caro. 1987. Insulin receptor binding and protein kinase activity in muscles of trained rats. *Am J Physiol* 252:E170-5.

Dolkas, C.B., K.J. Rodnick, and C.E. Mondon. 1990. Effect of body weight gain on insulin sensitivity after retirement from exercise training. *J Appl Physiol* 68:520-6.

Etgen Jr., G.J., J.T. Brozinick Jr., H.Y. Kang, and J.L. Ivy. 1993. Effects of exercise training on skeletal muscle glucose uptake and transport. *Am J Physiol* 264:C727-33.

Etgen Jr., G.J., J. Jensen, C.M. Wilson, D.G. Hunt, S.W. Cushman, and J.L. Ivy. 1997. Exercise training reverses insulin resistance in muscle by enhanced recruitment of GLUT-4 to the cell surface. *Am J Physiol* 272:E864-9.

Farese, R.V. 2002. Function and dysfunction of aPKC isoforms for glucose transport in insulin-sensitive and insulin-resistant states. *Am J Physiol Endocrinol Metab* 283:E1-11.

Fisher, J.S., J. Gao, D.H. Han, J.O. Holloszy, and L.A. Nolte. 2002. Activation of AMP kinase enhances sensitivity of muscle glucose transport to insulin. *Am J Physiol Endocrinol Metab* 282:E18-23.

Friedman, J.E., G.L. Dohm, N. Leggett-Frazier, C.W. Elton, E.B. Tapscott, W.P. Pories, and J.F. Caro. 1992. Restoration of insulin responsiveness in skeletal muscle of morbidly obese patients after weight loss. Effect on muscle glucose transport and glucose transporter GLUT4. *J Clin Invest* 89:701-5.

Gazdag, A.C., S. Sullivan, J.W. Kemnitz, and G.D. Cartee. 2000. Effect of long-term caloric restriction on GLUT4, phosphatidylinositol-3 kinase p85 subunit, and insulin receptor substrate-1 protein levels in rhesus monkey skeletal muscle. *J Gerontol A Biol Sci Med Sci* 55:B44-8.

Gazdag, A.C., T.J. Wetter, R.T. Davidson, K.A. Robinson, M.G. Buse, A.J. Yee, L.P. Turcotte, and G.D. Cartee. 2000. Lower calorie intake enhances muscle insulin action and reduces hexosamine levels. *Am J Physiol Regul Integr Comp Physiol* 278:R504-12.

Gonzalez, A.A., R. Kumar, J.D. Mulligan, A.J. Davis, R. Weindruch, and K.W. Saupe. 2004. Metabolic adaptations to fasting and chronic caloric restriction in heart, muscle, and liver do not include changes in AMPK activity. *Am J Physiol Endocrinol Metab* 287:E1032-7.

Grimditch, G.K., R.J. Barnard, S.A. Kaplan, and E. Sternlicht. 1986. Effect of training on insulin binding to rat skeletal muscle sarcolemmal vesicles. *Am J Physiol* 250:E570-5.

Gulve, E.A., G.D. Cartee, J.R. Zierath, V.M. Corpus, and J.O. Holloszy. 1990. Reversal of enhanced muscle glucose transport after exercise: Roles of insulin and glucose. *Am J Physiol* 259:E685-91.

Hamada, T., E.B. Arias, and G.D. Cartee. 2006. Increased submaximal insulin-stimulated glucose uptake in mouse skeletal muscle after treadmill exercise. *J Appl Physiol* 101: 1368-76.

Han, D.H., P.A. Hansen, M.M. Chen, and J.O. Holloszy. 1998. DHEA treatment reduces fat accumulation and protects against insulin resistance in male rats. *J Gerontol A Biol Sci Med Sci* 53:B19-24.

Hansen, P.A., L.A. Nolte, M.M. Chen, and J.O. Holloszy. 1998. Increased GLUT-4 translocation mediates enhanced insulin sensitivity of muscle glucose transport after exercise. *J Appl Physiol* 85:1218-22.

Hardie, D.G., and K. Sakamoto. 2006. AMPK: A key sensor of fuel and energy status in skeletal muscle. *Physiology (Bethesda)* 21:48-60.

Hevener, A.L., D. Reichart, and J. Olefsky. 2000. Exercise and thiazolidinedione therapy normalize insulin action in the obese Zucker fatty rat. *Diabetes* 49:2154-9.

Holloszy, J.O., and P.A. Hansen. 1996. Regulation of glucose transport into skeletal muscle. *Rev Physiol Biochem Pharmacol* 128:99-193.

Houmard, J.A., P.C. Egan, P.D. Neufer, J.E. Friedman, W.S. Wheeler, R.G. Israel, and G.L. Dohm. 1991. Elevated skeletal muscle glucose transporter levels in exercise-trained middle-aged men. *Am J Physiol* 261:E437-43.

Houmard, J.A., M.S. Hickey, G.L. Tyndall, K.E. Gavigan, and G.L. Dohm. 1995. Seven days of exercise increase GLUT-4 protein content in human skeletal muscle. *J Appl Physiol* 79:1936-8.

Houmard, J.A., C.D. Shaw, M.S. Hickey, and C.J. Tanner. 1999. Effect of short-term exercise training on insulin-stimulated PI 3-kinase activity in human skeletal muscle. *Am J Physiol* 277:E1055-60.

Houmard, J.A., M.H. Shinebarger, P.L. Dolan, N. Leggett-Frazier, R.K. Bruner, M.R. McCammon, R.G. Israel, and G.L. Dohm. 1993. Exercise training increases GLUT-4 protein concentration in previously sedentary middle-aged men. *Am J Physiol* 264:E896-901.

Howlett, K.F., K. Sakamoto, M.F. Hirshman, W.G. Aschenbach, M. Dow, M.F. White, and L.J. Goodyear. 2002. Insulin signaling after exercise in insulin receptor substrate-2-deficient mice. *Diabetes* 51:479-83.

Janssen, I., A. Fortier, R. Hudson, and R. Ross. 2002. Effects of an energy-restrictive diet with or without exercise on abdominal fat, intermuscular fat, and metabolic risk factors in obese women. *Diabetes Care* 25:431-8.

Kane, S., H. Sano, S.C. Liu, J.M. Asara, W.S. Lane, C.C. Garner, and G.E. Lienhard. 2002. A method to identify serine kinase substrates. Akt phosphorylates a novel adipocyte protein with a Rab GTPase-activating protein (GAP) domain. *J Biol Chem* 277:22115-8.

Kelley, D.E., R. Wing, C. Buonocore, J. Sturis, K. Polonsky, and M. Fitzsimmons. 1993. Relative effects of calorie restriction and weight loss in noninsulin-dependent diabetes mellitus. *J Clin Endocrinol Metab* 77:1287-93.

Kemppainen, J., H. Tsuchida, K. Stolen, H. Karlsson, M. Bjornholm, O.P. Heinonen, P. Nuutila, A. Krook, J. Knuuti, and J.R. Zierath. 2003. Insulin signalling and resistance in patients with chronic heart failure. *J Physiol* 550:305-15.

Kim, Y.B., K. Kotani, T.P. Ciaraldi, R.R. Henry, and B.B. Kahn. 2003. Insulin-stimulated protein kinase C lambda/zeta activity is reduced in skeletal muscle of humans with obesity and type 2 diabetes: Reversal with weight reduction. *Diabetes* 52:1935-42.

Kump, D.S., and F.W. Booth. 2005. Alterations in insulin receptor signalling in the rat epitrochlearis muscle upon cessation of voluntary exercise. *J Physiol* 562:829-38.

Lamarche, B., J.P. Despres, M.C. Pouliot, S. Moorjani, P.J. Lupien, G. Theriault, A. Tremblay, A. Nadeau, and C. Bouchard. 1992. Is body fat loss a determinant factor in the improvement of carbohydrate and lipid metabolism following aerobic exercise training in obese women? *Metabolism* 41:1249-56.

Lemieux, A.M., C.J. Diehl, J.A. Sloniger, and E.J. Henriksen. 2005. Voluntary exercise training enhances glucose trans-

port but not insulin signaling capacity in muscle of hypertensive TG(mREN2)27 rats. *J Appl Physiol* 99:357-62.

Luciano, E., E.M. Carneiro, C.R. Carvalho, J.B. Carvalheira, S.B. Peres, M.A. Reis, M.J. Saad, A.C. Boschero, and L.A. Velloso. 2002. Endurance training improves responsiveness to insulin and modulates insulin signal transduction through the phosphatidylinositol 3-kinase/Akt-1 pathway. *Eur J Endocrinol* 147:149-57.

McCurdy, C.E., and G.D. Cartee. 2005. Akt2 is essential for the full effect of calorie restriction on insulin-stimulated glucose uptake in skeletal muscle. *Diabetes* 54:1349-56.

McCurdy, C.E., R.T. Davidson, and G.D. Cartee. 2003. Brief calorie restriction increases Akt2 phosphorylation in insulin-stimulated rat skeletal muscle. *Am J Physiol Endocrinol Metab* 285:E693-700.

McCurdy, C.E., R.T. Davidson, and G.D. Cartee. 2005. Calorie restriction increases the ratio of phosphatidylinositol 3-kinase catalytic to regulatory subunits in rat skeletal muscle. *Am J Physiol Endocrinol Metab* 288:E996-1001.

Mikines, K.J., B. Sonne, P.A. Farrell, B. Tronier, and H. Galbo. 1988. Effect of physical exercise on sensitivity and responsiveness to insulin in humans. *Am J Physiol* 254:E248-59.

Mu, J., J.T. Brozinick Jr., O. Valladares, M. Bucan, and M.J. Birnbaum. 2001. A role for AMP-activated protein kinase in contraction- and hypoxia-regulated glucose transport in skeletal muscle. *Mol Cell* 7:1085-94.

Myers Jr., M.G., X.J. Sun, and M.F. White. 1994. The IRS-1 signaling system. *Trends Biochem Sci* 19:289-93.

Nassis, G.P., K. Papantakou, K. Skenderi, M. Triandafillopoulou, S.A. Kavouras, M. Yannakoulia, G.P. Chrousos, and L.S. Sidossis. 2005. Aerobic exercise training improves insulin sensitivity without changes in body weight, body fat, adiponectin, and inflammatory markers in overweight and obese girls. *Metabolism* 54:1472-9.

Perseghin, G., T.B. Price, K.F. Petersen, M. Roden, G.W. Cline, K. Gerow, D.L. Rothman, and G.I. Shulman. 1996. Increased glucose transport-phosphorylation and muscle glycogen synthesis after exercise training in insulin-resistant subjects. *New Engl J Med* 335:1357-62.

Reynolds, T.H.T., J.T. Brozinick Jr., M.A. Rogers, and S.W. Cushman. 1997. Effects of exercise training on glucose transport and cell surface GLUT-4 in isolated rat epitrochlearis muscle. *Am J Physiol* 272:E320-5.

Rice, B., I. Janssen, R. Hudson, and R. Ross. 1999. Effects of aerobic or resistance exercise and/or diet on glucose tolerance and plasma insulin levels in obese men. *Diabetes Care* 22:684-91.

Rodnick, K.J., J.O. Holloszy, C.E. Mondon, and D.E. James. 1990. Effects of exercise training on insulin-regulatable glucose-transporter protein levels in rat skeletal muscle. *Diabetes* 39:1425-9.

Ropelle, E.R., J.R. Pauli, P.O. Prada, C.T. de Souza, P.K. Picardi, M.C. Faria, D.E. Cintra, et al. 2006. Reversal of diet-induced insulin resistance with a single bout of exercise in the rat: the role of PTP1B and IRS-1 serine phosphorylation. *J Physiol* 577.3: 997-1007.

Sano, H., S. Kane, E. Sano, C.P. Miinea, J.M. Asara, W.S. Lane, C.W. Garner, and G.E. Lienhard. 2003. Insulin-stimulated phosphorylation of a Rab GTPase-activating protein regulates GLUT4 translocation. *J Biol Chem* 278:14599-602.

Santos, R.F., S. Azhar, C. Mondon, and E. Reaven. 1989. Prevention of insulin resistance by environmental manipulation as young rats mature. *Horm Metab Res* 21:55-8.

Shepherd, P.R., D.J. Withers, and K. Siddle. 1998. Phosphoinositide 3-kinase: The key switch mechanism in insulin signalling. *Biochem J* 333 (Pt. 3): 471-90.

Short, K.R., J.L. Vittone, M.L. Bigelow, D.N. Proctor, R.A. Rizza, J.M. Coenen-Schimke, and K.S. Nair. 2003. Impact of aerobic exercise training on age-related changes in insulin sensitivity and muscle oxidative capacity. *Diabetes* 52:1888-96.

Slentz, C.A., E.A. Gulve, K.J. Rodnick, E.J. Henriksen, J.H. Youn, and J.O. Holloszy. 1992. Glucose transporters and maximal transport are increased in endurance-trained rat soleus. *J Appl Physiol* 73:486-92.

Tamemoto, H., T. Kadowaki, K. Tobe, T. Yagi, H. Sakura, T. Hayakawa, Y. Terauchi, et al. 1994. Insulin resistance and growth retardation in mice lacking insulin receptor substrate-1. *Nature* 372:182-6.

Tan, M.H., and A. Bonen. 1987. Effect of exercise training on insulin binding and glucose metabolism in mouse soleus muscle. *Can J Physiol Pharmacol* 65:2231-4.

Tanner, C.J., T.R. Koves, R.L. Cortright, W.J. Pories, Y.B. Kim, B.B. Kahn, G.L. Dohm, and J.A. Houmard. 2002. Effect of short-term exercise training on insulin-stimulated PI 3-kinase activity in middle-aged men. *Am J Physiol Endocrinol Metab* 282:E147-53.

Thong, F.S., W. Derave, B. Kiens, T.E. Graham, B. Urso, J.F. Wojtaszewski, B.F. Hansen, and E.A. Richter. 2002. Caffeine-induced impairment of insulin action but not insulin signaling in human skeletal muscle is reduced by exercise. *Diabetes* 51:583-90.

Thorell, A., M.F. Hirshman, J. Nygren, L. Jorfeldt, J.F. Wojtaszewski, S.D. Dufresne, E.S. Horton, O. Ljungqvist, and L.J. Goodyear. 1999. Exercise and insulin cause GLUT-4 translocation in human skeletal muscle. *Am J Physiol* 277: E733-41.

Treadway, J.L., D.E. James, E. Burcel, and N.B. Ruderman. 1989. Effect of exercise on insulin receptor binding and kinase activity in skeletal muscle. *Am J Physiol* 256:E138-44.

van Loon, L.J., and B.H. Goodpaster. 2006. Increased intramuscular lipid storage in the insulin-resistant and endurance-trained state. *Pflugers Arch* 451:606-16.

Wallberg-Henriksson, H. 1987. Glucose transport into skeletal muscle. Influence of contractile activity, insulin, catecholamines and diabetes mellitus. *Acta Physiol Scand Suppl* 564:1-80.

Wang, Z.Q., A.D. Bell-Farrow, W. Sonntag, and W.T. Cefalu. 1997. Effect of age and caloric restriction on insulin receptor binding and glucose transporter levels in aging rats. *Exp Gerontol* 32:671-84.

Webster, B.A., S.R. Vigna, and T. Paquette. 1986. Acute exercise, epinephrine, and diabetes enhance insulin binding to skeletal muscle. *Am J Physiol* 250:E186-97.

Wetter, T.J., A.C. Gazdag, D.J. Dean, and G.D. Cartee. 1999. Effect of calorie restriction on in vivo glucose metabolism by individual tissues in rats. *Am J Physiol* 276:E728-38.

White, M.F. 2003. Insulin signaling in health and disease. *Science* 302:1710-1.

Whiteman, E.L., H. Cho, and M.J. Birnbaum. 2002. Role of Akt/protein kinase B in metabolism. *Trends Endocrinol Metab* 13:444-51.

Wilkes, J.J., and L.E. Nagy. 1996. Chronic ethanol feeding impairs glucose tolerance but does not produce skeletal muscle insulin resistance in rat epitrochlearis muscle. *Alcohol Clin Exp Res* 20:1016-22.

Willis, P.E., S.G. Chadan, V. Baracos, and W.S. Parkhouse. 1998. Restoration of insulin-like growth factor I action in skeletal muscle of old mice. *Am J Physiol* 275:E525-30.

Winder, W.W. 2001. Energy-sensing and signaling by AMP-activated protein kinase in skeletal muscle. *J Appl Physiol* 91:1017-28.

Wing, R.R., E. H. Blair, P. Bononi, M.D. Marcus, R. Watanabe, and R.N. Bergman. 1994. Caloric restriction per se is a significant factor in improvements in glycemic control and insulin sensitivity during weight loss in obese NIDDM patients. *Diabetes Care* 17:30-6.

Wojtaszewski, J.F., B.F. Hansen, J. Gade, B. Kiens, J.F. Markuns, L.J. Goodyear, and E.A. Richter. 2000. Insulin signaling and insulin sensitivity after exercise in human skeletal muscle. *Diabetes* 49:325-31.

Wojtaszewski, J.F., B.F. Hansen, B. Kiens, and E.A. Richter. 1997. Insulin signaling in human skeletal muscle: Time course and effect of exercise. *Diabetes* 46:1775-81.

Wojtaszewski, J.F., Y. Higaki, M.F. Hirshman, M.D. Michael, S.D. Dufresne, C.R. Kahn, and L.J. Goodyear. 1999. Exercise modulates postreceptor insulin signaling and glucose transport in muscle-specific insulin receptor knockout mice. *J Clin Invest* 104:1257-64.

Wright, D.C., K.A. Hucker, J.O. Holloszy, and D.H. Han. 2004. Ca(2+) and AMPK both mediate stimulation of glucose transport by muscle contractions. *Diabetes* 53:330-5.

Yeh, J.I., E.A. Gulve, L. Rameh, and M.J. Birnbaum. 1995. The effects of wortmannin on rat skeletal muscle. Dissociation of signaling pathways for insulin- and contraction-activated hexose transport. *J Biol Chem* 270:2107-11.

Youngren, J.F., S. Keen, J.L. Kulp, C.J. Tanner, J.A. Houmard, and I.D. Goldfine. 2001. Enhanced muscle insulin receptor autophosphorylation with short-term aerobic exercise training. *Am J Physiol Endocrinol Metab* 280:E528-33.

Zheng, D., P.S. MacLean, S.C. Pohnert, J.B. Knight, A.L. Olson, W.W. Winder, and G.L. Dohm. 2001. Regulation of muscle GLUT-4 transcription by AMP-activated protein kinase. *J Appl Physiol* 91:1073-83.

Zhou, Q., and G.L. Dohm. 1997. Treadmill running increases phosphatidylinositol 3-kinase activity in rat skeletal muscle. *Biochem Biophys Res Commun* 236:647-50.

Zhu, M., R. de Cabo, R.M. Anson, D.K. Ingram, and M.A. Lane. 2005. Caloric restriction modulates insulin receptor signaling in liver and skeletal muscle of rat. *Nutrition* 21:378-88.

Zhu, M., J. Miura, L.X. Lu, M. Bernier, R. DeCabo, M.A. Lane, G.S. Roth, and D.K. Ingram. 2004. Circulating adiponectin levels increase in rats on caloric restriction: The potential for insulin sensitization. *Exp Gerontol* 39:1049-59.

Ziel, F.H., N. Venkatesan, and M.B. Davidson. 1988. Glucose transport is rate limiting for skeletal muscle glucose metabolism in normal and STZ-induced diabetic rats. *Diabetes* 37:885-90.

Zorzano, A., T.W. Balon, L.P. Garetto, M.N. Goodman, and N.B. Ruderman. 1985. Muscle alpha-aminoisobutyric acid transport after exercise: Enhanced stimulation by insulin. *Am J Physiol* 248:E546-52.

Chapter 10

Mitochondrial Oxidative Capacity and Insulin Resistance

Kevin R. Short, PhD

Mitochondria are the biochemical powerhouses of most mammalian cells. Their primary role is the generation of chemical energy in the form of ATP. ATP is the main energy currency of cells; thus, mitochondria are vital for survival. In addition to producing ATP, mitochondria contain enzymes that interconvert carbohydrate, fat, and amino acids; help buffer intracellular calcium; and are increasingly appreciated for their role in cellular apoptosis, or cell death. The widespread effects of these functions have raised interest in the role of mitochondria in many areas of physiology and medicine.

The focus of this chapter is the association between insulin action and mitochondrial content and function. Over the past 10 to 15 y, several research groups have reported an intriguing connection between the markers of mitochondrial gene expression and ATP production capacity in skeletal muscle and the presence of insulin resistance and type 2 diabetes. This chapter reviews the evidence and rationale for the proposal that low mitochondrial function in skeletal muscle contributes to insulin resistance and presents the evidence that questions our understanding of this relationship.

Mitochondrial Structure and Function

Mitochondria are distinct organelles found within most cell types. Tissues with higher metabolic rate and oxygen consumption, such as the heart, brain, liver, and skeletal muscle, can contain hundreds of mitochondria per cell, whereas less active cells,

such as skin, bone, and white adipose cells, typically have lower mitochondrial content. Mitochondria are composed of at least 600 distinct proteins (Taylor et al. 2003), many of which are membrane transporters and enzymes that participate in the import and combustion of fat and carbohydrate in the beta-oxidation and tricarboxylic acid cycles and the electron transport chain (ETC).

The majority of the proteins required for mitochondrial assembly are encoded by the nuclear genome. These genes are transcribed in the nucleus, translated into protein in the endoplasmic reticulum, and then imported across the outer and inner mitochondrial membranes in a highly regulated process (Hood 2001). Additionally, mitochondria are unique among organelles in that they have their own DNA (mtDNA) that encodes for a small number of important proteins. The mtDNA is a circular genome of approximately 16,000 base pairs that encodes 13 subunits of the respiratory complexes in the ETC as well as 22 transfer RNAs and 2 ribosomal units that help synthesize proteins within the mitochondria. Ultimately, mitochondrial function depends on the coordinated expression and assembly of proteins from both the nuclear and the mitochondrial genomes.

Evidence of a Role for Mitochondria in Insulin Resistance and Diabetes

Several lines of evidence have emerged to support the possibility that reduced mitochondrial function may be associated with the development of

Dr. K.S. Nair, M.L. Bigelow, and numerous colleagues in the Endocrine Research Unit and the GCRC at Mayo Clinic Rochester made invaluable contributions to many of the studies described herein. Support was provided by the National Institutes of Health, Mayo Foundation, and American Federation of Aging Research.

insulin resistance in aging, obesity, and diabetes. Each of these areas are covered in the following sections.

Aging

The ability of muscle mitochondria to produce ATP has been reported to decline with age in humans in studies using in vitro methods on isolated mitochondria from muscle biopsies (Short et al. 2005; Tonkonogi et al. 2003; Trounce, Byrne, and Marzuki 1989) as well as in studies using in vivo measurements with magnetic resonance spectroscopy (Chilibeck et al. 1998; Conley, Jubrias, and Esselman 2000; Petersen et al. 2003). In some reports there was no loss of muscle mitochondrial ATP-generating capacity with age (Chretien et al. 1998; Kent-Braun and Ng 2000; Lanza, Befroy, and Kent-Braun 2005; Rasmussen et al. 2003). Discrepancies among these studies may be due to variations in sample size, location of the muscles tested, and medical and physical activity history of the subjects. Nevertheless, declining mitochondrial ATP production with age appears plausible since there are several reports that the activity of individual oxidative enzymes within mitochondria decrease with age (Coggan et al. 1993; Houmard et al. 1998; Rooyackers et al. 1996; Short et al. 2003, 2005). Furthermore, studies of aging human muscle found reduced mitochondrial protein content, which could be attributed to slower synthesis of mitochondrial proteins, lower content of mRNA transcripts that encode mitochondrial proteins, and reduced abundance or oxidative damage of mtDNA (Rooyackers et al. 1996; Short et al. 2003, 2005; Welle, Bhatt, and Thornton 2000; Welle et al. 2003).

Many of the genes responsible for the mitochondrial and oxidative phenotype of the cell are regulated by a family of nuclear transcription factors, the most studied of which has been PGC-1. In mice, PGC-1, particularly the alpha isoform, controls muscle mitochondrial biogenesis in response to cold stimulation, exercise, and transgenic overexpression (Baar et al. 2002; Lin et al. 2002; Pilegaard, Saltin, and Neufer 2002; Puigserver et al. 1998). Additionally, in studies of muscle cells in culture, PGC-1 has been shown to control the expression of the major insulin-sensitive glucose transporter protein, GLUT4 (Michael et al. 2001). Aging was first

reported to have no effect on the mRNA expression of the PGC-1α isoform in vastus lateralis muscle in healthy subjects aged 20 to 80 y, and there was also no change with age for two closely related transcription factors, NRF1 and TFAM (Short et al. 2003). This finding suggested that mitochondrial functional decline with age may be due to disrupted nuclear activity of these transcription factors rather than to a lower content. Subsequently, however, Ling and colleagues (2004) reported that the mRNA abundance of both the PGC-1α and the PGC-1β isoforms was reduced in vastus lateralis muscle from older subjects (62 y) compared with younger (28 y) subjects, so additional work is needed to confirm current findings at the levels of mRNA, protein content, and nuclear binding activity.

The decline seen in mitochondrial genes and functions with age likely contributes to the decline observed in peak oxygen uptake during exercise ($\dot{V}O_2$peak) and in muscle endurance (Haseler, Lin, and Richardson 2004; Hepple et al. 2002). It has also been proposed that age-related mitochondrial decline contributes to the lower insulin-mediated glucose uptake seen among older people (Ling et al. 2004; Petersen et al. 2003). Petersen and colleagues (2003) showed that mitochondrial ATP synthesis in the gastrocnemius muscle and insulin-stimulated glucose uptake, which were reduced in older people, were inversely correlated with intramuscular lipid content, which was elevated in older people. They reasoned that lipid accumulation is an intermediate step between reduced mitochondrial oxidative phosphorylative capacity and insulin action (depicted in figure 10.1). In agreement, Ling and colleagues (2004) reported that the expression of PGC-1 (alpha and beta isoforms) mRNA was reduced in older muscle and was significantly correlated with insulin-mediated glucose uptake and oxidation.

Obesity and Diabetes

Some of the changes in muscle oxidative capacity that are reported to occur with aging have also been observed in association with obesity and type 2 diabetes. Among the initial findings were that people with type 2 diabetes have lower $\dot{V}O_2$max and slower oxygen uptake kinetics during exercise when compared with control subjects (Regensteiner et al. 1995, 1998). This finding could be a result of both

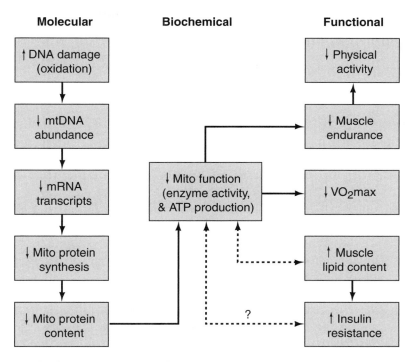

Figure 10.1 Proposed molecular changes in muscle mitochondrial gene and protein expression and the resulting effects on biochemical and physical functions. At present, there is moderately strong support that these changes occur with normal aging, whereas there is less agreement about whether these changes occur with obesity or diabetes alone. The proposed association and potential regulation among mitochondrial function, muscle lipid content, and insulin resistance are shown with the bidirectional dotted arrows since it is presently unresolved how these parameters are related. Mito, mitochondria.

cardiovascular and muscular changes. A study of Danes showed that people with a family history of diabetes were more likely to have an elevated content of Type IIb muscle fibers (Nyholm et al. 1997). Type IIb (fast-twitch, glycolytic) fibers are typically lower in mitochondrial content than Type I (slow-twitch, oxidative) or Type IIa (intermediate) fibers, so this fiber type shift was hypothesized to influence muscle metabolism and fuel selection.

In agreement with those findings, Simoneau and Kelley (1997; Simoneau et al. 1999) reported that activity of oxidative mitochondrial enzymes was reduced in people with obesity or type 2 diabetes when normalized to the activity of glycolytic enzymes. More recently, Kelley's group in Pittsburgh found that the activity of mitochondrial oxidative and ETC enzymes is reduced in the muscle of people with type 2 diabetes compared with lean controls without diabetes (Kelley et al. 2002; Ritov et al. 2005). The decline in ETC enzymes appears to be most prominent in subsarcolemmal mitochondria, the mitochondria that reside near the plasma membrane (Ritov et al. 2005), whereas intramyofibrillar mitochondria may be less affected.

This same research group has also shown that obesity and type 2 diabetes are associated with lower abundance of mtDNA and has found some evidence of mitochondrial structural damage in muscle from people with diabetes (Kelley et al. 2002; Ritov et al. 2005).

These measurements of mitochondrial enzymes and structure have all been performed on muscle biopsies. Supporting data from in vivo measurements of mitochondrial ATP synthesis were presented by Petersen and colleagues (2004), who showed that people with insulin resistance and a family history of diabetes had lower rates of ATP synthesis and elevated intramuscular lipids compared with a group of controls who were more insulin sensitive. This finding, along with their similar earlier results in elderly people (Petersen et al. 2003), suggested that reduced muscle mitochondrial content or function could lead to impaired fuel oxidation and energy phosphorylation potential (Lowell and Shulman 2005).

According to this hypothesis, reduced fuel oxidation contributes to the accumulation of intracellular lipids, which, in turn, is responsible for impaired insulin signaling and insulin-mediated glucose uptake (figure 10.1). Specifically, lipid metabolites, such as ceramides, DAG, or incompletely oxidized lipid chains, are thought to increase with accumulation of intracellular lipids and have been shown to impair insulin signaling pathways that lead to glucose uptake (Koves et al. 2005; Lowell and Shulman 2005). The accumulation of intracellular lipids may also result in increased lipid peroxidation by mitochondria-derived free radicals, and this has been suggested to cause further mitochondrial oxidative damage and functional decline (Schrauwen and Hesselink 2004). Work by Schrauwen and colleagues (2001) showed that people with type 2 diabetes have reduced content of uncoupling protein 3 (UCP3), a mitochondrial membrane protein that is thought to play a protective role against the generation of mitochondrial free radicals. Collectively, these findings support the possibility that a decline in mitochondrial oxidative capacity or fatty acid handling precedes or exacerbates insulin resistance.

Role of Genetics

The risk of developing diabetes varies greatly and is based on the complex interactions of many genes and the environment. Because of the potentially important role of mitochondria, several studies have attempted to identify mitochondrial genes with altered mRNA expression that may contribute to altered mitochondrial functions and insulin resistance (Asmann et al. 2006; Mootha et al. 2003, 2004; Patti et al. 2003; Sreekumar et al. 2002). Sreekumar and colleagues (2002) used microarray technology to measure mRNA expression of thousands of gene transcripts in muscle from people with type 2 diabetes compared with age-, sex-, and obesity-matched subjects without diabetes. They showed that mRNA abundance for several genes, including many encoding mitochondrial proteins, were reduced in people with type 2 diabetes during states of poor glycemic control. After 2 wk of tight glycemic control by insulin treatment expression level was reversed for many, but not all, of the altered genes. Subsequent work by other research groups (Mootha et al. 2003, 2004; Patti et al. 2003) confirmed that the expression of several mitochondrial genes was reduced in muscle from patients with type 2 diabetes and uncontrolled

hyperglycemia, as was the expression of PGC-1α mRNA and a related group of transcription factors regulating mitochondrial genes.

Recently it was reported that the expression of mitofusin 2 (MFN2), a gene involved in mitochondrial networking within the cell, was reduced in people with obesity and diabetes (Bach et al. 2005). This report strengthened the association between mitochondria and insulin action. It has since been shown that MFN2 is regulated by PGC-1α and related transcription factors (Soriano et al. 2006). Additional support comes from findings that single nucleotide polymorphisms in the PGC-1 gene were associated with diabetes in Danish and Pima Indian populations (Andersen et al. 2005; Ek et al. 2001; Muller et al. 2003). Thus, mitochondrial genes regulated by PGC-1α appear to have a potentially important role in insulin resistance and diabetes that should continue to be explored. It also appears that both quantitative and qualitative aspects of gene expression need to be considered.

Evidence That Mitochondria Are Not Responsible for Insulin Resistance

Despite the intriguing demonstrations that mitochondrial function may contribute to insulin resistance, solid experimental support for causation has yet to be presented. In addition, there are contradictory lines of evidence suggesting that parallel changes in mitochondrial function and insulin action are only moderately associated events or that insulin action or excess lipids may actually contribute to the regulation of mitochondrial function. In this section, support for these possibilities is considered.

Aging

As noted, several studies have demonstrated that muscle mitochondrial gene expression and function and peak aerobic exercise capacity decline with age (Coggan et al. 1993; Houmard et al. 1998; Rooyackers et al. 1996; Short et al. 2003, 2005; Tonkonogi et al. 2003; Trounce, Byrne, and Marzuki 1989; Welle, Bhatt, and Thornton 2000; Welle et al. 2003). The fact that most of these studies were performed in people without diabetes who had normal fasting glucose calls into question how essential a decline in mitochondrial content or function is for the development of insulin resistance. Given that mitochondrial decline with age is likely to be more common than the prevalence of diabetes, mitochondrial metabolism may make only a minor contribution to the etiology of insulin resistance in the elderly.

The capacity for muscle mitochondrial ATP production in younger (20-32 y) and older (65-80 y) people without diabetes was inversely related to the area under the plasma glucose curve following a mixed meal (Short et al. 2005; see figure 10.2). Although these data could support the view that mitochondrial function plays a role in glucose tolerance and insulin action, in these subjects there was an equally strong correlation between trunk fat mass and area under the glucose curve (figure 10.2). It is well established that abdominal adiposity is a major predictive factor for insulin resistance and diabetes (DeFronzo, Bonadonna, and Ferrannini 1992; Gabriely et al. 2002; Khort et al. 1993). Several studies have demonstrated that after accounting for variation in abdominal fat, most or all of the independent effect of age on insulin resistance is eliminated (Basu et al. 2003; Khort et al. 1993; Short et al. 2003). Thus, if age-related mitochondrial changes do have an effect on insulin resistance, this effect appears to be small, especially when compared with the effect of body fatness.

Exercise

Physical activity is clearly a key component for the prevention and treatment of diabetes, especially when used in conjunction with dietary modification and weight loss (Hu et al. 2001; Tuomilehto et al. 2001). Moderate to vigorous physical activity has potent stimulatory effects on mitochondrial biogenesis and insulin-mediated glucose metabolism, although it is not entirely clear how closely these outcomes are related. Endurance exercise increases mRNA expression and activity of mitochondrial oxidative enzymes in skeletal muscle, leading to increased capacity for mitochondrial ATP production (Holloszy 1967; Hood 2001; Short et al. 2003; Starritt, Angus, and Hargreaves 1999; Wibom et al. 1992). This adaptation appears to be regulated by elevated expression of PGC-1α and related nuclear transcription factors following both acute and chronic exercise training (Baar et al. 2002; Pilegaard, Saltin, and Neufer 2002; Short et al. 2003). Likewise, exercise or muscle contraction typically results in both acute and chronic improvements in GLUT4 content and insulin-mediated glucose uptake into skeletal muscle (Christ-Roberts et al. 2004; Cox et al. 1999; Dela et al. 1994; Houmard, Shinebarger et al. 1993; Ryder, Chibalin, and Zierath 2001; Short et al. 2003).

In many studies, exercise training responses are measured within 24 h after the last exercise session, when both insulin action and mitochondrial biogenesis markers are elevated and may even show

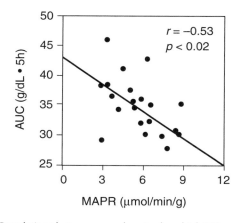
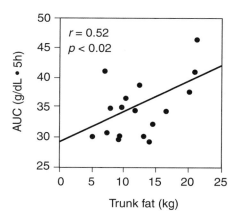

Figure 10.2 Correlations between muscle mitochondrial ATP production rate (MAPR, left panel) or trunk fat (measured by dual-energy X-ray absorptiometry, right panel) and meal glucose response. Subjects were 10 younger (20-30 y) and 10 older (65-80 y) people with normal fasting glucose who received a mixed meal. Area under the curve (AUC) for plasma glucose concentration was determined from repeated measurements taken for 5 h after the meal.

a correlated increase (Cox et al. 1999). However, when measured 4 to 5 d after the last exercise session, improvements in insulin action may not be evident in weight-stable subjects (Segal et al. 1991; Short et al. 2003), despite evidence of enhanced mitochondrial oxidative capacity (Short et al. 2003). This phenomenon may be due to a more transient effect of the last exercise bout on insulin action compared with mitochondrial function in some people, especially subjects who are older or overweight. It is well established that cessation of regular exercise training results in loss of glucose tolerance and insulin sensitivity within 5 to 7 d in both well-trained athletes and moderately trained, formerly sedentary people (Heath et al. 1983; Houmard et al. 1996; Nishida et al. 2004; Vukovich et al. 1996). This decline in insulin action is faster than the decline in activity of mitochondrial enzymes in skeletal muscle during detraining in endurance athletes (Coyle et al. 1984) and in swim-trained rats (Host et al. 1998).

Furthermore, Houmard and colleagues (Houmard, Hortobagyi et al. 1993) reported that glucose tolerance declined over 14 d of detraining in endurance- and strength-trained subjects, but this change was unrelated to changes in citrate synthase (CS), a mitochondrial marker enzyme, and GLUT4 content, which did not change. Studies of exercise training and detraining show that mitochondrial function does not always follow changes in insulin action, and therefore these adaptations do not appear to be directly related. Since exercise has strong effects on both mitochondrial biogenesis and insulin action, studies examining these outcomes among different subject groups must be careful to control for history of physical activity. Consideration of physical activity is often inadequate or ignored in many studies, so it is possible that variations in activity may contribute to discrepant findings among studies.

Genetic Factors

Several investigations have found lower abundance of mRNA transcripts for mitochondrial proteins in muscle from people with insulin resistance (Mootha et al. 2003, 2004; Patti et al. 2003; Sreekumar et al. 2002). Since exercise and diet strongly affect mRNA levels of many of these genes, it is not yet clear how much the reported differences are attributable to inheritance versus environment and lifestyle. These

microarray studies examined patients with type 2 diabetes during episodes of poor glycemic control. The study by Sreekumar and coworkers (2002) showed that many of the genes exhibiting altered transcript levels in subjects with poorly controlled diabetes were at least partly corrected following insulin treatment for 2 wk. Likewise, it was recently reported that many of the differences in muscle gene transcripts found between people with type 2 diabetes and carefully matched controls were partially or fully corrected following a 7 h infusion of insulin at postprandial levels (Asmann et al. 2006). In that study there were also no inherent differences in the basal amount of muscle mtDNA, intramuscular triglyceride, mitochondrial ATP production capacity or peak aerobic capacity ($\dot{V}O_2$peak) between the people with and without diabetes. Thus, many of the reported differences in mtDNA and mRNA levels of individual genes could be the result of a metabolic status arising from underappreciated differences in medical history, physical activity, and diet rather than underlying genetic mutations.

Furthermore, a report by Mohlke and colleagues (2005) casts some doubt on the possibility that common mutations in mtDNA play a role in diabetes development. These authors found no significant associations between single nucleotide polymorphisms or several haplotypes of mtDNA and type 2 diabetes or related metabolic traits in more than 1,600 Finnish people with and without diabetes. They argued that there was little support for a maternal or mtDNA-related inheritance pattern of risk for diabetes based on their results. This finding does not, of course, exclude important roles for involvement of the many mitochondrial genes encoded by the nucleus.

In agreement with the finding in Finns, another recent study demonstrated that a family history of type 2 diabetes does not appear to limit the adaptability of mitochondrial oxidative capacity. Østergård and colleagues (2006) studied healthy adult offspring of parents with type 2 diabetes and matched control subjects who did not have first-degree relatives with diabetes. In response to a supervised 10 wk aerobic exercise training program, improvements in insulin sensitivity, $\dot{V}O_2$max, and activity of mitochondrial enzymes was similar in the two groups. In contrast, Petersen and colleagues (Petersen et al. 2004; Peterson, Dufour, and Shulman 2005) reported that insulin-resistant offspring of parents with diabetes have lower mitochondrial ATP production during basal (fasting) and insulin-stimulated conditions compared with subjects who have normal insulin sensitivity. However, those studies were performed (Petersen et al. 2004; Petersen, Dufour, and Shulman 2005) only on offspring with very low glucose tolerance and insulin sensitivity, and family history of diabetes was not completely excluded in the control group. Thus, those studies would be more appropriately characterized as a comparison of insulin-resistant versus insulin-sensitive subjects, since only a defined segment of subjects with diabetes family history was tested.

Additional evidence for a disassociation between mitochondrial genes and insulin sensitivity comes from studies of transgenic mice. PGC-1, as described earlier, is a major regulator of mitochondrial biogenesis, GLUT4, and other oxidative genes. In animals in which PGC-1α protein content was suppressed, there was reduced mitochondrial protein content as might be expected, yet glucose metabolism remained largely unchanged, suggesting either a minor role for mitochondria in regulation of insulin sensitivity or the presence of compensatory mechanisms for lower oxidative capacity (Lin et al. 2004). Likewise, preliminary evidence gave no indication that overexpression of PGC-1α improved insulin sensitivity, despite a marked increase in muscle mitochondrial content (Choi et al. 2005). A slightly different outcome was discovered by Baar and coworkers (2003) in response to overexpression of NRF1, a nuclear transcription factor that works in conjunction with PGC-1α to control many oxidative phenotype genes. The increase in NRF1 led to increases in some, but not all, mitochondrial proteins (PGC-1α was unchanged) but to no change in mitochondrial respiration rate. Despite the lack of a clear functional improvement in mitochondria, there was an increase in insulin-stimulated glucose transport in muscle, and this was attributed to an increase in GLUT4 protein content. These reports illustrate that genetic manipulation of mitochondrial biogenesis and insulin sensitivity are not necessarily under coordinated control.

Regulatory Roles of Insulin and Lipid

Although it was proposed that reduced mitochondrial oxidative capacity results in intracellular lipid accumulation and insulin resistance (Lowell and Shulman 2005; Petersen et al. 2003), emerging data are suggesting that insulin and lipids may have important regulatory effects on mitochondria. Insulin infusion at physiological (i.e., postprandial) levels was shown to increase the abundance in skeletal muscle of several metabolic gene transcripts, including those encoding mitochondrial proteins (Ducluzeau et al. 2001; Huang et al. 1999; Stump et al. 2003). For some (Asmann et al. 2006; Ducluzeau et al. 2001) but not all (Asmann et al. 2006; Huang et al. 1999) of those genes, the effect of insulin was blunted in people with type 2 diabetes when compared with subjects without diabetes. Increased transcript availability may explain why insulin infusion stimulates the synthesis rate of mitochondrial proteins in skeletal muscle of healthy pigs and humans (Boirie et al. 2001; Stump et al. 2003).

More importantly, Stump and coworkers (2003) showed that insulin infusion resulted in increased mitochondrial ATP production rate in muscle biopsies from subjects without diabetes but that this stimulatory effect was blunted in patients with type 2 diabetes. A similar stimulatory effect of insulin on muscle ATP synthesis in healthy subjects was subsequently reported by Petersen and colleagues (Petersen, Dufour, and Shulman 2005) and Brehm and colleagues (2006) in studies using magnetic resonance spectroscopy in vivo. Petersen and coworkers (Petersen, Dufour, and Shulman 2005) also confirmed that the increase of ATP synthesis in response to insulin is inhibited in subjects who are insulin resistant. More recently, it was demonstrated that a short-term period (~9h) of insulin deprivation in patients with type 1 diabetes results in reductions in muscle mitochondrial ATP production and transcript levels of several genes involved in oxidative phosphorylation (Karakelides et al. 2007). These data suggest that insulin action is important for maintaining mitochondrial function and that lack of insulin action, either through insulin resistance or insulin deprivation, prevents this stimulatory effect. Insulin resistance may therefore be a cause, rather than an effect, of impaired mitochondrial oxidative function.

Lipid levels may also have an important regulatory effect on mitochondria. Intracellular lipid content and circulating fatty acid levels tend to be increased in people with insulin resistance

(Perseghin et al. 1999; Phillips et al. 1996). Intra-myocellular lipid content in itself is not predictive of insulin resistance, however, since well-trained endurance athletes with high insulin sensitivity can have muscle lipid levels that are similar to the high levels found in subjects who are obese or diabetic (Goodpaster et al. 2001). The difference in insulin sensitivity between these groups of people may be related to the size, number, or distribution of the lipid droplets in the cell (He, Goodpaster, and Kelley 2004; van Loon and Goodpaster 2006). It is not yet established whether or how mitochondrial oxidative capacity is related to the volume and distribution of intracellular lipids in muscle.

It is clear, however, that infusion of lipids can cause a rapid increase in insulin resistance (Dresner et al. 1999; Roden et al. 1996), and recent data show that lipid elevation also inhibits mitochondrial function and gene expression. Brehm and coworkers (2006) showed that insulin infusion resulted in a 60% increase of in vivo resting ATP synthesis in the calf muscles of healthy young men, but this response was inhibited by lipid infusion. In support of this finding, Richardson and colleagues (2005) reported that muscle mRNA levels of PGC-1α and some mitochondrial genes decreased following a 3 d infusion of lipids in subjects without diabetes. Likewise, mitochondrial fat oxidation capacity was impaired in rodents consuming a high-fat diet but could be preserved if animals performed regular exercise (Koves et al. 2005).

In another recent study, mitochondria were implicated in the development of insulin resistance in mice, but only when a high-fat diet was consumed by genetically altered mice. Pomplun and colleagues (2007) produced mice with markedly reduced amounts of frataxin, a protein involved in producing the iron-sulfur complexes of several mitochondrial enzymes. The experimental mice had a normal phenotype while on a standard chow diet, but they responded to a high-fat diet with greater increases in body fat and insulin resistance when compared with wild-type control mice. The authors proposed that their findings offer proof that a mitochondrial defect could predispose the mice to diabetes when the mice were eating a western-ized diet. This is an intriguing concept. Whether this proves to be the case in humans remains to be seen since a deficit in frataxin is typically associated with neurological disease. Viewed in another light, this study shows the potency of high fat consumption combined with a sedentary lifestyle since even wild-type mice had gains in body fat and diminished glucose tolerance.

Together, these findings provide evidence that increased circulating or intramuscular lipids may suppress mitochondrial genes and function as well as increase insulin resistance, particularly in sedentary people. It is therefore possible that lipid elevation from excess nutrient intake or sedentary lifestyle is an early event that contributes to the reduced mitochondrial oxidative capacity observed in obesity and diabetes.

Concluding Remarks

There is evidence for an association between muscle mitochondrial oxidative capacity and the occurrence of insulin resistance with aging, obesity, and diabetes. Findings in support of this association led to the proposal that reduced mito-chondrial function contributes to the development of insulin resistance, perhaps through the accumulation of intracellular lipid stores. However, several lines of evidence argue against this direction of causation. Recent data have shown that insulin stimulates muscle mitochondrial function in healthy people but this effect is blunted in people with insulin resistance. Additionally, elevation of circulating or tissue lipid concentration has inhibitory effects on mitochondrial function and gene expression in addition to the known inhibitory effect on insulin sensitivity. Thus, it is possible that lipid elevation results in insulin resistance and that the subsequent lack of insulin action on mitochondria contributes to the decrease in mitochondrial content and function observed in obesity and diabetes. Until more detailed experimental studies are performed, though, we cannot fully determine how and why mitochondrial function and insulin action are associated under several conditions. This relationship is clearly more complex than currently understood.

References

Andersen, G., L. Wegner, K. Yanagisawa, C.S. Rose, J. Lin, C. Glumer, T. Drivsholm, K. Borch-Johnsen, T. Jorgensen, T. Hansen, B.M. Spiegelman, and O. Pedersen. 2005. Evidence of an association between genetic variation of the coactivator PGC-1β and obesity. *J Med Genet* 42:402-07.

Asmann, Y.W., C.S. Stump, K.R. Short, J. M. Coenen-Schimke, Z.K. Guo, M.L. Bigelow, and K.S. Nair. 2006. Skeletal muscle mitochondrial functions, mitochondrial DNA copy numbers, and gene transcript profiles in type 2 diabetic and nondiabetic subjects at equal levels of low or high insulin and euglycemia. *Diabetes* 55:3309-19.

Baar, K., Z. Song, C. Semenkovich, T.E. Jones, D.H. Han, L.A. Nolte, E.O. Ojuka, M. Chen, and J.O. Holloszy. 2003. Skeletal muscle overexpression of nuclear respiratory factor 1 increases glucose transport capacity. *FASEB J* 17:1666-73.

Baar, K., A.R. Wende, T.E. Jones, M. Marison, L.A. Nolte, M. Chen, D.P. Kelly, and J.O. Holloszy. 2002. Adaptations of skeletal muscle to exercise: Rapid increase in the transcriptional coactivator PGC-1. *FASEB J* 16:1879-86.

Bach, D., D. Naon, S. Pich, F.X. Soriano, N. Vega, J. Rieusset, M. Laville, C. Guillet, Y. Boirie, H. Wallberg-Henriksson, M. Manco, M. Calvani, M. Castagneto, M. Palacin, G. Mingrone, J.R. Zierath, H. Vidal, and A. Zorzano. 2005. Expression of mfn2, the Charcot-Marie-Tooth neuropathy type 2A gene, in human skeletal muscle: Effects of type 2 diabetes, obesity, weight loss, and the regulatory role of tumor necrosis factor α and interleukin-6. *Diabetes* 54:2685-93.

Basu, R., E. Breda, A.L. Oberg, C.C. Powell, C. Dalla Man, A. Basu, J.L. Vittone, G.G. Klee, P. Arora, M.D. Jensen, G. Toffolo, C. Cobelli, and R.A. Rizza. 2003. Mechanisms of the age-associated deterioration in glucose tolerance: Contribution of alterations in insulin secretion, action, and clearance. *Diabetes* 52:1738-48.

Boirie, Y., K.R. Short, B. Ahlman, M. Charlton, and K.S. Nair. 2001. Tissue-specific regulation of mitochondrial and sarcoplasmic protein synthesis rates by insulin. *Diabetes* 50:2652-58.

Brehm, A., M. Krssak, A.I. Schmid, P. Nowotny, W. Waldhausl, and M. Roden. 2006. Increased lipid availability impairs insulin-stimulated ATP synthesis in human skeletal muscle. *Diabetes* 55:136-40.

Chilibeck, P.D., C.R. McCreary, G.D. Marsh, D.H. Paterson, E.G. Noble, A.W. Taylor, and R.T. Thompson. 1998. Evaluation of muscle oxidative potential by ^{31}P-MRS during incremental exercise in old and young humans. *Eur J Appl Physiol* 78:460-65.

Choi, C.S., R.M. Reznick, A. Kulkarni, S. Kim, J.K. Kim, J. Lin, B.M. Spiegelman, and G.I. Shulman. 2005. Diet induced insulin resistance is paradoxically prevented in PGC1α knockout mice but not altered in transgenic mice with muscle specific overexpression of PGC1α. *Diabetes* 54:A382-83.

Chretien, D., J. Gallego, A. Barrientos, J. Casademont, F. Cardellach, A. Munnich, A. Rotig, and P. Rustin. 1998. Biochemical parameters for the diagnosis of mitochondrial respiratory chain deficiency in humans, and their lack of age-related changes. *Biochem J* 329:249-54.

Christ-Roberts, C.Y., T. Pratipanawatr, W. Pratipanawatr, R. Berria, R. Belfort, S. Kashyap, and L.J. Mandarino. 2004. Exercise training increases glycogen synthase activity and GLUT4 expression but not insulin signaling in overweight nondiabetic and type 2 diabetic subjects. *Metabolism* 53:1233-42.

Coggan, A.R., A.M. Abduljalil, S.C. Swanson, M.S. Earle, J.W. Farris, L.A. Mendenhall, and P.M. Robitaille. 1993. Muscle metabolism during exercise in young and older untrained and endurance-trained men. *J Appl Physiol* 75:2125-33.

Conley, K.E., S.A. Jubrias, and P.C. Esselman. 2000. Oxidative capacity and ageing in human muscle. *J Physiol* 526:203-10.

Cox, J.H., R.N. Cortright, G.L. Dohm, and J.A. Houmard. 1999. Effect of aging on response to exercise training in humans: Skeletal muscle GLUT-4 and insulin sensitivity. *J Appl Physiol* 86:2019-25.

Coyle, E.F., W.H. Martin, D.R. Sinacore, M.J. Joyner, J.M. Hagberg, and J.O. Holloszy. 1984. Time course of loss of adaptations after stopping prolonged intense endurance training. *J Appl Physiol* 57:1857-64.

DeFronzo, R.A., R.C. Bonadonna, and E. Ferrannini. 1992. Pathogenesis of NIDDM. A balanced overview. *Diabetes Care* 15:318-68.

Dela, F., T. Ploug, A. Handberg, L.N. Petersen, J.J. Larsen, K.J. Mikines, and H. Galbo. 1994. Physical training increases muscle GLUT4 protein and mRNA in patients with NIDDM. *Diabetes* 43:862-65.

Dresner, A., D. Laurent, M. Marcucci, M.E. Griffin, S. Dufour, G.W. Cline, L.A. Slezak, D.K. Andersen, R.S. Hundal, D.L. Rothman, K.F. Petersen, and G.I. Shulman. 1999. Effects of free fatty acids on glucose transport and IRS-1-associated phosphatidylinositol 3-kinase activity. *J Clin Invest* 103:253-59.

Ducluzeau, P.H., N. Perretti, M. Laville, F. Andreelli, N. Vega, J.P. Riou, and H. Vidal. 2001. Regulation by insulin of gene expression in human skeletal muscle and adipose tissue. *Diabetes* 50:1134-42.

Ek, J., G. Andersen, S. Urhammer, P.H. Gæde, T. Drivsholm, K. Borch-Johnsen, T. Hansen, and O. Pedersen. 2001. Mutation analysis of peroxisome proliferator-activated receptor- coactivator-1 (PGC-1) and relationships of identified amino acid polymorphisms to type II diabetes mellitus. *Diabetologia* 44:2220-26.

Gabriely, I., X.H. Ma, X.M. Yang, G. Atzmon, M.W. Rajala, A.H. Berg, P. Scherer, L. Rossetti, and N. Barzilai. 2002. Removal of visceral fat prevents insulin resistance and glucose intolerance of aging. *Diabetes* 51:2951-58.

Goodpaster, B.H., J. He, S. Watkins, and D.E. Kelley. 2001. Skeletal muscle lipid content and insulin resistance: Evidence for a paradox in endurance-trained athletes. *J Clin Endocrinol Metab* 86:5755-61.

Haseler, L.J., A.P. Lin, and R.S. Richardson. 2004. Skeletal muscle oxidative metabolism in sedentary humans: ^{31}P-MRS assessment of O_2 supply and demand limitations. *J Appl Physiol* 97:1077-81.

He, J., B.H. Goodpaster, and D.E. Kelley. 2004. Effects of weight loss and physical activity on muscle lipid content and droplet size. *Obes Res* 12:761-9.

Heath, G.W., J.R. Gavin, J.M. Hinderliter, J.M. Hagberg, S.A. Bloomfield, and J.O. Holloszy. 1983. Effects of exercise and lack of exercise on glucose tolerance and insulin sensitivity. *J Appl Physiol* 55:512-7.

Hepple, R.T., J.L. Hagan, D.J. Krause, and C.C. Jackson. 2002. Aerobic power declines with aging in rat skeletal muscles perfused at matched convective O_2 delivery. *J Appl Physiol* 94:744-51.

Holloszy, J.O. 1967. Effects of exercise on mitochondrial oxygen uptake and respiratory enzyme activity in skeletal muscle. *J Biol Chem* 242:2278-82.

Hood, D.A. 2001. Contractile activity-induced mitochondrial biogenesis in skeletal muscle. *J Appl Physiol* 90:1137-57.

Host, H.H., P.A. Hansen, L.A. Nolte, M.M. Chen, and J.O. Holloszy. 1998. Rapid reversal of adaptive increases in muscle GLUT-4 and glucose transport capacity after training cessation. *J Appl Physiol* 84:798-802.

Houmard, J.A., T. Hortobagyi, P.D. Neufer, R.A. Johns, D.D. Fraser, R.G. Israel, and G.L. Dohm. 1993. Training cessation does not alter GLUT-4 protein levels in human skeletal muscle. *J Appl Physiol* 74:776-81.

Houmard, J.A., M.H. Shinebarger, P.L. Dolan, N. Leggett-Frazier, R.K. Bruner, M.R. McCammon, R.G. Isreal, and G.L. Dohm. 1993. Exercise training increases GLUT4 concentration in previously sedentary middle-aged men. *Am J Physiol Endocrinol Metab* 264:E896-901.

Houmard, J.A., G.L. Tyndall, J.B. Midyette, M.S. Hickey, P.L. Dolan, K.E. Gavigan, M.L. Weidner, and G.L. Dohm. 1996. Effect of reduced training and training cessation on insulin action and muscle GLUT-4. *J Appl Physiol* 81:1162-8.

Houmard, J.A., M.L. Weidner, K.E. Gavigan, G.L. Tyndall, M.S. Hickey, and A. Alshami. 1998. Fiber type and citrate synthase activity in the human gastrocnemius and vastus lateralis with aging. *J Appl Physiol* 85:1337-41.

Hu, F.B., J.E. Manson, J.S. Meir, G. Colditz, S. Liu, C.G. Solomon, and W.C. Willet. 2001. Diet, lifestyle, and the risk of type 2 diabetes mellitus in women. *New Engl J Med* 345:790-7.

Huang, X., K.F. Eriksson, A. Vaag, M. Lehtovirta, M. Hansson, E. Laurila, T. Kanninen, B.T. Olesen, I. Kurucz, L. Koranyi, and L. Groop. 1999. Insulin-regulated mitochondrial gene expression is associated with glucose flux in human skeletal muscle. *Diabetes* 48:1508-14.

Karakelides, H., Y.W. Asmann, M.L. Bigelow, K.R. Short, K. Dhatariya, J. Coenen-Schimke, J. Kahl, D. Mukhopadhyay, and K.S. Nair. Effect of insulin deprivation on muscle mitochondrial ATP production and gene transcript levels in type 1 diabetic subjects. *Diabetes* Published ahead on print, online July 2007. http://diabetes.diabetesjournals.org/cgi/reprint/db07-0378v1

Kelley, D.E., J. He, E.V. Menshikova, and V.B. Ritov. 2002. Dysfunction of mitochondria in human skeletal muscle in type 2 diabetes. *Diabetes* 51:2944-50.

Kent-Braun, J.A., and A.V. Ng. 2000. Skeletal muscle oxidative capacity in young and older women and men. *J Appl Physiol* 89:1072-8.

Khort, W.M., J.P. Kirwin, M.A. Staten, R.E. Bourey, D.S. King, and J.O. Holloszy. 1993. Insulin resistance in aging is related to abdominal obesity. *Diabetes* 42:273-81.

Koves, T.R., P. Li, J. An, T. Akimoto, D. Slentz, O. Ilkayeva, G.L. Dohm, Z. Yan, C.B. Newgard, and D.M. Muoio. 2005. Peroxisome proliferator-activated receptor- co-activator 1-mediated metabolic remodeling of skeletal myocytes mimics exercise training and reverses lipid-induced mitochondrial inefficiency. *J Biol Chem* 280:33588-98.

Lanza, I.R., D.E. Befroy, and J.A. Kent-Braun. 2005. Age-related changes in ATP-producing pathways in human skeletal muscle in vivo. *J Appl Physiol* 99:1736-44.

Lin, J., H. Wu, P.T. Tarr, C.Y. Zhang, Z. Wu, O. Boss, L.F. Michael, P. Puigserver, E. Isotani, E.N. Olson, B.B. Lowell, R. Bassel-Duby, and B.M. Spiegelman. 2002. Transcriptional co-activator PGC-1α drives the formation of slow-twitch muscle fibers. *Nature* 418:797-800.

Lin, J., P.-H. Wu, P.T. Tarr, K.S. Lindenberg, J. St. Pierre, C.-Y. Zhang, V.K. Mootha, S. Jager, C.R. Vianna, and R.M. Reznick. 2004. Defects in adaptive energy metabolism with CNS-linked hyperactivity in PGC-1α null mice. *Cell* 119:121-35.

Ling, C., P. Poulsen, E. Carlsson, M. Ridderstrale, P. Almgren, J. Wojtaszewski, H. Beck-Nielsen, L. Groop, and A. Vaag. 2004. Multiple environmental and genetic factors influence skeletal muscle PGC-1α and PGC-1β gene expression in twins. *J Clin Invest* 114:1518-26.

Lowell, B.B., and G.I. Shulman. 2005. Mitochondrial dysfunction and type 2 diabetes. *Science* 307:384-7.

Michael, L.F., Z. Wu, R.B. Cheatham, P. Puigserver, G. Adelmant, J.L. Lehman, D.P. Kelly, and B.M. Spiegelman. 2001. Restoration of insulin-sensitive glucose transporter (GLUT4) gene expression in muscle cells by the transcriptional coactivator PGC-1. *Proc Natl Acad Sci USA* 98:3820-5.

Mohlke, K., A. Jackson, L. Scott, E. Peck, Y. Suh, P. Chines, R. Watanabe, T. Buchanan, K. Conneely, M. Erdos, N. Narisu, S. Enloe, T. Valle, J. Tuomilehto, R. Bergman, M. Boehnke, and F. Collins. 2005. Mitochondrial polymorphisms and susceptibility to type 2 diabetes-related traits in Finns. *Hum Genet* 118:245-54.

Mootha, V.K., C. Handschin, D. Arlow, X. Xie, J. St. Pierre, S. Sihag, W. Yang, D. Altshuler, P. Puigserver, N. Patterson, P.J. Willy, I.G. Schulman, R.A. Heyman, E.S. Lander, and B.M. Spiegelman. 2004. Errα and Gabpa/b specify PGC-1α-dependent oxidative phosphorylation gene expression that is altered in diabetic muscle. *Proc Natl Acad Sci USA* 101:6570-5.

Mootha, V.K., C.M. Lindgren, K.F. Eriksson, A. Subramanian, S. Sihag, J. Lehar, P. Puigserver, E. Carlsson, M. Ridderstrale, E. Laurila, N. Houstis, M.J. Daly, N. Patterson, J.P. Mesirov, T.R. Golub, P. Tamayo, B. Spiegelman, E.S. Lander, J.N. Hirschhorn, D. Altshuler, and L.C. Groop. 2003. PGC-1α-responsive genes involved in oxidative phosphorylation are coordinately downregulated in human diabetes. *Nat Genet* 34:267-73.

Muller, Y.L., C. Bogardus, O. Pedersen, and L. Baier. 2003. A Gly482Ser missense mutation in the peroxisome proliferator-activated receptor gamma coactivator-1 is associated with altered lipid oxidation and early insulin secretion in Pima Indians. *Diabetes* 52:895-8.

Nishida, Y., K. Tokuyama, S. Nagasaka, Y. Higaki, Y. Shirai, A. Kiyonaga, M. Shindo, I. Kusaka, T. Nakamura, S. Ishibashi, and H. Tanaka. 2004. Effect of moderate exercise training on peripheral glucose effectiveness, insulin sensitivity, and endogenous glucose production in healthy humans estimated by a two-compartment-labeled minimal model. *Diabetes* 53:315-20.

Nyholm, B., Z. Qu, A. Kaal, S.B. Pedersen, C.H. Gravholt, J.L. Andersen, B. Saltin, and O. Schmitz. 1997. Evidence of an increased number of type IIb muscle fibers in insulin-resistant first-degree relatives of patients with NIDDM. *Diabetes* 46:1822-8.

Østergård, T., J.L. Andersen, B. Nyholm, S. Lund, K.S. Nair, B. Saltin, and O. Schmitz. 2006. Impact of exercise training on insulin sensitivity, physical fitness, and muscle oxidative capacity in first-degree relatives of type 2 diabetic patients. *Am J Physiol Endocrinol Metab* 290:E998-1005.

Patti, M.E., A.J. Butte, S. Crunkhorn, K. Cusi, R. Berria, S. Kashyap, Y. Miyazaki, I. Kohane, M. Costello, R. Saccone, E.J. Landaker, A.B. Goldfine, E. Mun, R. DeFronzo, J. Finlayson, C.R. Kahn, and L.J. Mandarino. 2003. Coordinated reduction of genes of oxidative metabolism in humans with insulin resistance and diabetes: Potential role of PGC1 and NRF1. *Proc Natl Acad Sci USA* 100:8466-71.

Perseghin, G., P. Scifo, F. De Cobelli, E. Pagliato, A. Battezzati, C. Arcelloni, A. Vanzulli, G. Testolin, G. Pozza, A. Del Maschio, and L. Luzi. 1999. Intramyocellular triglyceride content is a determinant of in vivo insulin resistance in humans: A ^{1}H-^{13}C nuclear magnetic resonance spectroscopy assessment in offspring of type 2 diabetic parents. *Diabetes* 48:1600-6.

Petersen, K.F., D. Befoy, S. Dufour, J. Dziura, C. Ariyan, D.L. Rothman, L. DiPietro, G.W. Cline, and G.I. Shulman. 2003. Mitochondrial dysfunction in the elderly: possible role in insulin resistance. *Science* 300:1140-2.

Petersen, K.F., S. Dufour, D. Befroy, R. Garcia, and G.I. Shulman. 2004. Impaired mitochondrial activity in the insulin-resistant offspring of patients with type 2 diabetes. *New Engl J Med* 350:664-71.

Petersen, K.F., S. Dufour, and G.I. Shulman. 2005. Decreased insulin-stimulated ATP synthesis and phosphate transport in muscle of insulin-resistant offspring of type 2 diabetic parents. *PLoS Med* 2:e233.

Phillips, D.I.W., S. Caddy, V. Ilic, B.A. Fielding, K.N. Frayn, A.C. Borthwick, and R. Taylor. 1996. Intramuscular triglyceride and muscle insulin sensitivity: Evidence for a relationship in nondiabetic subjects. *Metabolism* 45:947-50.

Pilegaard, H., B. Saltin, and P.D. Neufer. 2002. Exercise induces transient transcriptional activation of the PGC-1α gene in human skeletal muscle. *J Physiol* 340:831-8.

Pomplun, D., A. Voight, T.J. Schulz, R. Thierbach, A.F. Pfeiffer, and M. Ristow. 2007. Reduced expression of mitochondrial frataxin in mice exacerbates diet-induced obesity. *Proc Natl Acad Sci USA* 104:6377-81.

Puigserver, P., Z. Wu, C.W. Park, R. Graves, M. Wright, and B.M. Spiegelman. 1998. A cold-inducible coactivator of nuclear receptors linked to adaptive thermogenesis. *Cell* 92:829-39.

Rasmussen, U.F., P. Krustrup, M. Kjaer, and H.N. Rasmussen. 2003. Human skeletal muscle mitochondrial metabolism in youth and senescence: No signs of functional changes in ATP formation and mitochondrial oxidative capacity. *Pflugers Arch* 446:270-8.

Regensteiner, J.G., T.A. Bauer, J.E.B. Reusch, S.L. Brandenburg, J.M. Sippel, A.M. Vogelsong, S. Smith, E.E. Wolfel, R.H. Eckel, and W.R. Hiatt. 1998. Abnormal oxygen uptake kinetic responses in women with type II diabetes mellitus. *J Appl Physiol* 85:310-7.

Regensteiner, J.G., J. Sippel, E.T. McFarling, E.E. Wolfel, and W.R. Hiatt. 1995. Effects of non-insulin-dependent diabetes on oxygen consumption during treadmill exercise. *Med Sci Sports Exerc* 27:875-81.

Richardson, D.K., S. Kashyap, M. Bajaj, K. Cusi, S.J. Mandarino, J. Finlayson, R.A. DeFronzo, C.P. Jenkinson, and L.J. Mandarino. 2005. Lipid infusion decreases the expression of nuclear encoded mitochondrial genes and increases the expression of extracellular matrix genes in human skeletal muscle. *J Biol Chem* 280:10290-7.

Ritov, V.B., E.V. Menshikova, J. He, R.E. Ferrell, B.H. Goodpaster, and D.E. Kelley. 2005. Deficiency of subsarcolemmal mitochondria in obesity and type 2 diabetes. *Diabetes* 54:8-14.

Roden, M., T.B. Price, G. Perseghin, K.F. Petersen, D.L. Rothman, G.W. Cline, and G.I. Shulman. 1996. Mechanism of free fatty acid-induced insulin resistance in humans. *J Clin Invest* 97:2859-65.

Rooyackers, O.E., D.B. Adey, P.A. Ades, and K.S. Nair. 1996. Effect of age on in vivo rates of mitochondrial protein synthesis in human skeletal muscle. *Proc Natl Acad Sci USA* 93:15364-9.

Ryder, J.W., A.V. Chibalin, and J.R. Zierath. 2001. Intracellular mechanisms underlying increases in glucose uptake in response to insulin or exercise in skeletal muscle. *Acta Physiol Scand* 171:249-57.

Schrauwen, P., and K.C. Hesselink. 2004. Oxidative capacity, lipotoxicity, and mitochondrial damage in type 2 diabetes. *Diabetes* 53:1412-7.

Schrauwen, P., M.K. Hesselink, E.E. Blaak, L.B. Borghouts, G. Schaart, W.H. Saris, and H.A. Keizer. 2001. Uncoupling protein 3 content is decreased in skeletal muscle of patients with type 2 diabetes. *Diabetes* 50:2870-3.

Segal, K.R., A. Edano, A. Abalos, J. Albu, L. Blando, M.B. Tomas, and F.X. Pi-Sunyer. 1991. Effect of exercise training on insulin sensitivity and glucose metabolism in lean, obese, and diabetic men. *J Appl Physiol* 71:2402-11.

Short, K.R., M.L. Bigelow, J. Kahl, R. Singh, J. Coenen-Schimke, S. Raghavakaimal, and K.S. Nair. 2005. Decline in skeletal muscle mitochondrial function with aging in humans. *Proc Natl Acad Sci USA* 102:5618-23.

Short, K.R., J.L. Vittone, M.L. Bigelow, D.N. Proctor, R.R. Rizza, J.M. Coenen-Schimke, and K.S. Nair. 2003. Impact of aerobic training on age-related changes in insulin action and muscle oxidative capacity. *Diabetes* 52:1888-96.

Simoneau, J.A., and D.E. Kelley. 1997. Altered glycolytic and oxidative capacities of skeletal muscle contribute to insulin resistance in NIDDM. *J Appl Physiol* 83:166-71.

Simoneau, J.A., J.H. Veerkamp, L.P. Turcotte, and D.E. Kelley. 1999. Markers of capacity to utilize fatty acids in human skeletal muscle: Relation to insulin resistance and obesity and effects of weight loss. *FASEB J* 13:2051-60.

Soriano, F.X., M. Liesa, D. Bach, D.C. Chan, M. Palacin, and A. Zorzano. 2006. Evidence for a mitochondrial regulatory pathway defined by peroxisome proliferator-activated receptor-δ coactivator-1α, estrogen-related receptor-α, and mitofusin 2. *Diabetes* 55:1783-91.

Sreekumar, R., P. Halvatsiotis, J. Coenen-Schimke, and K.S. Nair. 2002. Gene expression profile in skeletal muscle of type 2 diabetes and the effect of insulin treatment. *Diabetes* 51:1913-20.

Starritt, E.C., D. Angus, and M. Hargreaves. 1999. Effect of short-term training on mitochondrial ATP production rate in human skeletal muscle. *J Appl Physiol* 86:450-4.

Stump, C.S., K.R. Short, M.L. Bigelow, J.C. Schimke, and K.S. Nair. 2003. Effect of insulin on human skeletal muscle mitochondrial ATP production, protein synthesis, and mRNA transcripts. *Proc Natl Acad Sci USA* 100:7996-8001.

Taylor, S.W., E. Fahy, B. Zhang, G.M. Glenn, D.E. Warnock, S. Wiley, A.N. Murphy, S.P. Gaucher, R.A. Capaldi, B.W. Gibson, and S.S. Ghosh. 2003. Characterization of the human heart mitochondrial proteome. *Nat Biotechnol* 21:281-6.

Tonkonogi, M., M. Fernstrom, B. Walsh, L.L. Ji, O. Rooyackers, F. Hammarqvist, J. Wernerman, and K. Sahlin. 2003. Reduced oxidative power but unchanged antioxidative capacity in skeletal muscle from aged humans. *Pflugers Arch* 446:261-9.

Trounce, I., E. Byrne, and S. Marzuki. 1989. Decline in skeletal muscle mitochondrial respiratory chain function: Possible factor in ageing. *Lancet* 25:637-9.

Tuomilehto, J., J. Lindstrom, J.G. Eriksson, T.T. Valle, H. Hamalainen, P. Ilanne-Parikka, S. Keinanen-Kiukaanniemi, M. Laakso, A. Louheranta, M. Rastas, V. Salminen, and M. Uusitupa. 2001. Prevention of type 2 diabetes mellitus by changes in lifestyle among subjects with impaired glucose tolerance. *New Engl J Med* 344:1343-50.

van Loon, L., and B. Goodpaster. 2006. Increased intramuscular lipid storage in the insulin-resistant and endurance-trained state. *Pflugers Arch* 451:606-16.

Vukovich, M.D., P.J. Arciero, W.M. Kohrt, S.B. Racette, P.A. Hansen, and J.O. Holloszy. 1996. Changes in insulin action and GLUT-4 with 6 days of inactivity in endurance runners. *J Appl Physiol* 80:240-4.

Welle, S., K. Bhatt, and C.A. Thornton. 2000. High-abundance mRNAs in human muscle: Comparison between young and old. *J Appl Physiol* 89:297-304.

Welle, S., A.I. Brooks, J.M. Delehanty, N. Needler, and C.A. Thornton. 2003. Gene expression profile of aging in human muscle. *Physiol Genomics* 14:149-59.

Wibom, R., E. Hultman, M. Johansson, K. Matherei, D. Constantin-Teodosiu, and P.G. Schantz. 1992. Adaptation of mitochondrial ATP production in human skeletal muscle to endurance training and detraining. *J Appl Physiol* 73:2004-10.

Chapter 11

Effects of Acute Exercise and Exercise Training on Insulin Action in Skeletal Muscle

Erik A. Richter, MD, DMSci; and Jørgen F.P. Wojtaszewski, PhD

During dynamic exercise, ATP turnover in skeletal muscle increases greatly. This increase is fuelled mainly by the catabolism of carbohydrate and lipid, while protein oxidation does not increase significantly unless muscle glycogen stores are depleted. During exercise in the postabsorptive state, the contribution of blood glucose to energy expenditure initially is relatively minor, but as exercise continues and muscle glycogen stores drop, the contribution of blood glucose becomes more substantial, reaching about 35% of leg oxidative metabolism and nearly 100% of muscle carbohydrate metabolism (for review, see Rose and Richter 2005). Following exercise, glucose uptake in muscle decreases rapidly and, depending on the duration and intensity of the preceding exercise, reaches resting levels within a few hours if no food is ingested. During this time fat oxidation is elevated (for review, see Kiens 2006). However, the sensitivity toward insulin stimulation is enhanced for several hours (up to 48 h) after exercise to facilitate rapid glycogen resynthesis (Mikines et al. 1988; Wojtaszewski et al. 1997; Braun, Zimmermann, and Kretchmer 1995; Bogardus et al. 1983; Devlin et al. 1987). Repeated exercise (physical training) leads to longer lasting increases in insulin sensitivity (Mikines et al. 1989; Bruce et al. 2004; Dela et al. 1992; Ebeling et al. 1993; Houmard et al. 2004; Youngren et al. 2001). The beneficial effects of both acute and chronic exercise on insulin sensitivity form the basis for recommending physical activity for the prevention and treatment of insulin resistance. In this chapter we discuss the possible mechanisms behind the increased metabolic effect of insulin that occurs following a single bout of exercise as well as with regular training, mainly from the perspective of insulin action on glucose transport and to a lesser extent on glycogen synthase (GS) activity. Other metabolic roles of insulin such as stimulation of amino acid transport and protein synthesis have been studied remarkably little.

Exercise and Contraction Signaling in Muscle

The factors thought to be involved in exercise-induced glucose transport were recently reviewed by Rose and Richter (2005). In brief, glucose uptake increases during exercise due to a coordinated increase in (a) muscle glucose delivery, which results from greatly enhanced capillary perfusion; (b) sarcolemmal and T-tubular transport of glucose from the interstitium to the muscle interior, which is aided by a greater membrane content of GLUT4; and (c) metabolism of the transported glucose. Whereas each of these events is important, in this chapter we focus on the exercise-activated molecular mechanisms that lead to GLUT4 translocation. The importance of GLUT4 has been shown in mice

Work from the authors' laboratory described in this chapter was supported by grants from the Danish Medical Research Council, the Danish Natural Science Research Council, the Novo Nordisk Foundation, the Danish Diabetes Association, the Copenhagen Muscle Research Centre, The Carlsberg Foundation and an Integrated Project funded by the European Union (#LSHM-CT-2004-005272) and the Lundbeck Foundation. J.F.P.W. was supported by a Hallas Møller Fellowship from the Novo Nordisk Foundation.

with systemic or muscle-specific (Zisman et al. 2000; Ryder et al. 1999) knockout of GLUT4. In these mice, muscle glucose transport during contractions in vitro is abolished or greatly diminished.

Translocation of GLUT4

The molecular events leading to the translocation of GLUT4 during muscle contraction are incompletely understood. It has been suggested that the regulation of glucose uptake during exercise can be divided into a Ca^{++}/calmodulin-dependent feed-forward mechanism that increases muscle glucose uptake at the onset of contraction and an AMPK-related feedback mechanism that adjusts glucose uptake to the energy needs of the muscle (Rose and Richter 2005; figure 11.1). However, recent data (Jensen et al. 2007a and b) suggest that AMPK may in fact to some extent be directly activated by Ca^{++}-signaling via CaMK kinase (CaMKK).

In figure 11.1, the higher glucose transport that accompanies exercise occurs mainly because of higher amounts of GLUT4 in surface membranes,

specifically, in sarcolemma and T-tubuli. This process may involve several kinases that sense changes in the intracellular environment during contraction (i.e., higher Ca^{++}, AMP concentrations) and transduce signals to other undefined proteins involved in GLUT4 movement and insertion into membranes.

Early evidence that intracellular Ca^{++} helps enhance glucose transport in muscle came from studies showing that pharmacologically elevated myoplasmic Ca^{++} concentrations increase glucose transport in resting muscle (Holloszy and Narahara 1967). More recently, in a series of studies based on inhibiting CaMK with KN62 and KN93, Wright and colleagues (2004, 2005) suggested that activation of CaMKII during muscle contraction plays an essential role in activating glucose transport, especially in slow-twitch fibers. They showed that the KN inhibitors did not affect AMPK phosphorylation with exercise (Wright et al. 2004), a finding suggesting that the Ca^{++} signaling pathway is distinct from that of AMPK. However, since it is likely that the KN substances are not selective toward CaMK, it cannot be excluded that the results obtained with these drugs

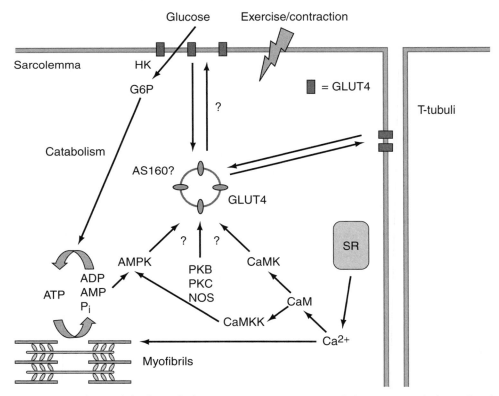

Figure 11.1 Contraction signaling in skeletal muscle during exercise contractions. T-tubuli, transverse tubuli; HK, hexokinase; G6P, glucose-6-phosphate; AS160, Akt substrate of 160 KDa; GLUT4, glucose transporter 4; SR, sarcoplasmic reticulum; AMPK, 5'AMP-activated protein kinase; CaM, calmodulin; CaMK, CA^{++}-calmodulin dependent protein kinase; PKB, protein kinase B (also known as Akt); PKC, protein kinase C; NOS, nitric oxide synthase.

are due to inhibition of enzymes other than CaMK and, as mentioned earlier, recent experiments in our lab suggest that AMPK in fact is activated by the Ca^{++} signaling pathway likely through CaMK kinase (CaMKK) (Jensen et al. 2007a and b). This means that the Ca^{++} signaling pathway and the AMPK pathway are not independent pathways.

Other downstream Ca^{++}-dependent kinases of possible importance for GLUT4 translocation include the conventional isoforms of PKC. Total PKC activity was found to translocate from the cytosol to the particulate fraction in electrically stimulated muscle (Richter et al. 1987). However, studies in humans did not show translocation of conventional PKC during exercise (Rose et al. 2004). Because the atypical isoforms of PKC are involved in activating glucose transport during insulin stimulation (Farese 2002; Sajan et al. 2006, Farese et al. 2007), activation of aPKC in muscle during exercise may suggest that it is involved in stimulating muscle glucose uptake (Richter et al. 2004; Perrini et al. 2004; Rose et al. 2004). However, at present this is unknown.

Activation of AMPK

The role of AMPK in contraction-induced glucose uptake is at present somewhat controversial. Pharmacological activation of AMPK with AICAR in resting muscle activates muscle glucose transport independently of insulin (for review, see Jorgensen, Richter, and Wojtaszewski 2006). Since AMPK is also activated in skeletal muscle during contraction or exercise in both rodent and human skeletal muscle (Jorgensen, Richter, and Wojtaszewski 2006), it has been speculated that AMPK may in fact be *the* contraction-activated kinase responsible for increasing glucose transport. More definitive mechanistic answers necessitate the use of genetically manipulated animals. In mice, overexpression of a dominant negative form of AMPK (a dead kinase) inhibits electrically induced muscle glucose transport by 30% to 40% in fast-twitch and even more in slow-twitch fibers of the kinase dead muscles compared with the wild-type muscles (Mu et al. 2001; Jensen et al. 2007a).

This finding suggests that contraction-induced glucose transport partly depends on AMPK. Interestingly, however, knockout of either the AMPK alpha2 catalytic subunit or the AMPK gamma3 regulatory subunit does not inhibit glucose transport in electrically stimulated incubated mouse muscle (Jorgensen et al. 2004; Barnes et al. 2004). In AMPKα1 KO mice, a small decrease in muscle glucose uptake was detected in slow-twitch but not in fast-twitch muscle (Jorgensen et al. 2004). A recent study has added weight to the role of AMPK in contraction-induced glucose uptake. In LKB1 KO mice, in which AMPK activation during electrical stimulation is virtually abolished, uptake of glucose was markedly inhibited (Sakamoto et al. 2004). Firm conclusions from these experiments are difficult, but the sum of the evidence seems to support a role for AMPK in contraction-induced glucose uptake. The role of AMPK likely varies in different kinds of contractions and possibly differs between in vivo exercise and electrically induced tetanic contractions, because the latter may lead to exaggerated activation of AMPK.

Recently, a new substrate of Akt, AS160, was described. AS160 apparently links insulin receptor signaling and GLUT4 trafficking (Watson and Pessin 2006). AS160 contains a GTPase activating protein (GAP) homology domain that has been shown to regulate the GTPase activity of certain Rab proteins in vitro. Phosphorylation of AS160 by Akt is likely to inhibit its GAP activity, such that as a consequence, the GTP form of Rab proteins is formed, in turn increasing GLUT4 vesicle movement to and fusion with the plasma membrane (see figure 11.2).

Interestingly, AS160 is phosphorylated (inactivated) during muscle contraction in a manner dependent on the alpha2 subunit but not the gamma3 regulatory subunit of AMPK (Treebak et al. 2006). However, since knockout of AMPKα2 does not inhibit contraction-induced glucose transport (Jorgensen et al. 2004), the role of AS160 in contraction-induced glucose transport remains unclear. In addition, we recently showed that AS160 phosphorylation during submaximal bicycle exercise does not increase until 60 min of exercise (Treebak et al. 2007), whereas muscle glucose uptake increases at the onset of exercise (Rose and Richter 2005). This finding further suggests that increased AS160 phosphorylation is not obligatory for enhanced glucose uptake during exercise. On the other hand, in mice in which AS160 is mutated so that it cannot be phosphorylated, electrically induced muscle contractions elicited a blunted muscle glucose uptake compared to uptake in WT mice (Kramer et al. 2006). This suggests that AS160 does play a role for contraction-induced muscle glucose uptake under some circumstances.

The regulation of GS activity during exercise has received less attention than glucose uptake. Nevertheless, during exercise GS is influenced by both stimulatory and inhibitory factors, and the consequent effect of exercise on GS activity is a result of the relative strength of the various stimuli. Several human studies have shown that muscle GS activity is higher in a glycogen-depleted state than in a glycogen-loaded state, and it has been suggested that exercise-induced GS activation is dependent on, and merely a result of, glycogen breakdown (Nielsen et al. 2001). The mechanism behind this dependency is unknown, but in rodents it seems to involve dephosphorylation of GS on sites 3a and 3b and changes in the subcellular localization of GS (Nielsen et al. 2001; Nielsen and Wojtaszewski 2004). Whether these two changes are linked is unknown, but cellular redistribution of GS induced by glycogen depletion could make GS more susceptible to dephosphorylation. Conversely, the covalent modifications of GS seen during conditions of high muscle glycogen content are only partly reversible by phosphatase treatment (Wojtaszewski et al. 2002). This finding indicates that unknown factors are involved in addition to phosphorylation-dependent regulatory mechanisms.

Insulin Signaling: A Web

The many biological effects of insulin are initiated when insulin binds to the alpha subunits of the transmembrane heterotetrameric insulin receptor, leading to the autophosphorylation and activation of the kinase associated with the beta subunits. A host of interrelated intracellular proteins then mediates the postreceptor signaling toward a diversity of both mitogenic and metabolic cellular responses. These signaling events are often depicted linearly, but over recent years evidence has gathered to suggest they form a much more complex weblike arrangement, within which inhibiting and promoting cross talk occurs (figure 11.2; Thong, Dugani, and Klip 2005).

The kinase of the insulin receptor phosphorylates tyrosine residues on other proteins. Such endogenous insulin receptor substrates (second messengers) are numerous and include IRS1 through IRS4, Cbl-associated protein (CAP), and APS, which are adapter proteins associated with pleckstrin homology (PH) and Src homology 2 (SH2) domains (Chang, Chiang, and Saltiel 2004). Of these receptors, the IRS family is by far the best studied, while the presence and function of the APS-Cbl-CAP complex in skeletal muscle are still unsolved. The IRS proteins do not by themselves have catalytic activity. However, phosphorylation on tyrosine residues within specific motifs allows them to bind to other proteins containing the SH2 domain and thus direct further signaling. The importance of IRS1 and to a lesser extent of IRS2 in insulin signaling to glucose transport in muscle tissues has been supported by studies that eliminated the protein by gene knockout or siRNA (White 2002).

When tyrosine-phosphorylated IRS interacts with the SH2 domain of the p85 regulatory subunit of the class IA PI3K, it activates the p110 catalytic subunit of the enzyme. This action subsequently increases the production of phosphatidylinositol 3-phosphate compounds (PIP_3, PIP_2, and PIP) within the plasma membrane. The production of these compounds also depends on the activation of additional lipid kinases that act on several of these sites. Many different techniques, including chemical poison and molecular inhibition, have been used to elucidate the role of class IA PI3K in glucose transport (Shepherd 2005). Interestingly, increased levels of phosphatidylinositol 3-phosphate compounds enhance GLUT4 translocation to the plasma membrane but do not increase glucose transport, raising the possibility that an additional signal is necessary for GLUT4 to fully function at the plasma membrane. Perhaps AS160 is involved. As mentioned earlier, AS160 apparently links insulin receptor signaling and GLUT4 trafficking (Watson and Pessin 2006). Phosphorylation of AS160 by Akt is likely to increase GLUT4 vesicle movement to the plasma membrane, vesicle fusion with the membrane, or both. Insulin increases AS160 phosphorylation, probably in a PI3K-dependent manner, in 3T3-L1 adipocytes and skeletal muscle.

Insulin stimulation leads to a PI3K-dependent dual phosphorylation and activation of the serine and threonine kinase Akt. Akt exists as two isoforms, Akt1 and Akt2, in skeletal muscle. Current evidence suggests that the insulin-induced increase in PIP_3 content helps recruit PDK1 to the plasma membrane as well as activates PDK1. Apparently, Akt is recruited to the plasma membrane in association with PDK1. Akt is subsequently phosphorylated by PDK1 on Thr-308. The full activation of Akt,

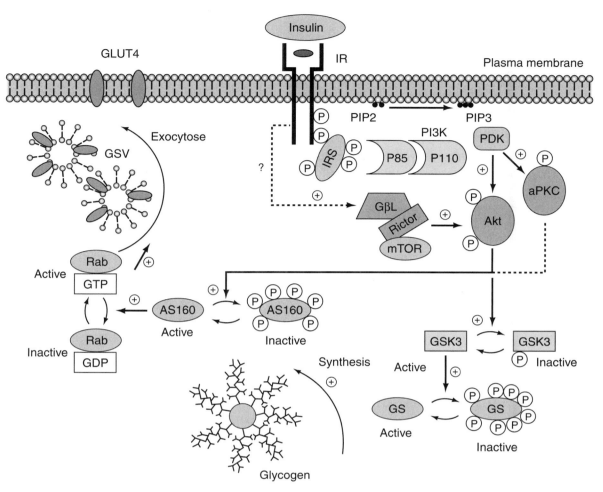

Figure 11.2 Depiction of some of the events from insulin binding to its receptor to the activation of the effector proteins GLUT4 and glycogen synthase. IR, insulin receptor, IRS, insulin receptor substrate, PI3K, phosphatidylinositol-3-kinase; PDK, phosphatidylinositol dependent kinase-3; GS, glycogen synthase; AS160, Akt substrate of 160 Kda; GSV, GLUT4 storage vesicles; GLUT4, glucose transporter 4.

however, requires additional phosphorylation on Ser-473 by PDK2, which may be the rictor-mTOR kinase complex. Transgenic approaches as well as KO and siRNA techniques have been used to confirm that Akt plays an important role in insulin-stimulated glucose transport. Akt2 has been suggested to be more important than Akt1 in this process (Whiteman, Cho, and Birnbaum 2002; Bouzakri et al. 2006).

PKCλ and PKCζ, atypical PKC isoforms, are also activated in a PI3K- and PDK-dependent manner in response to insulin stimulation. Phosphorylation by PDK1 induces autophosphorylation of PKCλ and PKCζ, and this is suggested to release autoinhibition of these atypical PKCs (Farese 2002). The role of PKCλ and PKCζ in insulin-stimulated glucose transport has also been verified by studies overexpressing both dead and active kinase constructs as

well as by studies using siRNA (Sajan et al. 2006). Recently, it was demonstrated that aPKC KO mice are severely insulin resistant and display symptoms of the metabolic syndrome (Farese et al. 2007).

Akt is at the crossroads at which insulin signals regulate glycogen synthesis by controlling GS activity. Upon insulin stimulation, Akt phosphorylates GSK3α and GSK3β in a PI3K-dependent manner, thus inactivating GSK3. Although GS action is regulated by multiple phosphorylation events, the sites assumed to exert the greatest effect on its activity are the N-terminal site 2 and the C-terminal site 3. Of these two sites, GSK3 directly phosphorylates site 3. The importance of GSK3 for insulin-induced GS activation was recently established by knock-in experiments of mutated GSK3 in murine muscle, in which insulin failed to regulate GS activity (McManus et al. 2005).

An important level of regulation in the insulin signaling cascade likely involves negative feedback loops. Over recent years it has become apparent that IRS1 is heavily serine phosphorylated at the time it becomes tyrosine phosphorylated during insulin stimulation. Multiple serine residues undergo phosphorylation by a variety of kinases (mTOR/JNK, GSK3, ERK, Akt, S6K1, AMPK). Although many aspects of these serine phosphorylation events are unresolved, some seem to downregulate whereas others upregulate the ability to signal through IRS1. Another level of regulation occurs through phosphatases, such as leukocyte antigen-related (LAR) phosphatase and protein tyrosine phosphatase 1B (PTP1B), acting on tyrosine residues within the insulin receptor and IRS. In addition, insulin signaling may also be modified by serine and threonine phosphatases (PP2A) acting on Akt, PKC, and GSK3 and by lipid phosphatases (PTEN, SHIP2) downregulating the levels of PIP_3.

Effects of a Single Exercise Bout on Insulin Sensitivity

Studies using the hyperinsulinemic-euglycemic clamp have shown that a single bout of cycling or stair-climbing improved the insulin sensitivity of glu-cose clearance at the whole-body level (for review, see Wojtaszewski, Nielsen, and Richter 2002). These changes have been observed as long as 2 d after the exercise bout (Mikines et al. 1988). The effect on whole body insulin sensitivity likely results from the increased insulin sensitivity localized to the muscles performing the work, as has been shown in rats (Richter et al. 1984) and humans (Richter et al. 1982; Richter et al. 1989; Wojtaszewski et al. 1997). Increased insulin action seems to be mainly a result of a leftward shift in the dose–response curve for insulin action on glucose uptake (figure 11.3). However, exercise does not always increase insulin action. For example, immediately after intense exercise, insulin action in vivo is impaired, possibly due to elevated concentrations of cat-echolamines and FFA (Kjær et al. 1990). Likewise, eccentric exercise or physical activity dominated by eccentric contractions elicits a prolonged decrease in insulin action, an effect that may be caused by altered protein expression and function brought on by muscle damage (Asp et al. 1996; Kristiansen, Asp, and Richter 1996; Del Aguila et al. 2000; Kirwan et al. 1992). Nevertheless, nondamaging dynamic exercise of a moderate intensity usually increases insulin sensitivity in the postexercise period.

What are the mechanisms behind this effect of exercise on insulin sensitivity? A simple hypothesis

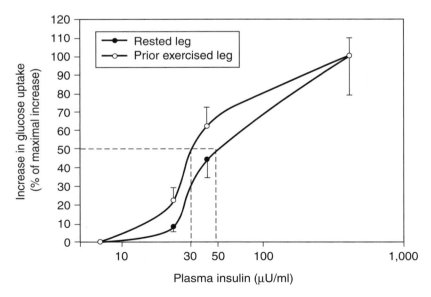

Figure 11.3 Exercise improves the insulin sensitivity of glucose uptake in human muscle. One-legged exercise enhances insulin sensi-tivity in the exercised compared to the rested leg in healthy human subjects. This finding is indicated by the decreased insulin concen-tration eliciting a half-maximal glucose uptake response when glucose uptake is given as a percent of maximal increase.

is that exercise somehow potentiates the insulin signaling cascade. It could theoretically do so via exercise-induced depletion of muscle glycogen, since a low glycogen concentration in rat skeletal muscle has been shown to enhance insulin signaling at the level of Akt (Derave et al. 2000). However, when the proximal insulin signaling cascade from the insulin receptor through IRS1-associated PI3K to Akt and GSK3 was analyzed in human skeletal muscle, exercise-induced increased insulin action was not accompanied by enhanced signaling (Wojtaszewski et al. 2000). On the contrary, IRS1-associated PI3K activity was stimulated slightly less in the exercised muscle than it was in the contralateral control muscle (Wojtaszewski et al. 1997). However, in human skeletal muscle two isoforms of IRS are expressed, and both contribute to PI3K activation during insulin stimulation (White 2002). The physiological role of IRS2-associated PI3K activity in human skeletal muscle is not well understood. Data from IRS KO mice seem to indicate that IRS1 is the major isoform mediating insulin-stimulated glucose uptake in skeletal muscle (Previs et al. 2000). Data from cultured human myotubes suggest that IRS1 and Akt2 are required for myoblast differentiation and glucose metabolism, whereas IRS2 and Akt1 are dispensable (Bouzakri et al. 2006). Still, knockout of IRS2 results in peripheral insulin resistance, although this effect could be to some extent a secondary consequence of prolonged exposure to hyperglycemia due to hepatic insulin resistance and beta-cell failure. Interestingly, immediately after in vivo exercise, activation of IRS2- but not IRS1-associated PI3K in response to a supraphysiological insulin stimulus is markedly increased in mouse skeletal muscle (Howlett et al. 2002). Furthermore, we have recently found that IRS2-associated PI3K activity is enhanced in humans 4 h after a single bout of knee-extensor exercise lasting for 60 min but that the effect of insulin stimulation on IRS2-associated PI3K activity is unchanged by previous exercise (Frosig et al. 2007b). This results in a generally heightened IRS2-associated PI3K activity in the exercised leg both at basal measurement and during insulin stimulation. Theoretically, this heightened activity should lead to a greater production of PIP_3. Downstream of PI3K, aPKC has emerged as an important signaling component stimulating GLUT4 translocation. One way to activate aPKC is via PIP_3. Thus, increased IRS2-associated PI3K activity in the exercised leg could be assumed to enhance allosteric activation

of aPKC if the increase in PIP_3 is restricted to areas in the cell in which aPKC is located. This suggestion is, however, speculative. Interestingly, while previous exercise does not enhance the ability of insulin to activate aPKC, exercise does increase the ability of PIP_3 to activate aPKC (Frosig et al. 2007b). Together, the increase in IRS2-associated PI3K activity and the increased ability of PIP_3 to activate aPKC in the exercised leg may be the first indication of direct molecular interaction between exercise and insulin signaling.

Exercise leads to AMPK activation in contracting muscles. It has been hypothesized that AMPK activation is involved in increased insulin sensitivity. In rats, pharmacological activation of AMPK by AICAR enhanced muscle insulin sensitivity without eliciting muscle contraction, suggesting that AMPK activation increases subsequent insulin sensitivity (Fisher et al. 2002). However, proof of this concept requires studies utilizing genetic manipulation of muscles such as knockout of AMPK.

Exercise expends energy and decreases muscle glycogen levels. In some rodent studies, the duration of enhanced insulin sensitivity after a single bout of exercise has been found to depend on the feeding status of the animals, such that enhanced insulin sensitivity is prolonged when carbohydrate intake is restricted after exercise. In humans, the amount of muscle glycogen broken down during the exercise bout correlates significantly with the insulin action 4 h after exercise (Wojtaszewski, Nielsen, and Richter 2002). This finding suggests that muscle glycogen plays a role in insulin sensitivity. The molecular mechanisms behind such an effect remain to be established.

Muscle contractions change gene expression. However, it is generally believed that the increased insulin sensitivity observed after a single bout of exercise is not caused by exercise-induced increased protein expression of signaling molecules or GLUT4 (Wojtaszewski et al. 1997).

Effects of Exercise Training on Insulin Action

While a single bout of exercise increases insulin action, chronic changes in muscle use or disuse cause more prolonged changes in insulin action. It should be realized that simply repeating a single bout of exercise regularly will maintain the muscle at a level of improved postexercise sensitivity. Still, exercise training also alters the gene expression of key proteins involved in glucose handling. Changing the gene expression enables the muscle to improve the reaction to insulin stimulation. Thus, the effects of exercise training cannot be discussed without taking into account the fact that the trained muscle more or less is chronically in a postexercise period.

Exercise training improves whole-body insulin sensitivity both in subjects with normal insulin sensitivity and in people experiencing insulin-resistant disease. This effect is largely attributable to enhanced insulin-induced glucose clearance in the peripheral tissues, particularly the trained skeletal muscle (Dela et al. 1995). Skeletal muscle is a heterogeneous tissue composed of fibers with different metabolic and contractile characteristics. Use or disuse—in other words, the contractile activity of the individual fibers—affects these characteristics. For example, use (endurance and resistance training) induces changes in myosin heavy-chain expression toward Type IIa (and away from Type IIx), whereas disuse induces changes in the opposite direction. Along the same lines but not necessarily linked to the changes in fiber type (Daugaard and Richter 2001, Daugaard et al. 2000), use or disuse changes the expression profile of a range of proteins. These proteins include GLUT4, a transporter known to be essential for insulin-induced glucose uptake into skeletal muscle (Zisman et al. 2000). In rodents, exercise training increases GLUT4 protein and mRNA expression in skeletal muscle, an effect that seems to onset rather early. Similar plasticity is observed in human muscle. Thus, resistance and endurance training increase and physical inactivity decreases GLUT4 mRNA and protein expression, and data suggest that the fibers recruited to perform the actual exercise are those in which the adaptations occur (Daugaard and Richter 2001). In addition, endurance training increases muscle capillarization and thus generally lowers the mean diffusion distance from capillary to muscle. In this way endurance training facilitates the delivery of insulin and glucose to the muscle. That this adaptation is important is indicated indirectly by the fact that insulin sensitivity has been shown to be related to capillary density in a large sample of subjects (Lillioja et al. 1987).

In rodent muscle, the ability of insulin to activate glucose transport is positively associated with GLUT4 expression. Among rodent muscles expressing variable amounts of GLUT4 (e.g., in genetically modified mice in which GLUT4 expression is manipulated by hetero- and homozygenous knockout of the GLUT4 gene), glucose uptake in response to insulin stimulation was graded according to GLUT4 protein expression. In animals treated with streptozotocin, in which muscle GLUT4 expression is decreased, a graded response was also observed. In humans, reductions in GLUT4 expression that occur with age correlate with decreased whole-body insulin sensitivity (Houmard et al. 1995). Still, in type 2 diabetes, muscle insulin resistance exists in the face of normal muscle GLUT4 expression (Andersen et al. 1993; Zierath et al. 1996), suggesting that other mechanisms may also be important for insulin action. Accordingly, several studies have addressed the idea that exercise training not only increases the amount of GLUT4 in the muscle but also changes insulin signaling capacity, enabling the muscle cell to respond to insulin with an enhanced sensitivity.

In rodents, exercise training increases mRNA coding for proteins within the signaling cascade (e.g., IR, IRS1, P85/P110) as well as mRNA coding for effector proteins (e.g., GLUT4 and GS). However, not all of these effects seem to translate into changes in actual protein levels. Although some studies have observed increased mRNA or protein levels of IR, p85, and IRS2 with exercise training (Chibalin et al. 2000; Kim et al. 1999; Saengsirisuwan et al. 2004), other studies in rodents did not see changes in insulin signaling protein expression (e.g., IR, IRS1, IRS2, AKT, and PKC) despite seeing increases in effector proteins like GS and GLUT4 as well as increases in insulin action (Krisan et al. 2004; Bernard et al. 2005; Lemieux et al. 2005; Hevener, Reichart, and Olefsky 2000; Luciano et al. 2002). Available data on effects in human skeletal muscle are sparse. Some exercise training studies have revealed no apparent changes in IR, IRS1, and IRS2 protein levels, whereas other investigations of endurance or resistance training have found increases in IR and AKT protein levels, GLUT4 and GS protein levels, and insulin action in muscle (Holten et al. 2004; Christ-Roberts et al. 2004; Frosig et al. 2007a). These responses to exercise training in muscle of both humans and rodents likely depend on training mode (intensity, duration) and animal species, yet the prevailing adaptive responses that were observed suggest that changes in the expression of the effector proteins GLUT4 and GS are more important than altered expression of the signaling proteins for the adaptive response to occur at the level of insulin action. Still, neither mRNA nor protein expression necessarily reflects adaptive responses in the signaling cascade upon activation. To address the question of whether exercise training changes signaling sensitivity, several studies have also investigated effects at the signaling level by measuring either activity or phosphorylation of different signaling elements.

Earlier studies investigating the insulin receptor function in vitro did not report any increases in insulin receptor tyrosine kinase activity after training in human skeletal muscle (Dela et al. 1993; Bak et al. 1989). However, a more recent study in humans (Youngren et al. 2001) indicates that in vitro insulin receptor autophosphorylation capacity is increased after 7 d of training. In none of the studies did exercise training change the amount of insulin receptor protein present. Similarly, some but not all studies of rodent muscle suggest that receptor signaling improves after training (Chibalin et al. 2000; Saengsirisuwan et al. 2004). Thus, from the available literature it can be concluded that exercise training improves receptor function under some conditions, and this enhanced receptor function may improve muscle insulin action. In some of the rodent studies, the improved insulin receptor function translated into increased activity at the postreceptor signaling level (IRS1, IRS2, and PI3K), whereas in other studies it did not (Zierath 2002). In human studies the picture is also unclear, as two studies observed improved signaling at the level of PI3K (Christ-Roberts et al. 2004; Tanner et al. 2002) and three did not (Houmard et al. 1999; Kirwan et al. 2000; Frosig et al. 2007a).

Likely this diversity of findings indicates that exercise training as a stimulus is complex and that our understanding of the intracellular signaling mechanism is limited. Endurance training usually enhances muscle glycogen levels. Because increased muscle glycogen content has been shown to be a negative regulator of insulin signaling, at least at the level of Akt phosphorylation (Derave et al. 2000), some of the variable signaling responses observed in different studies may

relate to differences in glycogen modulation by the training regime. Still, common to all of these training studies is, in addition to improved muscle insulin action to glucose handling, the discovery of elevated levels of effector proteins like GLUT4 and GS. One interpretation of this is that these adaptations are very important for the improved insulin action observed in trained muscle.

AMPK is activated during exercise and may play an important role in regulating metabolic events both acutely and chronically. AMPK activation may also to some extent improve muscle insulin sensitivity (Fisher et al. 2002; Oakes et al. 1997). Interestingly, we recently observed that AMPK activity in trained human muscle is increased at rest (Frosig et al. 2004). Chemical activation of AMPK regulates a variety of genes involved in mitochondria biogenesis as well as GLUT4 expression in resting muscle (Holmes, Kurth-Kraczek, and Winder 1999; Winder et al. 2000). Since chemical activation of AMPK increases muscle insulin sensitivity (Fisher et al. 2002) and expression of the GLUT4 gene, it seems intuitive that AMPK activation during the exercise bouts is responsible for the increased mitochondrial capacity and GLUT4 expression associated with training. However, this assumption has not been verified in studies of animal models with genetic deficiencies. For example, GLUT4 gene transcription induced by acute exercise is independent of the major alpha2 subunit of AMPK (Holmes et al. 2004; Jorgensen et al. 2005). In addition, training-induced increases in GLUT4 protein expression and mitochondrial enzyme activities in mouse muscle were found to be independent of the alpha2 AMPK subunit (Jorgensen et al. 2007). Still, in alpha2 AMPK KO mice the basal level of mitochondrial proteins was reduced by 20% to 25% when compared with levels in wild-type mice. This finding indicates that AMPK is important for the basal expression of mitochondrial proteins. Thus, the mechanisms behind the upregulation of insulin effector proteins during exercise training are unclear at present.

Interestingly, a study in which subjects exercised for 7 d showed that enhanced insulin sensitivity was found only if the subjects did not compensate for the training-induced increase in energy expenditure (Black et al. 2005). In contrast, when the calories expended during exercise training were carefully replaced by increased dietary intake, mainly of carbohydrate, no improvement in insulin sensitivity could be shown. Thus, decreased energy balance induced by exercise training may be important for enhancing insulin signaling—at least during short-term training.

Concluding Remarks

Because both a single bout of exercise and regular physical training enhance insulin sensitivity, physical activity is considered a cornerstone in the prevention and treatment of conditions of low insulin sensitivity such as metabolic syndrome and type 2 diabetes. Insulin and exercise each stimulate muscle glucose uptake via distinct molecular mechanisms that eventually converge on GLUT4 translocation to the plasma membrane. A single bout of exercise increases insulin sensitivity for several hours—up to 48 h—and the effect is mainly found in the muscles recruited during exercise. The molecular mechanism behind this effect is presently unresolved, but it is likely to be a complex phenomenon involving exercise-induced changes in several signaling parameters as well as changes in muscle energy stores. When exercise is repeated over time, adaptations to physical training occur. These include enhanced protein expression of GLUT4 and GS, changes in muscle fiber type, and increased capillarization. Changes in the expression or activation of signaling proteins are less consistently described.

References

Andersen, P.H., S. Lund, H. Vestergaard, S. Junker, B.B. Kahn, and O. Pedersen. 1993. Expression of the major insulin regulatable glucose transporter (GLUT4) in skeletal muscle of noninsulin-dependent diabetic patients and healthy subjects before and after insulin infusion. *J Clin Endocrinol Metab* 77:27-32.

Asp, S., J.R. Daugaard, S. Kristiansen, B. Kiens, and E.A. Richter. 1996. Eccentric exercise decreases maximal insulin action in humans: Muscle and systemic effects. *J Physiol* 494:891-8.

Bak, J., U. Jacobsen, F. Jørgensen, and O. Pedersen. 1989. Insulin receptor function and glycogen synthase activity in skeletal muscle biopsies from patients with insulin-dependent diabetes mellitus: Effects of physical training. *J Clin Endocrinol Metab* 69:158-64.

Barnes, B.R., S. Marklund, T.L. Steiler, M. Walter, G. Hjalm, V. Amarger, M. Mahlapuu, Y. Leng, C. Johansson, D. Galuska, K. Lindgren, M. Abrink, D. Stapleton, J.R. Zierath, and L. Andersson. 2004. The 5'-AMP-activated protein kinase gamma3 isoform has a key role in carbohydrate and lipid metabolism in glycolytic skeletal muscle. *J Biol Chem* 279:38441-7.

Bernard, J.R., A.M. Crain, D.A. Rivas, H.J. Herr, D.W. Reeder, and B.B. Yaspelkis III. 2005. Chronic aerobic exercise enhances components of the classical and novel insulin signalling cascades in Sprague-Dawley rat skeletal muscle. *Acta Physiol Scand* 183:357-66.

Black, S.E., E. Mitchell, P.S. Freedson, S.R. Chipkin, and B. Braun. 2005. Improved insulin action following short-term exercise training: Role of energy and carbohydrate balance. *J Appl Physiol* 99:2285-93.

Bogardus, C., P. Thuillez, E. Ravussin, B. Vasquez, M. Narimiga, and S. Azhar. 1983. Effect of muscle glycogen depletion on in vivo insulin action in man. *J Clin Invest* 72:1605-10.

Bouzakri, K., A. Zachrisson, L. Al-Khalili, B.B. Zhang, H.A. Koistinen, A. Krook, and J.R. Zierath. 2006. siRNA-based gene silencing reveals specialized roles of IRS-1/Akt2 and IRS-2/Akt1 in glucose and lipid metabolism in human skeletal muscle. *Cell Metab* 4:89-96.

Braun, B., M.B. Zimmermann, and N. Kretchmer. 1995. Effects of exercise intensity on insulin sensitivity in women with non-insulin-dependent diabetes mellitus. *J Appl Physiol* 78:300-6.

Bruce, C.R., A.D. Kriketos, G.J. Cooney, and J.A. Hawley. 2004. Disassociation of muscle triglyceride content and insulin sensitivity after exercise training in patients with type 2 diabetes. *Diabetologia* 47:23-30.

Chang, L., S.H. Chiang, and A.R. Saltiel. 2004. Insulin signaling and the regulation of glucose transport. *Mol Med* 10:65-71.

Chibalin, A.V., M. Yu, J.W. Ryder, X.M. Song, D. Galuska, A. Krook, H. Wallberg-Henriksson, and J.R. Zierath. 2000. Exercise-induced changes in expression and activity of proteins involved in insulin signal transduction in skeletal muscle: Differential effects on insulin-receptor substrates 1 and 2. *Proc Natl Acad Sci USA* 97:38-43.

Christ-Roberts, C.Y., T. Pratipanawatr, W. Pratipanawatr, R. Berria, R. Belfort, S. Kashyap, and L.J. Mandarino. 2004. Exercise training increases glycogen synthase activity and GLUT4 expression but not insulin signaling in overweight nondiabetic and type 2 diabetic subjects. *Metabolism* 53:1233-42.

Daugaard, J.R., J.N. Nielsen, S. Kristiansen, J.L. Andersen, M. Hargreaves, and E.A. Richter. 2000. Fiber type-specific expression of GLUT4 in human skeletal muscle: influence of exercise training. *Diabetes* 49: 1092-5,

Daugaard, J.R., and E.A. Richter. 2001. Relationship between muscle fibre composition, glucose transporter protein 4 and exercise training: Possible consequences in non-insulin-dependent diabetes mellitus. *Acta Physiol Scand* 171:267-76.

Del Aguila, L.F., R.K. Krishnan, J.S. Ulbrecht, P.A. Farrell, P.H. Correll, C.H. Lang, J.R. Zierath, and J.P. Kirwan. 2000. Muscle damage impairs insulin stimulation of IRS-1, PI 3-kinase, and Akt-kinase in human skeletal muscle. *Am J Physiol Endocrinol Metab* 279:E206-12.

Dela, F., A.A. Handberg, K.J. Mikines, J. Vinten, and H. Galbo. 1993. GLUT4 and insulin receptor binding and kinase activity in trained human muscle. *J Physiol* 469:615-24.

Dela, F., J.J. Larsen, K.J. Mikines, T. Ploug, L.N. Petersen, and H. Galbo. 1995. Insulin-stimulated muscle glucose clearance in patients with NIDDM. Effects of one-legged physical training. *Diabetes* 44:1010-20.

Dela, F., K.J. Mikines, M. von Linstow, N.H. Secher, and H. Galbo. 1992. Effect of training on insulin-mediated glucose uptake in human muscle. *Am J Physiol* 263:E1134-43.

Derave, W., B.F. Hansen, S. Lund, S. Kristiansen, and E.A. Richter. 2000. Muscle glycogen content affects insulin-stimulated glucose transport and protein kinase B activity. *Am J Physiol Endocrinol Metab* 279:E947-55.

Devlin, J.T., M. Hirshman, E.D. Horton, and E.S. Horton. 1987. Enhanced peripheral and splanchnic insulin sensitivity in NIDDM men after single bout of exercise. *Diabetes* 36:434-9.

Ebeling, P., R. Bourey, L. Koranyi, J.A. Tuominen, L.C. Groop, J. Henriksson, M. Mueckler, A. Sovijärvi, and V.A. Koivisto. 1993. Mechanism of enhanced insulin sensitivity in athletes. *J Clin Invest* 92:1623-31.

Farese, R.V. 2002. Function and dysfunction of aPKC isoforms for glucose transport in insulin-sensitive and insulin-resistant states. *Am J Physiol Endocrinol Metab* 283:E1-11.

Farese, R.V., M.P. Sajan, H. Yang, P. Li, S. Mastorides, W.R. Gower, S. Nimal, C.S. Choi, S. Kim, G.I. Shulman, C.R. Kahn, U. Braun, and M. Leitges. 2007. Muscle-specific knockout of PKC-lambda impairs glucose transport and induces metabolic and diabetic syndromes. *J Clin Invest* 117: 2289-2301.

Fisher, J.S., J. Gao, D.H. Han, J.O. Holloszy, and L.A. Nolte. 2002. Activation of AMP kinase enhances sensitivity of muscle glucose transport to insulin. *Am J Physiol Endocrinol Metab* 282:E18-23.

Frosig, C., A.J. Rose, J.T. Treebak, B. Kiens, E.A. Richter, and J.F.P. Wojtaszewski. 2007a. Effects of endurance exercise training on insulin signalling in human skeletal muscle-interactions at the level of PI3-K, Akt and AS160. *Diabetes* 56: 2093-2102.

Frosig, C., M.P. Sajan, S.J. Maarbjerg, N. Brandt, C. Roepstorff, J.F. Wojtaszewski, B. Kiens, R.V. Farese, and E.A. Richter. 2007. Exercise improves phosphatidylinositol-3,4,5-trisphosphate responsiveness of atypical protein kinase C and interacts with insulin signalling to peptide elongation in human skeletal muscle. *J Physiol* 582: 1289-1301.

Frosig, C., S.B. Jorgensen, D.G. Hardie, E.A. Richter, and J.F. Wojtaszewski. 2004. 5'-AMP-activated protein kinase activity and protein expression are regulated by endurance training in human skeletal muscle. *Am J Physiol Endocrinol Metab* 286:E411-7.

Hevener, A.L., D. Reichart, and J. Olefsky. 2000. Exercise and thiazolidinedione therapy normalize insulin action in the obese Zucker fatty rat. *Diabetes* 49:2154-9.

Holloszy, J., and H. Narahara. 1967. Enhanced permeability to sugar associated with muscle contraction. *J Gen Physiol* 50:551-62.

Holmes, B.F., E.J. Kurth-Kraczek, and W.W. Winder. 1999. Chronic activation of 5'-AMP-activated protein kinase increases GLUT-4, hexokinase, and glycogen in muscle. *J Appl Physiol* 87:1990-5.

Holmes, B.F., D.B. Lang, M.J. Birnbaum, J. Mu, and G.L. Dohm. 2004. AMP kinase is not required for the GLUT4 response to exercise and denervation in skeletal muscle. *Am J Physiol Endocrinol Metab* 287:E739-43.

Holten, M.K., M. Zacho, M. Gaster, C. Juel, J.F. Wojtaszewski, and F. Dela. 2004. Strength training increases insulin-mediated glucose uptake, GLUT4 content, and insulin signaling in skeletal muscle in patients with type 2 diabetes. *Diabetes* 53:294-305.

Houmard, J.A., C.D. Shaw, M.S. Hickey, and C.J. Tanner. 1999. Effect of short-term exercise training on insulin-stimulated PI 3-kinase activity in human skeletal muscle. *Am J Physiol* 277:E1055-60.

Houmard, J.A., M.D. Weidner, P.L. Dolan, N. Leggett-Frazier,

K.E. Gavigan, M.S. Hickey, G.L. Tyndall, D. Zheng, A. Alshami, and G.L. Dohm. 1995. Skeletal muscle GLUT 4 protein concentration and aging in humans. *Diabetes* 44:555-60.

Houmard, J.A., C.J. Tanner, C.A. Slentz, B.D. Duscha, J.S. McCartney, and W.E. Kraus. 2004. Effect of the volume and intensity of exercise training on insulin sensitivity. *J Appl Physiol* 96:101-6.

Howlett, K.F., K. Sakamoto, M.F. Hirshman, W.G. Aschenbach, M. Dow, M.F. White, and L.J. Goodyear. 2002. Insulin signaling after exercise in insulin receptor substrate-2-deficient mice. *Diabetes* 51:479-83.

Jensen, T.E., A.J. Rose, S.B. Jorgensen, N. Brandt, P. Schjerling, J.F. Wojtaszewski, and E.A. Richter. 2007a. Possible CaMKK-dependent regulation of AMPK phosphorylation and glucose uptake at the onset of mild tetanic skeletal muscle contraction. *Am J Physiol Endocrinol Metab* 292: E1308-1317.

Jensen, T.E., A.J. Rose, Y. Hellsten, J.F. Wojtaszewski, and E.A. Richter. 2007b. Caffeine-induced Ca^{2+} release increases AMPK-dependent glucose uptake in rodent soleus muscle. *Am J Physiol Endocrinol Metab* 293: E286-292.

Jorgensen, S.B., E.A. Richter, and J.F. Wojtaszewski. 2006. Role of AMPK in skeletal muscle metabolic regulation and adaptation in relation to exercise. *J Physiol* 574:17-31.

Jorgensen, S.B., J.T. Treebak, B. Viollet, P. Schjerling, S. Vaulont, J.F. Wojtaszewski, and E.A. Richter. 2007. Role of AMPKalpha2 in basal, training-, and AICAR-induced GLUT4, hexokinase II, and mitochondrial protein expression in mouse muscle. *Am J Physiol Endocrinol Metab* 292(1): E331-9.

Jorgensen, S.B., B. Viollet, F. Andreelli, C. Frosig, J.B. Birk, P. Schjerling, S. Vaulont, E.A. Richter, and J.F. Wojtaszewski. 2004. Knockout of the alpha2 but not alpha1 5'-AMP-activated protein kinase isoform abolishes 5-aminoimidazole-4-carboxamide-1-beta-4-ribofuranoside but not contraction-induced glucose uptake in skeletal muscle. *J Biol Chem* 279:1070-9.

Jorgensen, S.B., J.F. Wojtaszewski, B. Viollet, F. Andreelli, J.B. Birk, Y. Hellsten, P. Schjerling, S. Vaulont, P.D. Neufer, E.A. Richter, and H. Pilegaard. 2005. Effects of alpha-AMPK knockout on exercise-induced gene activation in mouse skeletal muscle. *FASEB J* 19: 1146-8.

Kiens, B. 2006. Skeletal muscle lipid metabolism in exercise and insulin resistance. *Physiol Rev* 86:205-43.

Kim, Y.B., T. Inoue, R. Nakajima, Y. Shirai-Morishita, K. Tokuyama, and M. Suzuki. 1999. Effect of long-term exercise on gene expression of insulin signaling pathway intermediates in skeletal muscle. *Biochem Biophys Res Commun* 254:720-7.

Kirwan, J.P., R.C. Hickner, K.E. Yarasheski, W.M. Kohrt, B.V. Wiethop, and J.O. Holloszy. 1992. Eccentric exercise induces transient insulin resistance in healthy individuals. *J Appl Physiol* 72:2197-202.

Kirwan, J.P., L.F. del Aguila, J.M. Hernandez, D.L. Williamson, D.J. O'Gorman, R. Lewis, and R.K. Krishnan. 2000. Regular exercise enhances insulin activation of IRS-1-associated PI3-kinase in human skeletal muscle. *J Appl Physiol* 88:797-803.

Kjær, M., C. Hollenbeck, B. Frey-Hewitt, H. Galbo, W. Haskell, and G. Reaven. 1990. Glucoregulation and hormonal responses to maximal exercise in non-insulin-dependent diabetes. *J Appl Physiol* 68:2067-74.

Kramer, H.F., C.A. Witczak, E.B. Taylor, N. Fujii, M.F. Hirshman, and L.J. Goodyear. 2006. AS160 regulates insulin- and contraction-stimulated glucose uptake in mouse skeletal muscle. *J Biol Chem* 281: 31478-85.

Krisan, A.D., D.E. Collins, A.M. Crain, C.C. Kwong, M.K. Singh, J.R. Bernard, and B.B. Yaspelkis III. 2004. Resistance training enhances components of the insulin signaling cascade in normal and high-fat-fed rodent skeletal muscle. *J Appl Physiol* 96:1691-700.

Kristiansen, S., S. Asp, and E.A. Richter. 1996. Decreased muscle GLUT-4 and contraction-induced glucose transport after eccentric contractions. *Am J Physiol* 271:R477-82.

Lemieux, A.M., C.J. Diehl, J.A. Sloniger, and E.J. Henriksen. 2005. Voluntary exercise training enhances glucose transport but not insulin signaling capacity in muscle of hypertensive TG(mREN2)27 rats. *J Appl Physiol* 99:357-62.

Lillioja, S., A. Young, C. Culter, J. Ivy, W.G. Abbott, J.K. Zawadzki, H. Yki-Järvinen, L. Christin, T.W. Secomb, and C. Bogardus. 1987. Skeletal muscle capillary density and fiber type are possible determinants of in vivo insulin resistance in man. *J Clin Invest* 80:415-24.

Luciano, E., E.M. Carneiro, C.R. Carvalho, J.B. Carvalheira, S.B. Peres, M.A. Reis, M.J. Saad, A.C. Boschero, and L.A. Velloso. 2002. Endurance training improves responsiveness to insulin and modulates insulin signal transduction through the phosphatidylinositol 3-kinase/Akt-1 pathway. *Eur J Endocrinol* 147:149-57.

McManus, E.J., K. Sakamoto, L.J. Armit, L. Ronaldson, N. Shpiro, R. Marquez, and D.R. Alessi. 2005. Role that phosphorylation of GSK3 plays in insulin and Wnt signalling defined by knockin analysis. *EMBO J* 24:1571-83.

Mikines, K., B. Sonne, P. Farrell, B. Tronier, and H. Galbo. 1988. Effect of physical exercise on sensitivity and responsiveness to insulin in humans. *Am J Physiol* 254:E248-59.

Mikines, K., B. Sonne, P. Farrell, B. Tronier, and H. Galbo. 1989. Effect of training on the dose-response relationship for insulin action in men. *J Appl Physiol* 66:695-703.

Mu, J., J.T. Brozinick Jr., O. Valladares, M. Bucan, and M.J. Birnbaum. 2001. A role for AMP-activated protein kinase in contraction- and hypoxia-regulated glucose transport in skeletal muscle. *Mol Cell* 7:1085-94.

Nielsen, J.N., W. Derave, S. Kristiansen, E. Ralston, T. Ploug, and E.A. Richter. 2001. Glycogen synthase localization and activity in rat skeletal muscle is strongly dependent on glycogen content. *J Physiol* 531:757-69.

Nielsen, J.N., and J.F. Wojtaszewski. 2004. Regulation of glycogen synthase activity and phosphorylation by exercise. *Proc Nutr Soc* 63:233-7.

Oakes, N.D., K.S. Bell, S.M. Furler, S. Camilleri, A.K. Saha, N.B. Ruderman, D.J. Chisholm, and E.W. Kraegen. 1997. Diet-induced muscle insulin resistance in rats is ameliorated by acute dietary lipid withdrawal or a single bout of exercise: Parallel relationship between insulin stimulation of glucose uptake and suppression of long-chain fatty acyl-CoA. *Diabetes* 46:2022-8.

Perrini, S., J. Henriksson, J.R. Zierath, and U. Widegren. 2004. Exercise-induced protein kinase C isoform-specific activation in human skeletal muscle. *Diabetes* 53:21-4.

Previs, S.F., D.J. Withers, J.M. Ren, M.F. White, and G.I. Shulman. 2000. Contrasting effects of IRS-1 versus IRS-2 gene disruption on carbohydrate and lipid metabolism in vivo. *J Biol Chem* 275:38990-4.

Richter, E.A., P.J.F. Cleland, S. Rattigan, and M.G. Clark. 1987. Contraction-associated translocation of protein kinase C in rat skeletal muscle. *FEBS Lett* 217:232-6.

Richter, E.A., L.P. Garetto, M.N. Goodman, and N.B. Ruderman. 1982. Muscle glucose metabolism following exercise in the rat. *J Clin Invest* 69:785-93.

Richter, E.A., L.P. Garetto, M.N. Goodman, and N.B. Ruderman. 1984. Enhanced muscle glucose metabolism after exercise: Modulation by local factors. *Am J Physiol* 246:E476-82.

Richter, E.A., K.J. Mikines, H. Galbo, and B. Kiens. 1989. Effect of exercise on insulin action in human skeletal muscle. *J Appl Physiol* 66:876-85.

Richter, E.A., B. Vistisen, S.J. Maarbjerg, M. Sajan, R.V. Farese, and B. Kiens. 2004. Differential effect of bicycling exercise intensity on activity and phosphorylation of atypical protein kinase C and extracellular signal-regulated protein kinase in skeletal muscle. *J Physiol* 560: 909-918.

Rose, A.J., B.J. Michell, B.E. Kemp, and M. Hargreaves. 2004. Effect of exercise on protein kinase C activity and localization in human skeletal muscle. *J Physiol* 561:861-70.

Rose, A.J., and E.A. Richter. 2005. Skeletal muscle glucose uptake during exercise: How is it regulated? *Physiology (Bethesda)* 20:260-70.

Ryder, J.W., Y. Kawano, D. Galuska, R. Fahlman, H. Wallberg-Henriksson, M.J. Charron, and J.R. Zierath. 1999. Postexercise glucose uptake and glycogen synthesis in skeletal muscle from GLUT4-deficient mice. *FASEB J* 13:2246-56.

Saengsirisuwan, V., F.R. Perez, J.A. Sloniger, T. Maier, and E.J. Henriksen. 2004. Interactions of exercise training and α-lipoic acid on insulin signaling in skeletal muscle of obese Zucker rats. *Am J Physiol Endocrinol Metab* 287: E529-36.

Sajan, M.P., J. Rivas, P. Li, M.L. Standaert, and R.V. Farese. 2006. Repletion of atypical protein kinase C following RNA interference-mediated depletion restores insulin-stimulated glucose transport. *J Biol Chem* 281:17466-73.

Sakamoto, K., O. Goransson, D.G. Hardie, and D.R. Alessi. 2004. Activity of LKB1 and AMPK-related kinases in skeletal muscle: Effects of contraction, phenformin, and AICAR. *Am J Physiol Endocrinol Metab* 287:E310-7.

Shepherd, P.R. 2005. Mechanisms regulating phosphoinositide 3-kinase signalling in insulin-sensitive tissues. *Acta Physiol Scand* 183:3-12.

Tanner C.J., T.R. Koves, R.L. Cortright, W.J. Pories, Y.B. Kim, B.B. Kahn, G.L. Dohm, and J.A. Houmard. 2002. Effect of short-term exercise training on insulin-stimulated PI 3-kinase activity in middle-aged men. *Am J Physiol Endocrinol Metab* 282:E147-53.

Thong, F.S., C.B. Dugani, and A. Klip. 2005. Turning signals on and off: GLUT4 traffic in the insulin-signaling highway. *Physiology (Bethesda)* 20:271-84.

Treebak, J.T., J.B. Birk, A.J. Rose, B. Kiens, E.A. Richter, and J.F. Wojtaszewski. 2007. AS160 phosphorylation is associated with activation of alpha2beta2gamma1- but not alpha2beta2gamma3-AMPK trimeric complex in skeletal muscle during exercise in humans. *Am J Physiol Endocrinol Metab* 292(3): E715-22.

Treebak, J.T., S. Glund, A. Deshmukh, D.K. Klein, Y.C. Long, T.E. Jensen, S.B. Jorgensen, B. Viollet, L. Andersson, D. Neumann, T. Wallimann, E.A. Richter, A.V. Chibalin, J.R. Zierath, and J.F. Wojtaszewski. 2006. AMPK-mediated AS160 phosphorylation in skeletal muscle is dependent on AMPK catalytic and regulatory subunits. *Diabetes* 55:2051-8.

Watson, R.T., and J.E. Pessin. 2006. Bridging the GAP between insulin signaling and GLUT4 translocation. *Trends Biochem Sci* 31:215-22.

White, M.F. 2002. IRS proteins and the common path to diabetes. *Am J Physiol Endocrinol Metab* 283:E413-22.

Whiteman, E.L., H. Cho, and M.J. Birnbaum. 2002. Role of Akt/protein kinase B in metabolism. *Trends Endocrinol Metab* 13:444-51.

Winder, W.W., B.F. Holmes, D.S. Rubink, E.B. Jensen, M. Chen, and J.O. Holloszy. 2000. Activation of AMP-activated protein kinase increases mitochondrial enzymes in skeletal muscle. *J Appl Physiol* 88:2219-26.

Wojtaszewski, J.F., B.F. Hansen, B. Kiens, and E.A. Richter. 1997. Insulin signaling in human skeletal muscle: Time course and effect of exercise. *Diabetes* 46:1775-81.

Wojtaszewski, J.F., S.B. Jorgensen, Y. Hellsten, D.G. Hardie, and E.A. Richter. 2002. Glycogen-dependent effects of 5-aminoimidazole-4-carboxamide (AICA)-riboside on AMP-activated protein kinase and glycogen synthase activities in rat skeletal muscle. *Diabetes* 51:284-92.

Wojtaszewski, J.F., J.N. Nielsen, and E.A. Richter. 2002. Invited review: Effect of acute exercise on insulin signaling and action in humans. *J Appl Physiol* 93:384-92.

Wojtaszewski, J.F.P., B.F. Hansen, J. Gade, B. Kiens, J.F. Markuns, L.J. Goodyear, and E.A. Richter. 2000. Insulin signaling and insulin sensitivity after exercise in human skeletal muscle. *Diabetes* 49:325-31.

Wright, D.C., P.C. Geiger, J.O. Holloszy, and D.H. Han. 2005. Contraction- and hypoxia-stimulated glucose transport is mediated by a Ca^{2+}-dependent mechanism in slow-twitch rat soleus muscle. *Am J Physiol Endocrinol Metab* 288:E1062-6.

Wright, D.C., K.A. Hucker, J.O. Holloszy, and D.H. Han. 2004. Ca^{2+} and AMPK both mediate stimulation of glucose transport by muscle contractions. *Diabetes* 53:330-5.

Youngren, J.F., S. Keen, J.L. Kulp, C.J. Tanner, J.A. Houmard, and I.D. Goldfine. 2001. Enhanced muscle insulin receptor autophosphorylation with short-term aerobic exercise training. *Am J Physiol Endocrinol Metab* 280:E528-33.

Zierath, J.R. 2002. Invited review: Exercise training-induced changes in insulin signaling in skeletal muscle. *J Appl Physiol* 93:773-81.

Zierath, J.R., L. He, A. Guma, E.O. Wahlström, A. Klip, and H. Wallberg-Henriksson. 1996. Insulin action on glucose transport and plasma membrane GLUT4 content in skeletal muscle from patients with NIDDM. *Diabetologia* 39:1180-9.

Zisman, A., O.D. Peroni, E.D. Abel, M.D. Michael, F. Mauvais-Jarvis, B.B. Lowell, J.F. Wojtaszewski, M.F. Hirshman, A. Virkamaki, L.J. Goodyear, C.R. Kahn, and B.B. Kahn. 2000. Targeted disruption of the glucose transporter 4 selectively in muscle causes insulin resistance and glucose intolerance. *Nat Med* 6:924-8.

Chapter 12

Resistance Exercise Training and the Management of Diabetes

Jørgen F.P. Wojtaszewski, PhD; Henriette Pilegaard, PhD; and Flemming Dela, MD

In spite of our current knowledge, many aspects of the effects of physical training on insulin resistance need to be addressed. On the molecular level it is yet not known how exercise leads to improved insulin action in skeletal muscle. The efficiency of different training regimens and different exercise intensities are poorly described. The necessary amount of exercise to enhance insulin action is also not well defined. Whether the exercise can be divided into smaller parts, or if it should be taken in one session, has not been investigated. The latter issues are not trivial, because, although somewhat disappointing, the clinical experience is that the patients do not ask "How much do I need to do?", but rather "How little can I get away with?"

Life style modifications, focusing on changing dietary habits and energy restriction, have proved to be difficult and are often perceived as a limitation (i.e., disallowing a behavior). In contrast, when prescribing exercise as a tool to prevent glucose intolerance from developing into overt type 2 diabetes or even to treat established insulin resistance, this is perceived as adding an element—not taking something away or prohibiting a behavior. However, as with any other treatment, a successful intervention requires that the individual is motivated and quickly experiences a solid effect of the personal efforts. Thus an important task for the health care professional is to visualize and explain such effects, e.g., by simple measurements of blood glucose, strength or aerobic capacity (Fritz and Rosenqvist, 2001). The concept of using physical training as a tool to prevent and/or treat diabetes is usually not that surprising to the patient, but most patients are quite unaware of the efficacy of physical training. Once the patient

has accepted and is motivated for this kind of treatment, it is very important that the health care system is capable of providing an organization and setting that can provide initial high quality supervision, guidance and support. In this respect, the role of physiotherapists cannot be underestimated. Given the high prevalence of inactivity-related diseases it is, however, unrealistic that the majority of the diseased people continue in a forever lasting health care system provided training program. At a certain time point, the patients ought to be able to continue by themselves and maintain their new lifestyle. This is the breakpoint, when the central part of "self-care" and "diabetes management" becomes apparent. Here all efforts should prove their worth, but unfortunately this is the point where most interventions are unsuccessful. An illustrative example of this is given in a recent study in which the authors clearly demonstrate that home-based training is not sufficient to maintain training induced improvements in glucose homeostasis (figure 12.1) (Dunstan et al., 2005). The situation resembles what is seen in many dietary intervention studies of obesity. Clearly, this lack of compliance is the one major obstacle of a successful introduction of exercise training for treatment of type 2 diabetes.

Apart from the effect on insulin resistance, exercise training also has positive effects on all of the other conditions associated with the metabolic syndrome (e.g., hypertension, dyslipidemia, obesity, atherosclerosis), which is also seen more frequently with advanced age. An increase in daily physical activity has been shown to prevent type 2 diabetes in pre-diabetic states such as impaired glucose tolerance, and in patients with diagnosed type 2 dia-

Part of the work mentioned in this chapter was supported by a European Union grant (6th framework LSHM-CT-2004-005272, EXGENESIS), the Novo Nordisk Foundation, and the Danish National Health Research Council.

JFPW was supported by a fellowship from the Novo Nordisk Foundation.

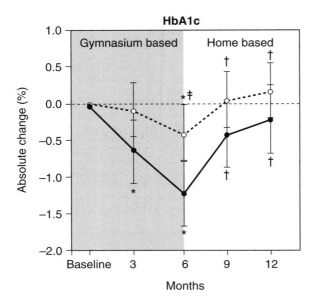

Figure 12.1 Thirty-six patients with type 2 diabetes were randomized to weight loss (open circle) or weight loss + resistance training (closed circle). The absolute change in HbA1C from baseline until 12 months is shown. The training was supervised the first six months (gymnasium-based; shaded area) and the following six months the training was conducted home-based. * P<0.05 within-group difference for the change from baseline; † P<0.05 within-group difference for the change from six months; ‡ P< 0.05 between-group difference for the change from baseline. Values are means and 95% CI.

betes it improves insulin mediated glucose uptake rates, glucose tolerance and the insulin response to glucose. The majority of intervention studies have used endurance / aerobic training programs and the effects of such programs on insulin sensitivity are undisputed (see chapter 2). However, aerobic training may be difficult for many of the patients, due to the existence of co-morbidities or simply because the majority of the patients are overweight if not severely obese. For these patients, resistance exercise may represent an attractive alternative (for a recent review see (Yaspelkis, 2006)).

The incidence of type 2 diabetes and other lifestyle related metabolic diseases increases with advancing age. A reduction in insulin sensitivity that plays a major role in the development of the metabolic syndrome and diabetes is often seen with aging, in addition to a reduction in the skeletal muscle mass that plays a major role for decrements in functional performance. For both the metabolic and strength impairment with aging it is not clear to what extent this can solely be explained by the age-associated reduction in activity level. Aging *per se*

has an influence upon skeletal muscle loss, but the metabolic impairment and functional losses can be largely counteracted by physical training, especially of a resistive nature. Resistance exercise results in a net increase in muscle protein synthesis and thus a rise in muscle fiber cross-sectional area.

Resistance Training and Insulin Sensitivity

Insulin resistance is not a natural development that inevitably follows aging. With "healthy" aging, (i.e., maintenance of a physically active lifestyle and avoidance of obesity), insulin mediated glucose uptake in skeletal muscle does not necessarily decrease with age *per se*. When insulin resistance is so often associated with aging, it is most of all due to a marked decrease in daily physical activity, decreased muscle mass, increased intramyocellular lipid deposits and accumulation of adipose tissue. For obvious reasons no longitudinal, randomized studies of this phenomenon have been carried out, but studies of glucose tolerance and skeletal muscle biochemistry in master athletes (Coggan et al., 1992; Rogers et al., 1990; Heath et al., 1981) as well as measurements of muscle glucose uptake rates in elderly and young healthy subjects (Dela et al., 1996) support this notion.

The number of studies on the effects of resistance training on glycemic control and insulin sensitivity in middle aged to elderly healthy subjects and patients with insulin resistance has increased substantially over the past 15 years (see table 12.1). In general, the interpretation and individual comparison of intervention studies using resistance training are to some extent difficult, because there is no uniform agreement in how the exercise intervention is described in terms of intensity and/or number of repetitions. Some studies merely report workload as RM (repetition maximum), while others report the number of repetitions at a given percentage of max workload. Furthermore, some studies describe in detail the use of progressive resistance training, while others do not give this information. Therefore, a clear dose-response relationship between the training intensity and the effect on strength is difficult to obtain from these studies. Furthermore, a clear dose-response relationship regarding insulin sensitivity cannot be deduced from the published studies, but from a biological point of view it is difficult to imagine that such a relationship should not exist – at least up to a yet undefined maximum. In those studies where insulin sensitivity has been measured, the majority has shown a significant positive effect (table 12.1). With some uncertainty the percentage of maximal workload can be calculated from given data on RM, and by doing so it appears that resistance training at an intensity range of 45-80% of maximal load results in ≈ 30% [range 10-75%] increase in strength and ≈ 28% [range 15-48%] increase in insulin sensitivity. The duration of the intervention is in most studies more than three months, but as little as six weeks of training have shown marked improvements in insulin sensitivity (Holten et al., 2004).

In the latter study (Holten et al., 2004), patients with type 2 diabetes and matched healthy controls performed one-leg resistance training three times per week, and only three different exercises were used. Sixteen hours after the last training session, insulin mediated glucose uptake rates were measured in both the trained and the untrained leg by isoglycemic, hyperinsulinemic clamp combined with femoral arterio-venous catheterization technique. Even with a moderate strength training program, in terms of both duration and number of exercise sessions, insulin mediated glucose uptake is significantly increased in insulin resistant skeletal muscle (figure 12.2).

As with pharmacological treatment of insulin resistance, the effect of training is lost when the treatment is stopped. This effect has been demonstrated in young, healthy subjects subjected to

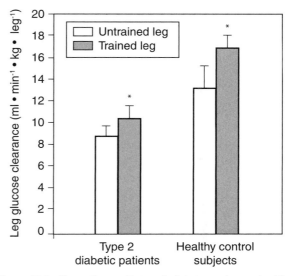

Figure 12.2 Ten patients with type 2 diabetes and seven healthy control subjects strength trained one leg for six weeks. Sixteen hours after the last training session glucose uptake in the leg was measured. Here presented as glucose clearance rate per kg leg at hyperinsulinemia (approximately 12000 pM). Skeletal muscles from patients with type 2 diabetes were always insulin resistant compared with the control, and in both groups trained muscles were more insulin sensitive than untrained muscle (* P<0.05).

Data taken from M.K. Holten et al., 2004, "Strength training increases insulin-mediated glucose uptake, GLUT4 content, and insulin signaling in skeletal muscle in patients with type 2 diabetes," *Diabetes* 53: 294-305.

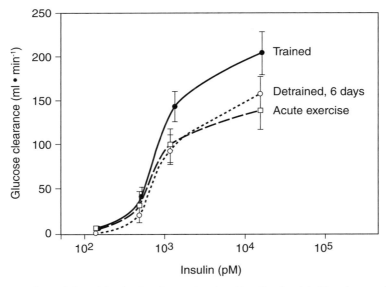

Figure 12.3 Effect of acute exercise and de-training for five days on basal and insulin-stimulated leg glucose clearance in the legs of seven patients with type 2 diabetes. The acute exercise (30 min ergometer one-legged bicycle exercise at 70 % $\dot{V}O_{2max}$) was performed with the previously non-trained leg 16 hours before the measurement. The detrained leg was studied six days after the last training session. The subjects had been training one leg for ten weeks, six days/wk at the above mentioned exercise intensity.

Data taken from F. Dela et al., 1995, "Insulin stimulated muscle glucose clearance in patients with NIDDM," *Diabetes* 44: 1010-1020.

Table 12.1 Effects of Supervised Resistance Training on Measures of Glycemic Control and Insulin Sensitivity

Year	Intervention/control (n) [a]	Health [b]	BMI [c]	Age (y) [d]	# weekly sessions [e]	# weeks [f]	# exerercises [g]
Trials with a sedentary control group							
1988 (Hurley et al. 1988)	10 M, 11 F	Healthy	OW	44 / 52	3-4	16	14
1993 (Smutok et al. 1993)	12 M, 10 F	CHD risk	OW	51 / 50	3	20	11
1997 (Honkola et al. 1997)	18 M, 20 F	T2DM	OW/OB	62 / 67	2	20	8-10
1998 (Dunstan et al. 1998)	8 M, 3 F; 5 M, 5 F	T2DM	OW/OB	50 / 51	3	8	10
1998 (Ishii et al. 1998)	9 M, 8 F	T2DM	NW	47 /52	5	4-6	9
2002 (Maiorana et al. 2000) [p]	14 M, 2 F (crossover)	T2DM	OW	52	3	8	7
2002 (Castaneda et al. 2002)	10 M, 21 F; 12 M, 19 F	T2DM	OB	66 / 66	3	16	5
2003 (Cuff et al. 2003) [p]	9 M, 10 F	T2DM	FAT / OB	60 / 63	3	16	5
2004 (Holten et al. 2004) [q]	17 M	T2DM + healthy	OW NW	62	3	6	3
2005 (Thomas et al. 2005)	60 M, 60 F	Healthy + T2DM, HT	NW	69 / 69	3	52	7
Trials without a sedentary control group							
1989 (Craig et al. 1989)	9 M	Healthy	NW	63	3	12	8
1994 (Miller et al. 1994)	11 M	Healthy	OW	58	3	16	14
1996 (Ryan et al. 1996) [s]	13 F	Healthy	NW and OW	58	3	16	14
1996 (Zachwieja et al. 1996)	9 M	Healthy	NW	60-75	4	16	9
1998 (Eriksson et al. 1998)	8 M	T2DM	NA	40	3	10	8
2001 (Ryan et al. 2001)	10 M, 8 F	Healthy + IGT	OW	69	3	24	11
2001 (Joseph et al. 2001) [s]	10 F	Healthy	OW/OB	65	3	4	5
2002 (Dunstan et al. 2002) [s]	16 M, 16 F	T2DM, HT	OW	68	3	24	9
2004 (Dionne et al. 2004)	12 F	Healthy	NW/OW	67	3	24	9
2004 (Fenicchia et al. 2004)	7 F, 8 F	T2DM, healthy	SO OW	50	3	6	9
2004 (Reynolds et al. 2004)	5 M, 6 F	HT	OW	67	3	16	8
2004 (Tokmakidis et al. 2004) [s]	9 F	T2DM	OB	55	2	16	6
2005 (Ibanez et al. 2005)	9 M	T2DM	OW	67	2	16	8
2005 (Cauza et al. 2005)	11 M, 11 F	T2DM	OB	56	3	16	10
2006 (Colberg et al. 2006)	6 M, 4 F; 5 M, 4 F	T2DM, healthy	OB OB	55 51	3	8	8

NA, data not available; ↔, no change; ↑ or ↓, significant increase or decrease.

[a] Number of subjects participating in the intervention and the control group. Gender is indicated by M for males and F for females.

[b] T2DM = type 2 diabetes; CHD = coronary heart disease; HT = hypertension; IGT = impaired glucose tolerance.

[c] NW = normal weight (BMI = 18.5-25 kg/m²); OW = overweight (BMI = 25-30 kg/m²); OB = obese (BMI = 30-35 kg/m²); SO = severely obese (BMI > 35 kg/m²).

[d] Values are mean or range.

[e] Number of training sessions per week.

[f] Number training weeks.

[g] Number of different exercises per session.

[h] Number of sets per training session

[i] Some studies report the training intensity as repetition max (RM), and others as number of repetitions (reps) at a certain percentage of 1RM. In parentheses ( ) are given the authors' rough recalculation of RM to % of 1RM.

[j] As reported in the paper or calculated from reported data.

# sets[h]	Intensity (RM or # of reps at % of 1RM)[i]	Effect on strength[j]	f- glucose[k]	f- insulin[k]	HbA1c[l]	Insulin sensitivity[m]	Body composition[n]
Trials with a sedentary control group							
1	8RM-20RM (52%-79%)	+33%-50%	↔	↓	NA	NA	Fat % ↔; BW ↔
2	12RM-15RM (62%-69%)	+36%-50%	↔°	↓°	NA	NA°	BW ↔; Fat % ↓
2	12RM-15RM (62%-69%)	+38% (excl. arms)	↔	↓	↓	NA	BW ↔; Fat % ↔
3	10-15 reps at 50%-55% of 1RM	+15%-43%	↔°	↔°	↔	NA°	BW ↔; Fat % ↔
2	10-12 reps at 40%-50% of 1RM	+16%	NA	NA	↔	+48 %	Fat % ↔; BW ↔
1-3	15 reps at 55%-65% of 1RM	+13%	↓	NA	↓	NA	BW ↔; Fat % ↓
3	8 reps at 70% of 1RM	+33%	↔	NA	↓	NA	BW ↔; FFM ↑
2	NA	+42%-49%	NA	NA	↔	+77%	BW ↓; Fat % ↓
4	20RM → 8RM (52%-79%)	+42%-75%	↔	↔	↔	+30%	—
NA	30 reps with an exercise band		↔	↔	↔	↑[r]	BW ↔; Fat % ↔
Trials without a sedentary control group							
3	8RM (79%)	+32%	↔	↓	NA	NA	BW ↑; Fat % ↓; FFM ↑
1-2	15RM (62%)	+50%	↔	↓	NA	+22%	BW ↔; Fat % ↓; FFM ↑
	15 reps at 90% of 3RM	+53%	↔	↔	NA	↔/↑[t]	BW ↔/↓; Fat % ↔/↓; FFM ↔/↓
4	4-10 reps at 75%-90% of 1RM	+14%-18%	↔	↔	NA	+24%[u]	BW ↔; Fat % ↓; FFM ↑
3	8-12 reps at 50%-60% of 1RM	NA	↔	↔	↔	+23%	BW ↔; Fat % ↔
1-2	8RM-10RM (74%-79%)	+26%	↔	↔	NA	↔	BW ↔; Fat % ↔; FFM ↑
3	8-12 reps at 80% of 1RM	+23%	NA	NA	NA	↔	BW ↓; Fat % ↓; FFM ↔
3	8-10 reps at 75%-85 % of 1RM	+28%-42%	↔	↔	↓	↔[v]	BW ↓; FFM ↑
3	10 reps at 80% of 1RM	+30%	NA	NA	NA	↔	BW ↔; Fat % ↓; FFM ↑
3	8-12 reps at 80% of 3RM	+19%-57%	↔	↔	NA	NA	BW ↔; Fat % ↔; FFM ↔
2	10RM-12RM (69%-74%)	+10%-24%	↓	↔	NA	+15%	BW ↔; Fat % ↔; FFM ↑
3	12 reps at 60% of 1RM	+31%-40%	↓	↓	↓	NA	BW ↔
3-4 → 5-6	10-15 reps at 50%-70% → 70%-80% of 1RM	+18%	↓	↔	NA	+40% [w]	BW ↔; Fat % ↓
3-6	10RM-15RM (62%-74%)	+22%-48%	↓	↔°	↓	+21%[v]	BW ↔; Fat % ↓; FFM ↑
3	8-12 reps at 50%, 60%, and 70% of 1RM	+19%-38%	↔	↔	↔	↔[v]	Fat % ↓/↔

[k] f denotes fasting value.
[l] Glycosylated hemoglobin.
[m] Insulin sensitivity measured by insulin clamp technique unless otherwise noted.
[n] Effect on body composition. BW = body weight; fat % = percentage of fat tissue; FFM = fat-free mass.
[o] Compared with the control group, an improvement in glycemic or insulin response to an oral glucose tolerance test was seen.
[p] Combined with aerobic exercise.
[q] Single-leg training; comparison between trained and untrained leg.
[r] Improvement as measured by the insulin tolerance test. No percentage given.
[s] Combined with diet.
[t] Only effect in the weight loss group.
[u] Measured by minimal model.
[v] Measured by HOMA index.
[w] Measured by intravenous glucose tolerance test.

a heavy resistance training program (Andersen et al., 2003), and in patients with type 2 diabetes after a one-legged endurance training program (figure 12.3). Thus, from the latter study, it is evident that the effect of training is quickly lost when training stops, and the insulin mediated glucose clearance becomes similar to values obtained with acute exercise (which, in turn, was only marginally and insignificantly increased compared with values obtained in an untrained leg (data not shown)).

Mechanisms Behind Resistance Training-Induced Improvements in Insulin Sensitivity

In response to aerobic exercise training, an increase in capillary density is a common finding in both healthy and insulin resistant human muscle. This increase facilitates diffusion of glucose from the capillaries into the muscle cells, and is therefore also a part of the mechanism behind the increase in insulin action seen after aerobic training. Capillary density is normally not increased by resistance training, so the mechanisms behind the enhancement of insulin action with the two different training regimens are not completely identical, but most likely the mechanisms differ only in minor details. Further studies may clarify this issue.

In contrast, resistance training induces a range of changes facilitating muscle hypertrophy and thus by mass increases, whole body insulin action. However, the mechanism by which resistance training improves insulin action on glucose metabolism is not merely a function of increased muscle mass (Holten et al., 2004; Andersen et al., 2003; Ferrara et al., 2006), but also a function of increased muscle quality. Thus, several proteins important for insulin signaling and/or glucose handling in muscle increase in abundance and/or activity. In contrast to endurance-like exercises, the available literature on effects of resistance exercise training on insulin signaling is sparse. However, those few studies seem to indicate that only minor differences may exist between effects of strength and endurance training (see chapter 11, see figure 12.4).

Strength Training Increases Effecter Proteins in Glucose Metabolism

A robust finding in both human and rodent muscle in response to resistance (and endurance) exercise

training is an increase in the abundance of the glucose transporter GLUT4 isoform mRNA and protein. This likely is of great importance for the glucose transport capacity of the sarcolemma during insulin stimulation as, at least in rodents, a close association between these two parameters has been reported (Kern et al., 1990). Once taken up by the cells glucose is rapidly phosphorylated and stored as glycogen, a process regulated by glycogen synthase, or oxidized, a process regulated by pyruvate dehydrogenase. The robust increase in glycogen synthase protein expression as well as activity in response to endurance exercise training is also found in response to strength training (Holten et al., 2004) likely facilitating glycogen super-compensation after exercise. Whether strength (like endurance) training increases hexokinase expression/activity is less clear, although it has been reported (Holten et al., 2004; Thibault et al., 1986). Thus, in general strength exercise training induces effects on protein expression/activity in glucose metabolism. These adaptations are likely important players in the associated enhanced muscle insulin action to stimulate glucose uptake.

Effects of Strength Training on Insulin Signaling

Whether these changes are sufficient in mediating the beneficial effects of training on insulin action is questionable. It could be argued that an improvement of the insulin signaling cascade is also necessary for this to occur. Again, this has been investigated in only a few studies. In the study by Holten et al. (2004) (figure 12.2) the expression level (but not the activation by insulin) of several key components in the classical insulin signal cascade through phosphatidylinositol 3 kinase (PI3K) was investigated in healthy subjects and subjects with type 2 diabetes undertaking one-legged resistance exercise training. It was reported that insulin receptor and Akt (1/2) proteins were upregulated, whereas the content of insulin receptor substrate 1 (IRS-1) and the regulatory subunit of PI3K (p85) was unchanged. Unfortunately, it was not possible from that study to evaluate whether these changes had any significant influence on the improved insulin stimulated muscle glucose uptake after training. This was, however, evaluated in high-fat fed rats subjected to resistance exercise training (Krisan et al., 2004). Rather than changes in expression

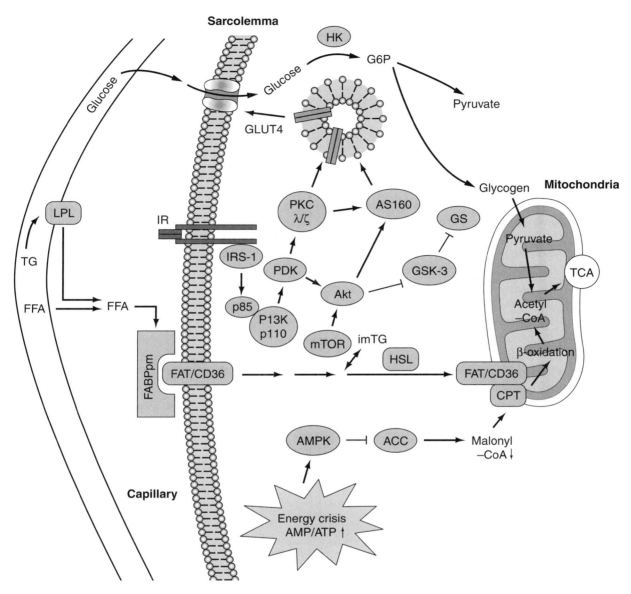

Figure 12.4   Key players in strength and/or endurance training induced improvements in fat and carbohydrate metabolism in skeletal muscle. The depicted signals and pathways are not representing the complete picture. Rather the most important proteins and enzymes in relation to physical training are shown. See text for details. TG, triglyceride; imTG, intramuscular TG; FFA, free fatty acid; G6P, glucose 6 phosphate; LPL, lipoprotein lipase; HSL, hormone sensitive lipase; HK, hexokinase; GS, glycogen synthase; FABP, fatty acid binding protein; FAT/CD36, fatty acid transport CD36; CPT, carnitine palmitoyltransferase; ACC, acetyl-CoA carboxylase; AMPK, 5'AMP activated protein kinase; AS160, Akt substrate of 160 kDa; GSK3, glycogen synthase kinase 3; PKCλ/ζ, atypical protein kinase C isoform λ/ζ; mTOR, mammalian target of rapamycin; PDK, phosphatidylinositol dependent kinase; IR, insulin receptor; IRS1, IR substrate 1.

levels of insulin signal components, it was found that the response to insulin in perfused hind-limb muscles of normal rats was elevated by training at the level of PI3K, Akt and aPKC activities and that the decreased signaling through these elements in the high-fat fed rat was fully or partly reversed by the resistance exercise training. These adaptations occurred mainly in muscle represented by red fibers and occurred without measurable muscle hypertrophy. The reader is referred to chapter 11 for more detailed analyses of the responses in the insulin signaling cascade to endurance type of exercise training, but in general similar findings have been observed with resistance exercise. Still, the responses likely depend on the species, mode and duration of the training as well as the pre-training levels, likely explaining the large inconsistency in responses among different studies.

Strength Training Increases Effecter Proteins in Lipid Metabolism

In addition to the above mentioned training effects on carbohydrate metabolism in skeletal muscle, it is well-known that physical training (in particular endurance training) increases the β-oxidative capacity (e.g., by enhancing the activity of hydroxy-acyl-CoA-dehydrogenase (HAD)). More recently, it has been found that the maximal activity of carnitine palmitoyltransferase I (CPT I), which is the rate-limiting step for mitochondrial oxidation of long-chain fatty acids, is increased with training (Bruce et al., 2006). In addition, it was found that training diminished the sensitivity of CPT I to inhibiton by malonyl-CoA (Bruce et al., 2006). The fatty acid translocase, FAT/CD36, has now been found to be present in mitochondria (Bezaire et al., 2006), and furthermore, it co-immunoprecipitates with CPT I and this association increases with training (Schenk and Horowitz, 2006). Further up-stream of this pathway (figure 12.4), the contents of FAT/CD36 and the fatty acid binding protein in the plasma membrane (FABPpm) have been shown to increase with training (Kiens et al., 1997; Kiens et al., 2004; Tunstall et al., 2002). All of these effects on proteins involved in the transport of fatty acids to the mitochondria form the basis for the increased capacity for fatty acid oxidation. One might also expect that the activity of the hormone sensitive lipase (HSL) is increased with training, but so far this has not been found (Helge et al., 2006). However, lipoprotein lipase (LPL) activity increases with training (Kiens et al., 2004; Kiens and Lithell, 1989; Nikkila et al., 1978) – additionally contributing to improving the overall capacity for trained muscle to metabolize fatty acids at the expense of carbohydrates.

AMPK—A Potential Regulator of Exercise Training Induced Protein Expression

AMP-activated protein kinase (AMPK) plays a key role in regulating fuel combustion in skeletal muscle (Hardie and Sakamoto, 2006). Thus, in resting muscle pharmacological AMPK activation increases both glucose uptake and lipid oxidation (Jorgensen et al., 2006). The latter via inactivation

of ACC causing a drop in malonyl-CoA content. However, as to whether the activation of AMPK seen during both endurance (Wojtaszewski et al., 2000; Fujii et al., 2000) and strength (Dreyer et al., 2006; Koopman et al., 2006) exercise is a key regulator of these acute metabolic changes is questionable (Hardie and Sakamoto, 2006). In accordance, the physiological significance of the changes in AMPK subunit expression observed both in response to endurance (Frosig et al., 2004) and strength (Wojtaszewski et al., 2005) training and the subsequent lower activation of AMPK signaling during an acute bout of exercise after training is still unclear (McConell et al., 2005; Nielsen et al., 2003). Repeated AMPK activation by pharmacological means or by genetic modification is sufficient to drive an enhanced expression of a range of both cytosolic, membrane bound as well as mitochondrial enzymes in muscle (Jorgensen et al., 2005; Zheng et al., 2001; Holmes et al., 1999; Winder et al., 2000; Bergeron et al., 2001; Zong et al., 2002; Nilsson et al., 2006). Thus, the repeated AMPK activation during exercise training may be involved in upregulation of mRNA and/or protein levels. However, genetic interference with the AMPK system has not given a clear picture as to whether AMPK is involved in this response to exercise as it has been reported in some (Barnes et al., 2005) but not other models (Jorgensen et al., 2005; Holmes et al., 2004).

AMPK activation may induce protein expression and mitochondrial biogenesis via upregulation of the transcriptional co-activator peroxisome proliferator activated receptor γ coactivator 1α (PGC-1α), and thus likely via activation expression of the transcription factors nuclear respiratory factor (NRF)1/2, mitochondrial transcription factor A (mtTFA) and peroxisome proliferator activated receptor α (PPARα). This is evident from the observation that the PGC-1α mRNA and protein is upregulated upon repeated pharmacological activation of AMPK in resting muscle (Zong et al., 2002; Jorgensen et al., 2005)—an effect dismissed in the absences of AMPK. Still, evidence from transgenic animal models does not support the view that AMPK is necessary for linking exercise training, PGC-1α and gene/protein expression in muscle (Jorgensen et al., 2005; Holmes et al., 2004). Some of these important roles of PGC-1α are discussed below.

Training-Induced Gene Expression

Training-induced changes in the protein content of a given protein may be the result of changes at any of the steps involved in synthesis and/or degradation of the protein. But regulation of both translation and transcription may be involved with the relative importance of these likely depending on the type of exercise and the gene/protein. Adaptive cellular responses to exercise training are thought at least in part to involve the cumulative effects of transient increases in expression of genes encoding proteins in metabolism leading to an increased oxidative capacity and a more efficient metabolism of skeletal muscle (Williams and Neufer, 1996; Pilegaard et al., 2003). These changes are thought to contribute to the positive health effects of regular exercise.

An important role of PGC-1α in regulating exercise-induced adaptive responses in human skeletal muscle has been suggested (Pilegaard et al., 2003). PGC-1α is a master regulator of mitochondrial biogenesis (Lin et al., 2002; Lin et al., 2005) and has the potential to exert a coordinating role in adaptive responses through regulation of a broad range of transcription factors (Lin et al., 2005). As an acute endurance type of exercise induces an upregulation of PGC-1α transcription and mRNA in human (Pilegaard et al., 2003) and mRNA and protein in rat skeletal muscle (Goto et al., 2000; Baar et al., 2002), it is possible that exercise-induced transient inductions of PGC-1α to each exercise session serve a central role in regulating other metabolic genes in response to exercise.

The drastic effects of overexpression of PGC-1α in mice with marked increases in several oxidative metabolism proteins and white muscles turning red (Lin et al., 2002) demonstrate the impact of PGC-1α in regulating the expression of metabolic genes and suggest that PGC-1α dysregulation could be important in metabolically related diseases like type 2 diabetes. In support of lowered PGC-1α expression potentially playing a role in insulin resistance and type 2 diabetes are the findings of reduced PGC-1α mRNA content in both patients with type 2 diabetes and family history-positive nondiabetics insulin resistance subjects (Patti et al. 2003). Moreover, type 2 diabetes was found to be associated with lowered mRNA content of the transcription factor NRF-1 (Patti et al., 2003), known to be regulated by PGC-1α (Lin et al., 2005), and many NRF-1 regulated genes encoding proteins in oxidative metabolism (Patti et al., 2003).

Additionally, specific variations in the sequence of the PGC-1α gene have been shown to be associated with type 2 diabetes. Thus, a frequent single nucleotide polymorphism leading to a change in an amino-acid in the PGC-1α protein was found to be associated with increased risk of type 2 diabetes (Ek et al., 2001) and also PGC-1α promoter polymorphisms have been shown to be coupled with early onset type 2 diabetes (Kim et al., 2005). It may be suggested that such polymorphisms can lead to reduced PGC-1α expression and influence the function of the PGC-1α protein. Taken together it may therefore be speculated that transient exercise-induced up-regulations of PGC-1α expression can contribute to the positive health effects of regular physical exercise.

As mentioned above, the beneficial effects of resistance exercise training may include both increased muscle mass and the concomitant increased muscle strength as well as improved insulin sensitivity of skeletal muscles. The improved insulin sensitivity with resistance exercise and the apparent link between PGC-1α and type 2 diabetes may suggest that resistance exercise also regulates the PGC-1α gene as endurance exercise has been shown to do (Goto et al., 2000; Baar et al., 2002; Pilegaard et al., 2003).

In support of this possibility is our recent finding that a single heavy resistance exercise bout increases the PGC-1α mRNA content 2.5 fold 6 h after the end of the exercise and having returned to pre-levels 24 h into recovery (figure 12.5) (Pilegaard, Speerschneider and Saltin, unpublished results). However, resistance exercise training 3 days per week for 6 weeks did not affect the PGC-1α mRNA content neither in type 2 diabetes patients with apparent reduced PGC-1α mRNA levels nor in healthy controls 16 hours after the last exercise bout (Pilegaard and Dela, unpublished results). But this does not rule out that transient resistance-exercise induced increases in PGC-1α protein may regulate metabolic genes and thus influence insulin sensitivity without the need for an accumulation in PGC-1α mRNA or protein similar to what has been shown with endurance exercise training 5 times per week (Pilegaard et al., 2003).

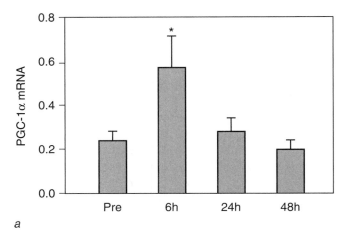
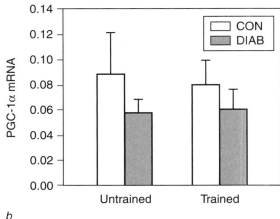

a

b

Figure 12.5 *(a)* PGC-1α mRNA content in vastus lateralis of healthy young male subjects in response to a single heavy resistance exercise bout of the knee extensor muscles. Muscle biopsies were obtained before exercise and after 6, 24 and 48 h of recovery. Values are mean ± SE, n=8 *: significantly different from before exercise, P<0.05. *(b)* PGC-1α mRNA content in vastus lateralis of patients with type 2 diabetes (DIAB) and aged-matched controls (CON) before and after 6 weeks of resistance exercise training. Values are mean ± SE, n=6 for controls and n=9 for patients with type 2 diabetes.

Data from Pilegaard and Dela, unpublished results.

Concluding Remarks

The decreased muscle metabolic capacity and reduction of muscle mass, which is seen in inactivity-related diseases, can be counteracted by physical activity, including resistance exercise training. Thus, resistance exercise training represents an alternative to aerobic / endurance exercise training and it is advantageous to the latter when focus is on increasing muscle mass. Resistance training also improves insulin sensitivity beyond the effect of just adding muscle mass. Although many aspects of the mechanisms behind both endurance- and strength training induced improvements in glucose homeostasis are still unresolved, there is ample evidence to recommend both training modalities for the prevention and treatment of insulin resistance. The major challenge in the future is to translate the biological evidence for effect into sustainable changes of behavior in a hitherto sedentary population, characterized by increasing prevalence of obesity and inactivity-related disorders.

The evidence that aerobic exercise training improves insulin mediated glucose uptake and reduces insulin resistance in patients with type 2 diabetes is substantial (see chapters 2, 4 and 12), and need not a repetition here. The fact that a sedentary lifestyle—for both healthy and diseased people—is detrimental to health is also a universally accepted standpoint. The current major challenge is how these facts can be translated into the community, e.g., how we can convince patients with type 2 diabetes and even the public in general to take up exercise and use it as their treatment of choice. Preferably exercise should be incorporated as a part of the everyday life and clearly this involves much more than just putting forward the biological scientific evidence for the effect. Even though much can still be achieved on the individual level, Public Health approaches consisting of not only national/regional guidelines but also involving city- and landscape planning, architecture, and the labour market are called for, if the burden of the inactivity-related diseases should be reduced.

References

Andersen, J.L., P. Schjerling, L.L. Andersen, and F. Dela. 2003. Resistance training and insulin action in humans: effects of de-training. *J Physiol* 551: 1049-1058.

Baar, K., A.R. Wende, T.E. Jones, M. Marison, L.A. Nolte, M. Chen, D.P. Kelly, and J.O. Holloszy. 2002. Adaptations of skeletal muscle to exercise: rapid increase in the transcriptional coactivator PGC-1. *FASEB J* 16: 1879-1886.

Barnes, B.R., Y.C. Long, T.L. Steiler, Y. Leng, D. Galuska, J.F. Wojtaszewski, L. Andersson, and J.R. Zierath. 2005. Changes in exercise-induced gene expression in 5'-AMP-activated protein kinase gamma3-null and gamma3 R225Q transgenic mice. *Diabetes* 54: 3484-3489.

Bergeron, R., J.M. Ren, K.S. Cadman, I.K. Moore, P. Perret, M. Pypaert, L.H. Young, C.F. Semenkovich, and G.I. Shulman. 2001. Chronic activation of AMP kinase results in NRF-1 activation and mitochondrial biogenesis. *Am J Physiol Endocrinol Metab* 281: E1340-E1346.

Bezaire, V., C.R. Bruce, G.J. Heigenhauser, N.N. Tandon, J.F. Glatz, J.J. Luiken, A. Bonen, and L.L. Spriet. 2006. Identification of fatty acid translocase on human skeletal muscle mitochondrial membranes: essential role in fatty acid oxidation. *Am J Physiol Endocrinol Metab* 290: E509-E515.

Bruce, C.R., A.B. Thrush, V.A. Mertz, V. Bezaire, A. Chabowski, G.J. Heigenhauser, and D.J. Dyck 2006. Endurance training in obese humans improves glucose tolerance and mitochondrial fatty acid oxidation and alters muscle lipid content. *Am J Physiol Endocrinol Metab* 291:E99-E107.

Castaneda, C., J.E. Layne, L. Munoz-Orians, P.L. Gordon, J. Walsmith, M. Foldvari, R. Roubenoff, K.L. Tucker, and M.E. Nelson. 2002. A randomized controlled trial of resistance exercise training to improve glycemic control in older adults with type 2 diabetes. *Diabetes Care* 25: 2335-2341.

Cauza, E., U. Hanusch-Enserer, B. Strasser, K. Kostner, A. Dunky, and P. Haber. 2005. Strength and endurance training lead to different post exercise glucose profiles in diabetic participants using a continuous subcutaneous glucose monitoring system. *Eur J Clin Invest* 35: 745-751.

Coggan, A.R., R.J. Spina, D.S. King, M.A. Rogers, M. Brown, P.M. Nemeth, and J.O. Holloszy. 1992. Histochemical and enzymatic comparison of the gastrocnemius muscle of young and elderly men and women. *J Gerontol* 47: B71-B76.

Colberg, S.R., H.K. Parson, T. Nunnold, M.T. Herriott, and A.I. Vinik. 2006. Effect of an 8-week resistance training program on cutaneous perfusion in type 2 diabetes. *Microvasc Res* 71: 121-127.

Craig, B.W., J. Everhart, and R. Brown. 1989. The influence of high-resistance training on glucose tolerance in young and elderly subjects. *Mech Ageing Dev* 49: 147-157.

Cuff, D.J., G.S. Meneilly, A. Martin, A. Ignaszewski, H.D .Tildesley, and J.J. Frohlich. 2003. Effective exercise modality to reduce insulin resistance in women with type 2 diabetes. *Diabetes Care* 26: 2977-2982.

Dela, F., J.J. Larsen, K.J. Mikines, T. Ploug, L.N. Petersen, and H. Galbo. 1995. Insulin stimulated muscle glucose clearance in patients with NIDDM. *Diabetes* 44: 1010-1020.

Dela, F., K.J. Mikines, J.J. Larsen, and H. Galbo. 1996. Training-induced enhancement of insulin action in human skeletal muscle: the influence of aging. *J Gerontol A Biol Sci Med Sci* 51: B247-B252.

Dionne, I.J., M.O. Melancon, M. Brochu, P.A. Ades, and E.T. Poelhman. 2004. Age-related differences in metabolic adaptations following resistance training in women. *Exp Gerontol* 39: 133-138.

Dreyer, H.C., S. Fujita, J.G. Cadenas, D.L. Chinkes, E. Volpi, and B.B. Rasmussen. 2006. Resistance exercise increases AMPK activity and reduces 4E-BP1 phosphorylation and protein synthesis in human skeletal muscle. *J Physiol*.

Dunstan, D.W., R.M. Daly, N. Owen, D. Jolley, M. De Courten, J. Shaw, and P. Zimmet. 2002. High-intensity resistance training improves glycemic control in older patients with type 2 diabetes. *Diabetes Care* 25: 1729-1736.

Dunstan, D.W., R.M. Daly, N. Owen, D. Jolley, E. Vulikh, .J Shaw, and P. Zimmet. 2005. Home-based resistance training is not sufficient to maintain improved glycemic control following supervised training in older individuals with type 2 diabetes. *Diabetes Care* 28: 3-9.

Dunstan, D.W., I.B. Puddey, L.J. Beilin, V. Burke, A.R. Morton, and K.G. Stanton. 1998. Effects of a short-term circuit weight training program on glycaemic control in NIDDM. *Diabetes Res Clin Pract* 40: 53-61.

Ek, J., G. Andersen, S.A. Urhammer, P.H. Gaede, T. Drivsholm, K. Borch-Johnsen, T. Hansen, and O. Pedersen. 2001. Mutation analysis of peroxisome proliferator-activated receptor-gamma coactivator-1 (PGC-1) and relationships of identified amino acid polymorphisms to Type II diabetes mellitus. *Diabetologia* 44(12): 2220-2226.

Ek, J., G. Andersen, S.A. Urhammer, P.H. Gaede, T. Drivsholm, K. Borch-Johnsen, T. Hansen, and O. Pedersen. 2001. Related Articles, Links Mutation analysis of peroxisome proliferator-activated receptor-gamma coactivator-1 (PGC-1) and relationships of identified amino acid polymorphisms to Type II diabetes mellitus. *Diabetologia* 44(12):2220-6.

Eriksson, J., J. Tuominen, T. Valle, S. Sundberg, A. Sovijarvi, H. Lindholm, J. Tuomilehto, and V. Koivisto. 1998. Aerobic endurance exercise or circuit-type resistance training for individuals with impaired glucose tolerance? *Horm Metab Res* 30: 37-41.

Fenicchia, L.M., J.A. Kanaley, J.L. Azevedo Jr., C.S. Miller, R.S. Weinstock, R.L. Carhart, and L.L. Ploutz-Snyder. 2004. Influence of resistance exercise training on glucose control in women with type 2 diabetes. *Metabolism* 53: 284-289.

Ferrara, C.M., A.P. Goldberg, H.K. Ortmeyer, and A.S. Ryan. 2006. Effects of aerobic and resistive exercise training on glucose disposal and skeletal muscle metabolism in older men. *J Gerontol A Biol Sci Med Sci* 61: 480-487.

Fritz, T. and U. Rosenqvist. 2001. Walking for exercise - immediate effect on blood glucose levels in type 2 diabetes. *Scand J Prim Health Care* 19: 31-33.

Frosig, C., S.B. Jorgensen, D.G. Hardie, E.A. Richter, and J.F. Wojtaszewski. 2004. 5'-AMP-activated protein kinase activity and protein expression are regulated by endurance training in human skeletal muscle. *Am J Physiol Endocrinol Metab* 286: E411-E417.

Fujii, N., T. Hayashi, M.F. Hirshman, T.J. Smith, S.A. Habinowski, L. Kaijser, J. Mu, O. Ljungqvist, M.J. Birnbaum, L.A. Witters, A. Thorell, and L.J. Goodyear. 2000. Exercise induces isoform specific increase in 5´AMP-activated protein kinase activity in human skeletal muscle. *Biochem Biophys Res Commun* 273: 1150-1155.

Goto, M., S. Terada, M. Kato, M. Katoh, T. Yokozeki, I. Tabata, and T. Shimokawa. 2000. cDNA Cloning and mRNA analysis of PGC-1 in epitrochlearis muscle in swimming-exercised rats. *Biochem Biophys Res Commun* 274: 350-354.

Hardie, D.G. and K. Sakamoto. 2006. AMPK: a key sensor of fuel and energy status in skeletal muscle. *Physiology (Bethesda)* 21: 48-60.

Heath, G.W., J.M. Hagberg, A.A. Ehsani, and J.O. Holloszy. 1981. A physiological comparison of young and older endurance athletes. *J Appl Physiol* 51: 634-640.

Helge, J.W., T.O. Biba, H. Galbo, M. Gaster, and M. Donsmark. 2006. Muscle triacylglycerol and hormone-sensitive lipase activity in untrained and trained human muscles. *Eur J Appl Physiol* 97: 566-572.

Holmes, B.F., E.J. Kurth-Kraczek, and W.W. Winder. 1999. Chronic activation of 5'-AMP-activated protein kinase increases GLUT-4, hexokinase, and glycogen in muscle. *J Appl Physiol* 87: 1990-1995.

Holmes, B.F., D.B. Lang, M.J. Birnbaum, J. Mu, and G.L. Dohm. 2004. AMP kinase is not required for the GLUT4 response to exercise and denervation in skeletal muscle. *Am J Physiol Endocrinol Metab* 287: E739-E743.

Holten, M.K., M. Zacho, M. Gaster, C. Juel, J.F. Wojtaszewski, and F. Dela. 2004. Strength training increases insulin-mediated glucose uptake, GLUT4 content, and insulin signaling in skeletal muscle in patients with type 2 diabetes. *Diabetes* 53: 294-305.

Honkola, A., T. Forsen, and J. Eriksson. 1997. Resistance training improves the metabolic profile in individuals with type 2 diabetes. *Acta Diabetol* 34: 245-248.

Hurley, B.F., J.M. Hagberg, A.P. Goldberg, D.R. Seals, A.A. Ehsani, R.E. Brennan, and J.O. Holloszy. 1988. Resistive training can reduce coronary risk factors without altering VO2max or percent body fat. *Med Sci Sports Exerc* 20: 150-154.

Ibanez, J., M. Izquierdo, I. Arguelles, L. Forga, J.L. Larrion, M. Garcia-Unciti, F. Idoate, and E.M. Gorostiaga. 2005. Twice-weekly progressive resistance training decreases abdominal fat and improves insulin sensitivity in older men with type 2 diabetes. *Diabetes Care* 28: 662-667.

Ishii, T., T. Yamakita, T. Sato, S. Tanaka, and S. Fujii. 1998. Resistance training improves insulin sensitivity in NIDDM subjects without altering maximal oxygen uptake. *Diabetes Care* 21: 1353-1355.

Jorgensen, S.B., E.A. Richter, and J.F. Wojtaszewski. 2006. Role of AMPK in skeletal muscle metabolic regulation and adaptation in relation to exercise. *J Physiol* 574: 17-31.

Jorgensen, S.B., J.F. Wojtaszewski, B. Viollet, F. Andreelli, J.B. Birk, Y. Hellsten, P. Schjerling, S. Vaulont, P.D. Neufer, E.A. Richter, and H. Pilegaard. 2005. Effects of alpha-AMPK knockout on exercise-induced gene activation in mouse skeletal muscle. *FASEB J* 19: 1146-1148.

Joseph, L.J., T.A. Trappe, P.A. Farrell, W.W. Campbell, K.E. Yarasheski, C.P. Lambert, and W.J. Evans. 2001. Short-term moderate weight loss and resistance training do not affect insulin-stimulated glucose disposal in postmenopausal women. *Diabetes Care* 24: 1863-1869.

Kern, M., J.A. Wells, J.M. Stephens, C.W. Elton, J.E. Friedman, E.B. Tapscott, P.H. Pekala, and G.L. Dohm. 1990. Insulin responsiveness in skeletal muscle is determined by glucose transporter (Glut4) protein level. *Biochem J* 270: 397-400.

Kiens, B., S. Kristiansen, P. Jensen, E.A. Richter, and L.P. Turcotte. 1997. Membrane associated fatty acid binding protein (FABPpm) in human skeletal muscle is increased by endurance training. *Biochem Biophys Res Commun* 231: 463-465.

Kiens, B. and H. Lithell. 1989. Lipoprotein metabolism influenced by training-induced changes in human skeletal muscle. *J Clin Invest* 83: 558-564.

Kiens, B., C. Roepstorff, J.F. Glatz, A. Bonen, P. Schjerling, J. Knudsen, and J.N. Nielsen. 2004. Lipid-binding proteins and lipoprotein lipase activity in human skeletal muscle: influence of physical activity and gender. *J Appl Physiol* 97: 1209-1218.

Kim, J.H., H.D. Shin, B.L. Park, Y.M. Cho, S.Y. Kim, H.K. Lee, and K.S. Park. 2005. Peroxisome proliferator-activated receptor gamma coactivator 1 alpha promoter polymophisms are associated with early-onset type 2 diabetes mellitus in the Korean population. *Diabetologia* 48(7): 1323-30.

Koopman, R., A.H. Zorenc, R.J. Gransier, D. Cameron-Smith, and L.J. van Loon. 2006. Increase in S6K1 phosphorylation in human skeletal muscle following resistance exercise occurs mainly in type II muscle fibers. *Am J Physiol Endocrinol Metab* 290: E1245-E1252.

Krisan, A.D., D.E. Collins, A.M. Crain, C.C. Kwong, M.K. Singh, J.R. Bernard, and B.B. Yaspelkis III. 2004. Resistance training enhances components of the insulin signaling cascade in normal and high-fat-fed rodent skeletal muscle. *J Appl Physiol* 96: 1691-1700.

Lin, J., C. Handschin, and B.M. Spiegelman. 2005. Metabolic control through the PGC-1 family of transcription coactivators. *Cell Metab* 1: 361-370.

Lin, J., H. Wu, P.T. Tarr, C.Y. Zhang, Z. Wu, O. Boss, L.F. Michael, P. Puigserver, E. Isotani, E.N. Olson, B.B. Lowell, R. Bassel-Duby, and B.M. Spiegelman. 2002. Transcriptional co-activator PGC-1 alpha drives the formation of slow-twitch muscle fibres. *Nature* 418: 797-801.

Maiorana, A., G. O'driscoll, C. Cheetham, J. Collis, C. Goodman, S. Rankin, R. Taylor, and D. Green. 2000. Combined aerobic and resistance exercise training improves functional capacity and strength in CHF. *J Appl Physiol* 88: 1565-1570.

McConell, G.K., R.S. Lee-Young, Z.P. Chen, N.K. Stepto, N.N. Huynh, T.J. Stephens, B.J. Canny, and B.E. Kemp. 2005. Short-term exercise training in humans reduces AMPK signalling during prolonged exercise independent of muscle glycogen. *J Physiol* 568: 665-676.

Miller, J.P., R.E. Pratley, A.P. Goldberg, P. Gordon, M. Rubin, M.S. Treuth, A.S. Ryan, and B.F. Hurley. 1994. Strength training increases insulin action in healthy 50- to 65-yr-old men. *J Appl Physiol* 77: 1122-1127.

Nielsen, J.N., K.J. Mustard, D.A. Graham, H. Yu, C.S. MacDonald, H. Pilegaard, L.J. Goodyear, D.G. Hardie, E.A. Richter, and J.F. Wojtaszewski. 2003. 5'-AMP-activated protein kinase activity and subunit expression in exercise-trained human skeletal muscle. *J Appl Physiol* 94: 631-641.

Nikkila, E.A., M.R. Taskinen, S. Rehunen, and M. Harkonen. 1978. Lipoprotein lipase activity in adipose tissue and skeletal muscle of runners: relation to serum lipoproteins. *Metabolism* 27: 1661-1667.

Nilsson, E.C., Y.C. Long, S. Martinsson, S. Glund, P. Garcia-Roves, L.T. Svensson, L. Andersson, J.R. Zierath, and M. Mahlapuu. 2006. Opposite transcriptional regulation in skeletal muscle of AMP-activated protein kinase gamma3 R225Q transgenic versus knock-out mice. *J Biol Chem* 281: 7244-7252.

Patti, M.E., A.J. Butte, S. Crunkhorn, K. Cusio, R. Berriak, S. Kashyap, Y. Miyazaki, I. Kohane, M. Costello, R. Saccone, E.J. Landaker, A.B. Goldfine, E. Mun, R. DeFronzo, J. Finlayson, C.R. Kahn, and L.J. Mandarino. 2003. Coordinated reduction of genes of oxidative metabolism in humans with insulin resistance and diabetes: Potential role of PGC1 and NRF1. *Proc Natl Acad Sci USA* 100(14): 8466-8471.

Pilegaard, H., B. Saltin, and P.D. Neufer. 2003. Exercise induces transient transcriptional activation of the PGC-1alpha gene in human skeletal muscle. *J Physiol* 546: 851-858.

Reynolds, T.H., M.A. Supiano, and D.R. Dengel. 2004. Resistance training enhances insulin-mediated glucose disposal with minimal effect on the tumor necrosis factor-alpha system in older hypertensives. *Metabolism* 53: 397-402.

Rogers, M.A., D.S. King, J.M. Hagberg, A.A. Ehsani, and J.O. Holloszy. 1990. Effect of 10 days of physical inactivity on glucose tolerance in master athletes. *J Appl Physiol* 68: 1833-1837.

Ryan, A.S., D.E. Hurlbut, M.E. Lott, F.M. Ivey, J. Fleg, B.F. Hurley, and A.P. Goldberg. 2001. Insulin action after resistive training in insulin resistant older men and women. *J Am Geriatr Soc* 49: 247-253.

Ryan, A.S., R.E. Pratley, A.P. Goldberg, and D. Elahi. 1996. Resistive training increases insulin action in postmenopausal women. *J Gerontol A Biol Sci Med Sci* 51: M199-M205.

Schenk, S. and J.F. Horowitz. 2006. Coimmunoprecipitation of FAT/CD36 and CPT I in skeletal muscle increases proportionally with fat oxidation after endurance exercise training. *Am J Physiol Endocrinol Metab* 291: E254-E260.

Smutok, M.A., C. Reece, P.F. Kokkinos, C. Farmer, P. Dawson, R. Shulman, J. DeVane-Bell, J. Patterson, C. Charabogos, A.P. Goldberg, et al. 1993. Aerobic versus strength training for risk factor intervention in middle-aged men at high risk for coronary heart disease. *Metabolism* 42: 177-184.

Thibault, M.C., J.A. Simoneau, C. Cote, M.R. Boulay, P. Lagasse, M. Marcotte, and C. Bouchard. 1986. Inheritance of human muscle enzyme adaptation to isokinetic strength training. *Hum Hered* 36: 341-347.

Thomas, G.N., A.W. Hong, B. Tomlinson, E. Lau, C.W. Lam, J.E. Sanderson, and J. Woo. 2005. Effects of Tai Chi and resistance training on cardiovascular risk factors in elderly Chinese subjects: a 12-month longitudinal, randomized, controlled intervention study. *Clin Endocrinol (Oxf)* 63: 663-669.

Tokmakidis, S.P., C.E. Zois, K.A. Volaklis, K. Kotsa, and A.M. Touvra. 2004. The effects of a combined strength and aerobic exercise program on glucose control and insulin action in women with type 2 diabetes. *Eur J Appl Physiol* 92: 437-442.

Tunstall, R.J., K.A. Mehan, G.D. Wadley, G.R. Collier, A. Bonen, M. Hargreaves, and D. Cameron-Smith. 2002. Exercise training increases lipid metabolism gene expression in human skeletal muscle. *Am J Physiol Endocrinol Metab* 283, E66-E72.

Williams, R.S. and P.D. Neufer. 1996. Regulation of gene expression in skeletal muscle by contractile activity. In *The handbook of physiology,* eds. Rowell LB and Shepherd JT, pp. 1124-1150. Amercian Physiological Society, Bethesda.

Winder, W.W., B.F. Holmes, D.S. Rubink, E.B. Jensen, M. Chen, and J.O. Holloszy. 2000. Chronic chemical activation of AMP-activated protein kinase increases mitochondrial enzymes and GLUT4 in skeletal muscle of resting rats. *Diabetes* 49: A10.

Wojtaszewski, J.F., J.B. Birk, C. Frosig, M. Holten, H. Pilegaard, and F. Dela. 2005. 5'AMP activated protein kinase expression in human skeletal muscle: effects of strength training and type 2 diabetes. *J Physiol* 564: 563-573.

Wojtaszewski, J.F.P., P. Nielsen, B.F. Hansen, E.A. Richter, and B. Kiens. 2000. Isoform specific and exercise intensity dependent activation of 5'AMP-activated protein kinase in human skeletal muscle. *J Physiol* 528: 221-226.

Yaspelkis, B.B., 3rd. 2006. Resistance training improves insulin signaling and action in skeletal muscle. *Exerc Sport Sci Rev* 34: 42-46.

Zachwieja, J.J., G. Toffolo, C. Cobelli, D.M. Bier, and K.E. Yarasheski. 1996. Resistance exercise and growth hormone administration in older men: effects on insulin sensitivity and secretion during a stable-label intravenous glucose tolerance test. *Metabolism* 45: 254-260.

Zheng, D., P.S. MacLean, S.C. Pohnert, J.B. Knight, A.L. Olson, W.W. Winder, and G.L. Dohm 2001. Regulation of muscle GLUT-4 transcription by AMP-activated protein kinase. *J Appl Physiol* 91: 1073-1083.

Zong, H., J.M. Ren, L.H. Young, M. Pypaert, J. Mu, M.J. Birnbaum, and G.I. Shulman. 2002. AMP kinase is required for mitochondrial biogenesis in skeletal muscle in response to chronic energy deprivation. *Proc Natl Acad Sci USA* 99: 15983-15987.

Part IV

Prevention of Type 2 Diabetes

Identification of Novel Molecular Targets and Pathways

Chapter 13

AMPK

The Master Switch for Type 2 Diabetes?

W.W. Winder, PhD; and D.M. Thomson, PhD

AMP-activated protein kinase (AMPK) is a heterotrimeric protein consisting of three subunits: alpha, beta, and gamma. Two alpha and beta subunit isoforms and three gamma subunit isoforms are expressed. The kinase domain is located on the alpha subunit along with the Thr-172 residue that must be phosphorylated by an upstream kinase for activation (Hardie and Carling 1997). AMP sensitivity is thought to reside in two Bateman domains of the gamma subunit (Daniel and Carling 2002). ATP and AMP compete for binding to these sites, with AMP activating the complex and ATP inhibiting it. As muscle contracts, the concentration of AMP increases as a result of the adenylate kinase reaction, 2 ADP → ATP + AMP. The rise in the ratio of AMP to ATP activates AMPK, and therefore AMPK can be considered an energy-sensing protein of the muscle. The binding of AMP to the gamma subunit makes AMPK a poorer substrate for protein phosphase 2C (PP2C). In cells in culture, the upstream kinase has been identified to be the protein complex LKB1-MO25-STRAD (Hardie, 2004, 2005; Hardie and Sakamoto, 2006; Hawley et al. 2003; Shaw et al. 2004; Witters, Kemp, and Means, 2006; Woods et al. 2003). In skeletal muscle of mice deficient in muscle LKB1 (Serine, Threonine Kinase 11 or STK11), AMPK is not phosphorylated or activated in response to contraction (Sakamoto et al. 2005). CaMK kinase (CaMKK) is another upstream kinase, but its expression, distribution, and role in skeletal muscle fibers are not well defined (Hawley et al. 2005).

Discoveries Suggesting AMPK Could Be Important for Prevention and Treatment of Type 2 Diabetes

Several key discoveries suggest that AMPK signaling is important for the prevention and treatment of type 2 diabetes. First, AMPK is activated in response to muscle contraction (Hutber, Hardie, and Winder 1997; Vavvas et al. 1997; Winder and Hardie 1996; Winder et al. 1997). Second, in resting muscle AMPK can be activated chemically by the drug AICAR (Merrill et al. 1997). AICAR can be used to probe for downstream actions of AMPK in muscle and other tissues. In perfused rat hind limb, AICAR triggers not only an increase in fatty acid oxidation but also an increase in glucose uptake across the hind limb (Merrill et al. 1997).

This study was followed up with data showing that the activation of AMPK with AICAR in incubated epitrochlearis triggered an increase in the transport of a nonmetabolizable glucose analogue into the muscle (Hayashi et al. 1998). The increase in glucose transport was not blocked by wortmannin, a characteristic shared with contraction-induced glucose uptake. Other stimuli that decreased the creatine phosphate content of the muscle (hypoxia, use of metabolic inhibitors that block the electron transport chain) also activated AMPK and increased glucose transport (Hayashi et al. 2000). The increase in glucose transport caused by perfusing muscles with AICAR was later

Studies performed in the author's lab were supported by NIH grants RO1 AR41438 and RO1 AR51928.

attributed to an increase in GLUT4 translocation from the microsome fraction to the sarcolemma (Kurth-Kraczek et al. 1999). These data suggest that AMPK could mediate the stimulation of glucose uptake by muscle contraction. Chronic activation of AMPK using daily injections of AICAR increased the expression of GLUT4, hexokinase (HK), and several mitochondrial oxidative enzymes (Barnes et al. 2005; Holmes, Kurth-Kraczek, and Winder 1999; Winder et al. 2000). These and other studies have led to the idea that AMPK should be targeted for developing pharmaceuticals to prevent and treat type 2 diabetes (Musi and Goodyear 2002; Pold et al. 2005; Winder 2000; Winder and Hardie 1999).

Could Type 2 Diabetes Result From a Deficiency in AMPK Signaling?

First, it should be emphasized that a large number of proteins are involved in insulin signaling and metabolism regulation. Type 2 diabetes does not result from the deficiency of just one protein, but likely has multiple etiologies, all resulting in decreased insulin sensitivity progressing to hyperglycemia in late stages. Because of the demonstrated effects of AMPK on glucose uptake, GLUT4 and HK expression, insulin sensitivity, and fatty acid metabolism, it was suggested that defects in AMPK signaling could produce symptoms of type 2 diabetes (Winder 2000; Winder and Hardie 1999). Data available on a small number of human subjects with type 2 diabetes, however, indicate that muscle AMPK subunit expression is not defective and that it can be activated in response to exercise (Hojlund et al. 2004; Musi et al. 2001). Deficiency of AMPK signaling in other tissues and its effects have not been addressed in human studies.

Several studies have addressed deficiencies of the alpha subunit isoforms of AMPK. Insulin-stimulated glucose uptake into incubated muscle has been reported to be unimpaired in the dominant negative alpha AMPK mouse (Fryer et al. 2002). Knockout of the alpha2 AMPK isoform resulted in a higher blood glucose concentration and lower plasma insulin during the glucose tolerance test (Viollet et al. 2003). The high blood glucose and low insulin were attributed to increased sympathetic activity. Insulin-stimulated glucose uptake

by isolated muscle was not influenced by the deficiency in alpha2 AMPK. Specific knockout of the alpha2 isoform has also been reported to increase fat deposition in response to a high-fat diet, but it was reported to have no effect on glucose tolerance or insulin tolerance in that study (Villena et al. 2004). In another study, knockout of the alpha2 isoform but not the alpha1 isoform resulted in glucose intolerance and insulin resistance (Jorgensen et al. 2004). Studies on knockout of the alpha1 and alpha2 isoforms are complicated by the fact that when one isoform is knocked out, the presence of the alternate isoform may compensate for its absence. In the alpha2 KO mouse, alpha1 protein in muscle is approximately three times as high compared with the wild type (Jorgensen et al. 2004).

A recent study demonstrated that insulin resistance induced by high-fat feeding was accompanied by a 60% decrease in the gastrocnemius alpha2 subunit for AMPK (Liu et al. 2006). Phospho-AMPK was reduced 77% compared with normal control muscles. When rats on a high-fat diet were treated with metformin, AMPK was activated and insulin sensitivity was restored.

The skeletal muscle–specific LKB1 KO mouse was developed to determine the roles of LKB1 in muscle. In these mice, contraction-stimulated glucose uptake by muscle is for the most part inhibited, but insulin stimulation (100 nM) of glucose uptake in isolated muscle is not (Sakamoto et al. 2005). It should be remembered that LKB1 is upstream of many other protein kinases besides AMPK (Alessi, Sakamoto, and Bayascas 2006; Sakamoto et al. 2004), including BR serine/threonine kinase 1 and 2 (BRSK1, BRSK2), NUAK family SNF1-like kinase 1 and 2 (NUAK1, NAUK2), salt-inducible kinase (SIK), qin-induced kinase (QIK), serine/threonine kinase QSK (QSK), MAP/microtubule affinity-regulating kinase 1-4 (MARK1 - MARK4, and SNF-related serine/threonine protein kinase (SNRK). Although LKB1 deficiency eliminates activation of skeletal muscle AMPK in response to contraction, AICAR, and metformin, the effects of LKB1 deficiency and consequent effects on these other downstream kinases are yet undefined. As with all KO models, failure to see a phenotypic deficiency must be interpreted with caution if redundant signaling mechanisms may compensate for the defective or missing protein.

Two studies have now demonstrated reduced skeletal muscle AMPK activity or phospho-AMPK in the ZDF rat (deficient in leptin) model of type 2 diabetes (Sriwijitkamol et al. 2006; Yu et al. 2004). AMPK subunit expression was not altered, but protein expression of LKB1 was reduced in the obese rats by 43% (Sriwijitkamol et al. 2006). In this same study, endurance exercise increased muscle LKB1 expression by 2.8-fold and PCG-1α protein by 2.5-fold. Thus, at least in this model of type 2 diabetes, there appears to be some deficiency in AMPK signaling. Endurance training has previously been demonstrated to increase skeletal muscle GLUT4 expression and to compensate for insulin resistance of ZDF rats (Etgen Jr. et al. 1996; Ivy 2004).

In a recent landmark study, liver LKB1 deficiency was induced in mice bred to be homozygous in the conditional floxed allele of LKB1. Mice were injected with adenovirus carrying the Cre recombinase gene which, after being expressed in cells, cleaved the floxed LKB1 gene, resulting in a marked decrease in LKB1 expression in the liver. Mice with deficiency of LKB1 in the liver exhibited hyperglycemia (+22 mM; Carling 2006; Shaw et al. 2005). The LKB1 deficiency resulted in a marked decline in liver phospho-AMPK, indicating that LKB1 is the major upstream kinase for AMPK in liver and that phosphorylation and activation of AMPK are responsible in part for regulation of glucose production. The LKB1-deficient mice had impaired glucose tolerance. Results of the insulin tolerance test were similar to those seen in wild-type mice. This implies that increased hepatic glucose production and not altered insulin sensitivity of target tissues was responsible for the increase in blood glucose. The loss of LKB1 resulted in increased gluconeogenic and lipogenic enzyme expression and the hyperglycemia and glucose intolerance were likely due to enhanced hepatic gluconeogenesis.

Liver-specific alpha2 AMPK KO mice also exhibit hyperglycemia and glucose intolerance (Andreelli et al. 2006). Hepatic glucose production was elevated 50% compared with controls. The increase in glucose production was not inhibited by insulin. In liver, the alpha2 isoform contributes approximately 50% of the total AMPK activity.

How Can AMPK Activation Help Prevent Type 2 Diabetes?

Strong evidence is accumulating that suggests regular exercise delays or even prevents the onset of type 2 diabetes (Hawley 2004; Knowler et al.

2002). Although the mechanisms are not clear, it does seem probable that the activation of AMPK via muscle contraction during exercise may be responsible in part for these beneficial effects (Musi and Goodyear 2002, 2006). It is clear that muscle contraction has beneficial effects independent of AMPK activation and that other signaling pathways are involved. In this review we focus on those effects attributed to AMPK, summarized in figure 13.1.

The insulin resistance of muscle in type 2 diabetes can be attributed to a reduction in insulin receptor phosphorylation, a reduction in IRS1 phosphorylation, a reduction in PI3K activation, and a reduction in Akt activation (Musi and Goodyear 2006). Although AMPK has been shown to phosphorylate IRS1, the physiological significance of that phosphorylation is not well understood (Jakobsen et al. 2001). As was noted earlier, GLUT4 translocation and stimulation of glucose transport can be blocked by wortmannin in incubated epitrochlearis muscles, but contraction-induced glucose transport and AICAR-induced glucose transport cannot (Hayashi et al. 1998). Thus, one or more insulin-independent pathways exist that bypass the PI3K/Akt step of insulin signaling (Brozinick and Birnbaum 1998; Yeh et al. 1995). This implies that muscle contraction and AMPK activation might be an alternative way to stimulate glucose transport in insulin-resistant muscle. Strong evidence from several labs supports this concept (Bergeron et al. 1999; Fryer et al. 2000, 2002; Fujii et al. 2004; Hayashi et al. 1998, 2000; Jessen et al. 2003; Kurth-Kraczek et al. 1999; Merrill et al. 1997; Nielsen et al. 2003; Russell et al. 1999; Wright et al. 2004). Thus, acute stimulation of glucose transport into muscle by AMPK activation is expected to be helpful in people with insulin resistance.

A second way AMPK activation might be helpful is to stimulate an increase in insulin sensitivity; that is, to repair the defect in insulin signaling. Evidence has accumulated from studies on the isolated incubated epitrochlearis muscle to support this hypothesis (Fisher et al. 2002). Muscles were incubated in serum for 1 h with AICAR with consequent activation of AMPK and increase in stimulation of glucose transport by a submaximal insulin concentration in the post-AICAR treatment time frame. This approximate twofold increase in insulin-stimulated glucose transport was not accompanied by an increase in

Akt phosphorylation. This effect was not observed in primary cultures of human myocytes, however (Al-Khalili et al. 2004).

Additional evidence that AMPK activation increases insulin sensitivity was obtained using muscle cells in culture. C2C12 myotubes exhibit an approximate twofold increase in 2-deoxyglucose transport in response to insulin (Smith, Patil, and Fisher 2005). This increase in glucose transport can be blocked with an inhibitor of GLUT4 translocation and with an AMPK inhibitor. If AICAR (2 mM) is included in the medium for 1 h, followed by 2 h of recovery in the absence of AICAR, insulin-stimulated glucose uptake is significantly increased by a submaximal insulin concentration (10 nM) and markedly increased (threefold) by a 100 nM insulin concentration (Smith, Patil, and Fisher 2005). Hyperosmolarity, another way to activate AMPK, also stimulated an increase in insulin sensitivity in these cultured myotubes.

One possible mechanism by which AMPK could improve insulin sensitivity is downregulation of the mTOR signaling pathway. In liver (Reiter et al. 2005) and skeletal muscle (Bolster et al. 2002), AICAR stimulation of AMPK inhibits mTOR signaling, as evidenced by suppression of mTOR phosphorylation and other signaling events downstream of mTOR at the level of ribosomal protein S6K, eIF4E, and 4EBP1. AMPK can regulate the mTOR pathway by at least two mechanisms. First, it phosphorylates and activates TSC2 (tuberin), which inhibits mTOR via decreased Rheb activity (Inoki, Zhu, and Guan 2003). Second, AMPK itself can directly phosphorylate and inhibit mTOR (Cheng et al. 2004).

Activity of mTOR or its downstream target, S6K, appears to mediate the induction of insulin resistance due to various stimuli, including high-fat diet and obesity (Um et al. 2004), hyperinsulinemia (Berg, Lavan, and Rondinone 2002; Ueno et al. 2005), and amino acids (Tremblay and Marette 2001). Hyperactivation of mTOR/S6K by deletion of TSC2 in mouse embryonic fibroblast cells or by overexpression of Rheb in HEK 293 cells results in insulin resistance due to downregulation of IRS1 and IRS2 (Shah et al. 2004). The inhibitory effect of mTOR/S6K activity on insulin signaling is due, at least in part, to serine phosphorylation of IRS1 at Ser-307 and Ser-636/Ser-639 by mTOR/S6K (Harrington et al. 2004; Shah et al. 2004; Um et al. 2004), which disrupts its ability to interact with the insulin receptor and PI3K. On the other hand, rapamycin (an inhibitor of mTOR) reverses hyperinsulinemia-induced IRS1 phosphorylation and insulin resistance in rat skeletal and liver tissues (Khamzina et al. 2005; Tzatsos and Kandror 2006; Ueno et al. 2005). Given its inhibitory effect on mTOR signaling, it is reasonable to expect that AMPK activation can enhance insulin sensitivity. Accordingly, activation of AMPK via LKB1 overexpression enhanced insulin-stimulated Akt phosphorylation (Jessen et al. 2003; Tzatsos and Kandror 2006), whereas AMPK activation via metformin and 2-deoxyglucose reduced mTOR activity toward IRS1 in hepatic and muscle cultures (Tzatsos and Kandror 2006).

In cultured endothelial cells, metformin as well as constitutively active AMPK expression resulted in phosphorylation (activation) of endothelial nitric oxide synthase (eNOS) and elevated bioactive NO

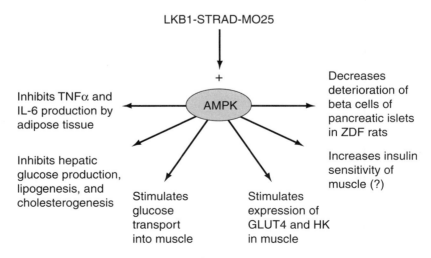

Figure 13.1 Putative roles of AMPK after activation by LKB1 in liver, adipose tissue, muscle, and pancreatic islets.

levels (Davis et al. 2006). Activation of AMPK in that study also attenuated the suppression of NO bioactivity and endothelial cell apoptosis induced by high glucose levels, suggesting that a beneficial effect of AMPK in diabetes may be improved vascular endothelial cell performance and viability. AMPK has also been shown to inhibit iNOS in skeletal muscle (Pilon, Dallaire, and Marette 2004), which is elevated in models of diabetes and insulin resistance and protects against insulin resistance when disrupted (Perreault and Marette 2001).

Finally, activation of AMPK in adipose tissue increases adiponectin production and reduces TNFα and IL-6 production (Lihn et al. 2004). These changes might be expected to increase whole-body insulin sensitivity.

Can Chemical AMPK Activation Prevent Diabetes?

Evidence is available from a few studies on animal models showing that chronic chemical activation of AMPK may be useful in preventing and treating type 2 diabetes. Administration of AICAR (0.5 mg/g body weight) for 7 wk to obese ZDF rats reduced plasma triglycerides (3.6 mM versus 10.4 mM) and plasma FFAs (0.54 mM versus 1.05 mM). The AICAR treatment increased both plasma high-density lipoprotein (HDL) cholesterol (2.7 mM versus 1.6 mM) and plasma total cholesterol (5.3 mM versus 3.4 mM). Plasma glucose was normalized and plasma insulin reduced from 1,119 pM to 282 pM in response to AICAR. The results of the oral glucose tolerance test administered 24 h following the last AICAR injection were also indistinguishable from the results obtained from lean rats. The total membrane-associated muscle GLUT4 was increased and the stimulation of 3-methylglucose uptake into isolated epitrochlearis and extensor digitorum longus muscles by 60 nM insulin (maximally effective concentration) was significantly improved in the AICAR-treated rats (Buhl et al. 2002).

A more recent study provides evidence that chronic chemical activation with AICAR prevents the spontaneous rise in plasma glucose (from 6 mM up to 14 mM) that occurs at about 9 wk of age in this strain. Exercise training has a similar effect. The normalization of plasma glucose is due in part to increased glucose disposal and reduced glucose production (Pold et al. 2005). Interestingly, the deterioration of islet cell morphology seen in these rats

was for the most part prevented in both the chronic AICAR-treated rats and in the endurance-trained rats (Pold et al. 2005). AMPK activity was found to be reduced in muscle and liver of the ZDF rat. Administration of AICAR 3 times/wk resulted in a marked decrease in blood glucose and also in less triglyceride accumulation in the liver, muscle, and pancreatic islets (Yu et al. 2004).

The KKAy-CETP mouse is a transgenic animal that overexpresses cholesteryl ester transfer protein (CETP). This overexpression results in increased plasma very low-density lipoprotein (VLDL), decreased plasma HDL, and increased plasma low-density lipoprotein (LDL). These mice exhibit hyperphagia, obesity, and hyperinsulinemia and provide a model of type 2 diabetes (Castle et al. 1998). Chronic daily subcutaneous injection with AICAR (0.5 mg/g) resulted in significantly lower blood glucose and reduction in plasma insulin in these mice (Fiedler et al. 2001). Euglycemic clamp studies demonstrated the reduction in blood glucose to be due to a significant decrease in endogenous glucose production. No significant difference was noted in either in vivo glucose uptake or uptake of 2-deoxyglucose by isolated soleus or extensor digitorum longus (EDL). In this diabetes model, plasma FFAs increased 1.9-fold and triglycerides increased 1.3-fold in response to the chronic AICAR treatment. Thus, some beneficial effects on glucoregulation were observed, but the dyslipidemia was compounded.

Mice deficient in the leptin receptor (ob/ob) also develop hyperglycemia (18 mM) and symptoms of type 2 diabetes. Daily injection of these mice with AICAR (1 mg/g body weight) over 7 d was reported to cause a progressive decrease in blood glucose to levels seen in the lean mice. Glucose tolerance was improved and the mice had higher levels of GLUT4 and HK2 compared with saline-injected controls. Plasma FFAs and triglycerides were actually increased by the AICAR treatment (Song et al. 2002). Impairment in insulin-induced PI3K and glucose transport in isolated muscle was not repaired by AICAR treatment.

Feasibility of Using AMPK Activators

Although AICAR has been administered to human subjects in clinical trials for other applications, it is not likely to be the compound of choice for treating type 2 diabetes (Musi and Goodyear 2002; Winder 2000; Winder and Hardie 1999; Winder et al. 2000). Doses required to activate muscle AMPK in animals are high (500-1,000 mg/kg) and have been reported to cause liver hypertrophy and lactacidemia (Winder et al. 2000). In addition, the muscles appear to become somewhat refractory to chronic treatment with AICAR. An intermittent dosing pattern may be more effective (Winder et al. 2000). Nevertheless, it seems that it will be essential to develop more potent AMPK activators.

At least two types of hypoglycemic drugs that have been reported to activate AMPK, metformin and TZDs, are currently being used to treat type 2 diabetes (Fryer, Parbu-Patel, and Carling 2002; Hawley et al. 2002; Lebrasseur et al. 2006; Musi et al. 2002; Saha et al. 2004; Sakamoto et al. 2004; Shaw et al. 2005; Zhou et al. 2001). Both activate AMPK in isolated liver and muscle cells, albeit at high concentration (Hawley et al. 2002; Lessard et al. 2006; Zhou et al. 2001). In hepatocyte culture, 0.5 mM metformin was required to activate AMPK after 3 h. A lower concentration (0.05 mM) was effective after 7 h incubation. In skeletal muscle, 2 mM metformin increased AMPK activity in incubated skeletal muscle after 3 h. Rosiglitazone (3 mg/kg) has recently been reported to restore AMPKα2 signaling in muscle of ZDF rats (Lessard et al. 2006). There remains the question of whether these concentrations would be achieved in patients at therapeutic doses. One study shows that patients who receive metformin show higher muscle AMPK activity after 4 wk and 10 wk of treatment (Musi et al. 2002). Activity of ACC, one of the downstream targets of AMPK, was not reduced after 4 wk, but it was reduced after 10 wk of treatment.

Liver LKB1 has recently been shown to be required in mice to mediate the ability of metformin (250 mg/kg) to lower blood glucose (Shaw et al. 2005). This was true in mice that are hyperglycemic in response to a high-fat diet or in mice that are deficient in the leptin receptor (ob/ob).

The inhibitory effect of AMPK on mTOR signaling and protein synthesis could present a potential negative side effect of prolonged AMPK therapy. For example, it has been hypothesized that chronic AMPK activation in skeletal muscle could lead to atrophy or to an impaired ability for growth in response to a hypertrophic stimulus (Atherton et al. 2005; Thomson and Gordon 2005, 2006). This would be of particular concern for the elderly, among whom age-related muscle loss and insulin resistance are prevalent. Alternatively, improved muscle energetics due to elevated AMPK activity could likewise lead to increased muscle loading with physical activity, thus counteracting any potential catabolic effects of AMPK. Such speculation can only be resolved by further research. Although the potential inhibition of protein synthesis by AMPK might be undesirable in muscle and some other tissues, data suggest that the same phenomenon along with a related suppression of cell proliferation by AMPK may have a protective effect against the development of atherosclerosis and cancer (Motoshima et al. 2006). The risk of both is elevated in patients with diabetes (Ahmed et al. 2006).

Future Directions

It seems clear that the LKB1/AMPK signaling pathway is intimately involved in regulating carbohydrate and fat metabolism in the major insulin-sensitive tissues, skeletal muscle, liver, and heart. Since type 2 diabetes involves dysregulation of metabolism and since beneficial effects have been observed in animal models of type 2 diabetes, it seems reasonable to continue targeting these proteins in order to develop pharmaceuticals for prevention and treatment of the disease. The development of more potent and specific AMPK activators will be essential. Although the hypoglycemic agents do activate AMPK, large doses are required and it is somewhat uncertain whether therapeutic doses are effective in increasing AMPK activity in target tissues. It also seems clear that none of the pharmacologic agents currently in use is completely effective in preventing long-term complications of diabetes.

Concluding Remarks

AMPK activation enhances glucose uptake by muscle, increases fatty acid oxidation, and increases insulin sensitivity. Chronic activation increases expression of GLUT4 and HK, thus increasing capacity for glucose transport into muscle. Because of these beneficial actions, this signaling system remains an active target for developing pharmaceuticals that can prevent and treat type 2 diabetes. Regardless of slow progress in the development of potent AMPK activators, it seems clear that muscle contraction rapidly activates AMPK, resulting in enhanced insulin sensitivity and reduced tendency to progress to full-blown type 2 diabetes. Regular exercise still remains the most effective way to specifically activate AMPK in skeletal muscle.

References

Ahmed, R.L., K.H. Schmitz, K.E. Anderson, W.D. Rosamond, and A.R. Folsom. 2006. The metabolic syndrome and risk of incident colorectal cancer. *Cancer* 107:28-36.

Al-Khalili, L., A. Krook, J.R. Zierath, and G.D. Cartee. 2004. Prior serum- and AICAR-induced AMPK activation in primary human myocytes does not lead to subsequent increase in insulin-stimulated glucose uptake. *Am J Physiol Endocrinol Metab* 287:E553-7.

Alessi, D.R., K. Sakamoto, and J.R. Bayascas. 2006. LKB1-dependent signaling pathways. *Annu Rev Biochem* 75:137-63.

Andreelli, F., M. Foretz, C. Knauf, P.D. Cani, C. Perrin, M.A. Iglesias et al. 2006. Liver adenosine monophosphate-activated kinase-alpha2 catalytic subunit is a key target for the control of hepatic glucose production by adiponectin and leptin but not insulin. *Endocrinology* 147:2432-41.

Atherton, P.J., J. Babraj, K. Smith, J. Singh, M.J. Rennie, and H. Wackerhage. 2005. Selective activation of AMPK-PGC-1alpha or PKB-TSC2-mTOR signaling can explain specific adaptive responses to endurance or resistance training-like electrical muscle stimulation. *FASEB J* 19:786-8.

Barnes, B.R., S. Glund, Y.C. Long, G. Hjalm, L. Andersson, and J.R. Zierath. 2005. 5'-AMP-activated protein kinase regulates skeletal muscle glycogen content and ergogenics. *FASEB J* 19:773-9.

Berg, C.E., B.E. Lavan, and C.M. Rondinone. 2002. Rapamycin partially prevents insulin resistance induced by chronic insulin treatment. *Biochem Biophys Res Commun* 293:1021-7.

Bergeron, R., R.R. Russell III, L.H. Young, J.M. Ren, M. Marcucci, A. Lee et al. 1999. Effect of AMPK activation on muscle glucose metabolism in conscious rats. *Am J Physiol* 276:E938-44.

Bolster, D.R., S.J. Crozier, S.R. Kimball, and L.S. Jefferson. 2002. AMP-activated protein kinase suppresses protein synthesis in rat skeletal muscle through downregulated mammalian target of rapamycin (mTOR) signaling. *J Biol Chem* 277:23977-80.

Brozinick Jr., J.T., and M.J. Birnbaum. 1998. Insulin, but not contraction, activates Akt/PKB in isolated rat skeletal muscle. *J Biol Chem* 273:14679-82.

Buhl, E.S., N. Jessen, R. Pold, T. Ledet, A. Flyvbjerg, S.B. Pedersen et al. 2002. Long-term AICAR administration reduces metabolic disturbances and lowers blood pressure in rats displaying features of the insulin resistance syndrome. *Diabetes* 51:2199-206.

Carling, D. 2006. LKB1: A sweet side to Peutz-Jeghers syndrome? *Trends Mol Med* 12:144-7.

Castle, C.K., S.L. Kuiper, W.L. Blake, B. Paigen, K.R. Marotti, and G.W. Melchior. 1998. Remodeling of the HDL in NIDDM: A fundamental role for cholesteryl ester transfer protein. *Am J Physiol* 274:E1091-8.

Cheng, S.W., L.G. Fryer, D. Carling, and P.R. Shepherd. 2004. Thr2446 is a novel mammalian target of rapamycin (mTOR) phosphorylation site regulated by nutrient status. *J Biol Chem* 279:15719-22.

Daniel, T., and D. Carling. 2002. Functional analysis of mutations in the gamma 2 subunit of AMP-activated protein kinase associated with cardiac hypertrophy and Wolff-Parkinson-White syndrome. *J Biol Chem* 277:51017-24.

Davis, B.J., Z. Xie, B. Viollet, and M.H. Zou. 2006. Activation of the AMP-activated kinase by antidiabetes drug metformin stimulates nitric oxide synthesis in vivo by promoting the association of heat shock protein 90 and endothelial nitric oxide synthase. *Diabetes* 55:496-505.

Etgen Jr., G.J., C.M. Wilson, J. Jensen, S.W. Cushman, and J.L. Ivy. 1996. Glucose transport and cell surface GLUT-4 protein in skeletal muscle of the obese Zucker rat. *Am J Physiol* 271:E294-301.

Fiedler, M., J.R. Zierath, G. Selen, H. Wallberg-Henriksson, Y. Liang, and K.S. Sakariassen. 2001. 5-aminoimidazole-4-carboxy-amide-1-beta-D-ribofuranoside treatment ameliorates hyperglycaemia and hyperinsulinaemia but not dyslipidaemia in KKAy-CETP mice. *Diabetologia* 44:2180-6.

Fisher, J.S., J. Gao, D.H. Han, J.O. Holloszy, and L.A. Nolte. 2002. Activation of AMP kinase enhances sensitivity of muscle glucose transport to insulin. *Am J Physiol Endocrinol Metab* 282:E18-23.

Fryer, L.G., F. Foufelle, K. Barnes, S.A. Baldwin, A. Woods, and D. Carling. 2002. Characterization of the role of the AMP-activated protein kinase in the stimulation of glucose transport in skeletal muscle cells. *Biochem J* 363:167-74.

Fryer, L.G., E. Hajduch, F. Rencurel, I.P. Salt, H.S. Hundal, D.G. Hardie et al. 2000. Activation of glucose transport by AMP-activated protein kinase via stimulation of nitric oxide synthase. *Diabetes* 49:1978-85.

Fryer, L.G., A. Parbu-Patel, and D. Carling. 2002. The antidiabetic drugs rosiglitazone and metformin stimulate AMP-activated protein kinase through distinct signaling pathways. *J Biol Chem* 277:25226-32.

Fujii, N., W.G. Aschenbach, N. Musi, M.F. Hirshman, and L.J. Goodyear. 2004. Regulation of glucose transport by the AMP-activated protein kinase. *Proc Nutr Soc* 63:205-10.

Hardie, D.G. 2004. The AMP-activated protein kinase pathway—new players upstream and downstream. *J Cell Sci* 117:5479-87.

Hardie, D.G. 2005. New roles for the LKB1-->AMPK pathway. *Curr Opin Cell Biol* 17:167-73.

Hardie, D.G., and D. Carling. 1997. The AMP-activated protein kinase—fuel gauge of the mammalian cell? *Eur J Biochem* 246:259-73.

Hardie, D.G., and K. Sakamoto. 2006. AMPK: A key sensor of fuel and energy status in skeletal muscle. *Physiology (Bethesda)* 21:48-60.

Harrington, L.S., G.M. Findlay, A. Gray, T. Tolkacheva, S. Wigfield, H. Rebholz et al. 2004. The TSC1-2 tumor suppressor controls insulin-PI3K signaling via regulation of IRS proteins. *J Cell Biol* 166:213-23.

Hawley, J.A. 2004. Exercise as a therapeutic intervention for the prevention and treatment of insulin resistance. *Diabetes Metab Res Rev* 20:383-93.

Hawley, S.A., J. Boudeau, J.L. Reid, K.J. Mustard, L. Udd, T.P. Makela et al. 2003. Complexes between the LKB1 tumor suppressor, STRAD alpha/beta and MO25 alpha/beta are upstream kinases in the AMP-activated protein kinase cascade. *J Biol* 2:28.

Hawley, S.A., A.E. Gadalla, G.S. Olsen, and D.G. Hardie. 2002. The antidiabetic drug metformin activates the AMP-activated protein kinase cascade via an adenine nucleotide-independent mechanism. *Diabetes* 51:2420-5.

Hawley, S.A., D.A. Pan, K.J. Mustard, L. Ross, J. Bain, A.M. Edelman et al. 2005. Calmodulin-dependent protein kinase kinase-beta is an alternative upstream kinase for AMP-activated protein kinase. *Cell Metab* 2:9-19.

Hayashi, T., M.F. Hirshman, N. Fujii, S.A. Habinowski, L.A. Witters, and L.J. Goodyear. 2000. Metabolic stress and altered glucose transport: Activation of AMP-activated protein kinase as a unifying coupling mechanism. *Diabetes* 49:527-31.

Hayashi, T., M.F. Hirshman, E.J. Kurth, W.W. Winder, and L.J. Goodyear. 1998. Evidence for 5' AMP-activated protein kinase mediation of the effect of muscle contraction on glucose transport. *Diabetes* 47:1369-73.

Hojlund, K., K.J. Mustard, P. Staehr, D.G. Hardie, H. Beck-Nielsen, E.A. Richter et al. 2004. AMPK activity and isoform protein expression are similar in muscle of obese subjects with and without type 2 diabetes. *Am J Physiol Endocrinol Metab* 286:E239-44.

Holmes, B.F., E.J. Kurth-Kraczek, and W.W. Winder. 1999. Chronic activation of 5'-AMP-activated protein kinase increases GLUT-4, hexokinase, and glycogen in muscle. *J Appl Physiol* 87:1990-5.

Hutber, C.A., D.G. Hardie, and W.W. Winder. 1997. Electrical stimulation inactivates muscle acetyl-CoA carboxylase and increases AMP-activated protein kinase. *Am J Physiol* 272:E262-6.

Inoki, K., T. Zhu, and K.L. Guan. 2003. TSC2 mediates cellular energy response to control cell growth and survival. *Cell* 115:577-90.

Ivy, J.L. 2004. Muscle insulin resistance amended with exercise training: Role of GLUT4 expression. *Med Sci Sports Exerc* 36:1207-11.

Jakobsen, S.N., D.G. Hardie, N. Morrice, and H.E. Tornqvist. 2001. 5'-AMP-activated protein kinase phosphorylates IRS-1 on Ser-789 in mouse C2C12 myotubes in response to 5-aminoimidazole-4-carboxamide riboside. *J Biol Chem* 276:46912-6.

Jessen, N., R. Pold, E.S. Buhl, L.S. Jensen, O. Schmitz, and S. Lund. 2003. Effects of AICAR and exercise on insulin-stimulated glucose uptake, signaling, and GLUT-4 content in rat muscles. *J Appl Physiol* 94:1373-9.

Jorgensen, S.B., B. Viollet, F. Andreelli, C. Frosig, J.B. Birk, P. Schjerling et al. 2004. Knockout of the alpha2 but not alpha1 5'-AMP-activated protein kinase isoform abolishes 5-aminoimidazole-4-carboxamide-1-beta-4-ribofuranoside but not contraction-induced glucose uptake in skeletal muscle. *J Biol Chem* 279:1070-9.

Khamzina, L., A. Veilleux, S. Bergeron, and A. Marette. 2005. Increased activation of the mammalian target of rapamycin pathway in liver and skeletal muscle of obese rats: Possible involvement in obesity-linked insulin resistance. *Endocrinology* 146:1473-81.

Knowler, W.C., E. Barrett-Connor, S.E. Fowler, R.F. Hamman, J.M. Lachin, E.A. Walker et al. 2002. Reduction in the incidence of type 2 diabetes with lifestyle intervention or metformin. *New Engl J Med* 346:393-403.

Kurth-Kraczek, E.J., M.F. Hirshman, L.J. Goodyear, and W.W. Winder. 1999. 5' AMP-activated protein kinase activation causes GLUT4 translocation in skeletal muscle. *Diabetes* 48:1667-71.

Lebrasseur, N.K., M. Kelly, T.S. Tsao, S.R. Farmer, A.K. Saha, N.B. Ruderman et al. 2006. Thiazolidinediones can rapidly activate AMP-activated protein kinase (AMPK) in mammalian tissues. *Am J Physiol Endocrinol Metab* 291:E175-81.

Lessard, S.J., Z.P. Chen, M.J. Watt, M. Hashem, J.J. Reid, M.A. Febbraio et al. 2006. Chronic rosiglitazone treatment restores AMPKalpha2 activity in insulin-resistant rat skeletal muscle. *Am J Physiol Endocrinol Metab* 290:E251-7.

Lihn, A.S., N. Jessen, S.B. Pedersen, S. Lund, and B. Richelsen. 2004. AICAR stimulates adiponectin and inhibits cytokines in adipose tissue. *Biochem Biophys Res Commun* 316:853-8.

Liu, Y., Q. Wan, Q. Guan, L. Gao, and J. Zhao. 2006. High-fat diet feeding impairs both the expression and activity of AMPKa in rats' skeletal muscle. *Biochem Biophys Res Commun* 339:701-7.

Merrill, G.F., E.J. Kurth, D.G. Hardie, and W.W. Winder. 1997. AICA riboside increases AMP-activated protein kinase, fatty acid oxidation, and glucose uptake in rat muscle. *Am J Physiol* 273:E1107-12.

Motoshima, H., B.J. Goldstein, M. Igata, and E. Araki. 2006. AMPK and cell proliferation—AMPK as a therapeutic target for atherosclerosis and cancer. *J Physiol* 574:63-71.

Musi, N., N. Fujii, M.F. Hirshman, I. Ekberg, S. Froberg, O. Ljungqvist et al. 2001. AMP-activated protein kinase (AMPK) is activated in muscle of subjects with type 2 diabetes during exercise. *Diabetes* 50:921-7.

Musi, N., and L.J. Goodyear. 2002. Targeting the AMP-activated protein kinase for the treatment of type 2 diabetes. *Curr Drug Targets Immune Endocr Metabol Disord* 2:119-27.

Musi, N., and L.J. Goodyear. 2006. Insulin resistance and improvements in signal transduction. *Endocrine* 29:73-80.

Musi, N., M.F. Hirshman, J. Nygren, M. Svanfeldt, P. Bavenholm, O. Rooyackers et al. 2002. Metformin increases AMP-activated protein kinase activity in skeletal muscle of subjects with type 2 diabetes. *Diabetes* 51:2074-81.

Nielsen, J.N., S.B. Jorgensen, C. Frosig, B. Viollet, F. Andreelli, S. Vaulont et al. 2003. A possible role for AMP-activated protein kinase in exercise-induced glucose utilization: Insights from humans and transgenic animals. *Biochem Soc Trans* 31:186-90.

Perreault, M., and A. Marette. 2001. Targeted disruption of inducible nitric oxide synthase protects against obesity-linked insulin resistance in muscle. *Nat Med* 7:1138-43.

Pilon, G., P. Dallaire, and A. Marette. 2004. Inhibition of inducible nitric-oxide synthase by activators of AMP-activated protein kinase: A new mechanism of action of insulin-sensitizing drugs. *J Biol Chem* 279:20767-74.

Pold, R., L.S. Jensen, N. Jessen, E.S. Buhl, O. Schmitz, A. Flyvbjerg et al. 2005. Long-term AICAR administration and exercise prevents diabetes in ZDF rats. *Diabetes* 54:928-34.

Reiter, A.K., D.R. Bolster, S.J. Crozier, S.R. Kimball, and L.S. Jefferson. 2005. Repression of protein synthesis and mTOR signaling in rat liver mediated by the AMPK activator aminoimidazole carboxamide ribonucleoside. *Am J Physiol Endocrinol Metab* 288:E980-8.

Russell III, R.R., R. Bergeron, G.I. Shulman, and L.H. Young. 1999. Translocation of myocardial GLUT-4 and increased glucose uptake through activation of AMPK by AICAR. *Am J Physiol* 277:H643-9.

Saha, A.K., P.R. Avilucea, J.M. Ye, M.M. Assifi, E.W. Kraegen, and N.B. Ruderman. 2004. Pioglitazone treatment activates AMP-activated protein kinase in rat liver and adipose tissue in vivo. *Biochem Biophys Res Commun* 314:580-5.

Sakamoto, K., O. Goransson, D.G. Hardie, and D.R. Alessi. 2004. Activity of LKB1 and AMPK-related kinases in skeletal muscle: Effects of contraction, phenformin, and AICAR. *Am J Physiol Endocrinol Metab* 287:E310-7.

Sakamoto, K., A. McCarthy, D. Smith, K.A. Green, D. Grahame Hardie, A. Ashworth et al. 2005. Deficiency of LKB1 in skeletal muscle prevents AMPK activation and glucose uptake during contraction. *EMBO J* 24:1810-20.

Shah, O.J., Z. Wang, and T. Hunter. 2004. Inappropriate activation of the TSC/Rheb/mTOR/S6K cassette induces IRS1/2 depletion, insulin resistance, and cell survival deficiencies. *Curr Biol* 14:1650-6.

Shaw, R.J., M. Kosmatka, N. Bardeesy, R.L. Hurley, L.A. Witters, R.A. DePinho et al. 2004. The tumor suppressor LKB1 kinase directly activates AMP-activated kinase and regulates apoptosis in response to energy stress. *Proc Natl Acad Sci USA* 101:3329-35.

Shaw, R.J., K.A. Lamia, D. Vasquez, S.H. Koo, N. Bardeesy, R.A. Depinho et al. 2005. The kinase LKB1 mediates glucose homeostasis in liver and therapeutic effects of metformin. *Science* 310:1642-6.

Smith, J.L., P.B. Patil, and J.S. Fisher. 2005. AICAR and hyperosmotic stress increase insulin-stimulated glucose transport. *J Appl Physiol* 99:877-83.

Song, X.M., M. Fiedler, D. Galuska, J.W. Ryder, M. Fernstrom, A.V. Chibalin et al. 2002. 5-Aminoimidazole-4-carboxamide ribonucleoside treatment improves glucose homeostasis in insulin-resistant diabetic (ob/ob) mice. *Diabetologia* 45:56-65.

Sriwijitkamol, A., J.L. Ivy, C. Christ-Roberts, R.A. DeFronzo, L.J. Mandarino, and N. Musi. 2006. LKB1-AMPK signaling in muscle from obese insulin-resistant Zucker rats and effects of training. *Am J Physiol Endocrinol Metab* 290:E925-32.

Thomson, D.M., and S.E. Gordon. 2005. Diminished overload-induced hypertrophy in aged fast-twitch skeletal muscle is associated with AMPK hyperphosphorylation. *J Appl Physiol* 98:557-64.

Thomson, D.M., and S.E. Gordon. 2006. Impaired overload-induced muscle growth is associated with diminished translational signaling in aged rat fast-twitch skeletal muscle. *J Physiol* 574:291-305.

Tremblay, F., and A. Marette. 2001. Amino acid and insulin signaling via the mTOR/p70 S6 kinase pathway. A negative feedback mechanism leading to insulin resistance in skeletal muscle cells. *J Biol Chem* 276:38052-60.

Tzatsos, A., and K.V. Kandror. 2006. Nutrients suppress phosphatidylinositol 3-kinase/Akt signaling via raptor-dependent mTOR-mediated insulin receptor substrate 1 phosphorylation. *Mol Cell Biol* 26:63-76.

Ueno, M., J.B. Carvalheira, R.C. Tambascia, R.M. Bezerra, M.E. Amaral, E.M. Carneiro et al. 2005. Regulation of insulin signalling by hyperinsulinaemia: Role of IRS-1/2 serine phosphorylation and the mTOR/p70 S6K pathway. *Diabetologia* 48:506-18.

Um, S.H., F. Frigerio, M. Watanabe, F. Picard, M. Joaquin, M. Sticker et al. 2004. Absence of S6K1 protects against age- and diet-induced obesity while enhancing insulin sensitivity. *Nature* 431:200-5.

Vavvas, D., A. Apazidis, A.K. Saha, J. Gamble, A. Patel, B.E. Kemp et al. 1997. Contraction-induced changes in acetyl-CoA carboxylase and 5'-AMP-activated kinase in skeletal muscle. *J Biol Chem* 272:13255-61.

Villena, J.A., B. Viollet, F. Andreelli, A. Kahn, S. Vaulont, and H.S. Sul. 2004. Induced adiposity and adipocyte hypertrophy in mice lacking the AMP-activated protein kinase-alpha2 subunit. *Diabetes* 53:2242-9.

Viollet, B., F. Andreelli, S.B. Jorgensen, C. Perrin, A. Geloen, D. Flamez et al. 2003. The AMP-activated protein kinase alpha2 catalytic subunit controls whole-body insulin sensitivity. *J Clin Invest* 111:91-8.

Winder, W.W. 2000. AMP-activated protein kinase: Possible target for treatment of type 2 diabetes. *Diabetes Technol Ther* 2:441-8.

Winder, W.W., and D.G. Hardie. 1996. Inactivation of acetyl-CoA carboxylase and activation of AMP-activated protein kinase in muscle during exercise. *Am J Physiol* 270:E299-304.

Winder, W.W., and D.G. Hardie. 1999. AMP-activated protein kinase, a metabolic master switch: Possible roles in type 2 diabetes. *Am J Physiol* 277:E1-10.

Winder, W.W., B.F. Holmes, D.S. Rubink, E.B. Jensen, M. Chen, and J.O. Holloszy. 2000. Activation of AMP-activated protein kinase increases mitochondrial enzymes in skeletal muscle. *J Appl Physiol* 88:2219-26.

Winder, W.W., H.A. Wilson, D.G. Hardie, B.B. Rasmussen, C.A. Hutber, G.B. Call et al. 1997. Phosphorylation of rat muscle acetyl-CoA carboxylase by AMP-activated protein kinase and protein kinase A. *J Appl Physiol* 82:219-25.

Witters, L.A., B.E. Kemp, and A.R. Means. 2006. Chutes and ladders: The search for protein kinases that act on AMPK. *Trends Biochem Sci* 31:13-16.

Woods, A., S.R. Johnstone, K. Dickerson, F.C. Leiper, L.G. Fryer, D. Neumann et al. 2003. LKB1 is the upstream kinase in the AMP-activated protein kinase cascade. *Curr Biol* 13:2004-8.

Wright, D.C., K.A. Hucker, J.O. Holloszy, and D.H. Han. 2004. Ca²⁺ and AMPK both mediate stimulation of glucose transport by muscle contractions. *Diabetes* 53:330-5.

Yeh, J.I., E.A. Gulve, L. Rameh, and M.J. Birnbaum. 1995. The effects of wortmannin on rat skeletal muscle. Dissociation of signaling pathways for insulin- and contraction-activated hexose transport. *J Biol Chem* 270:2107-11.

Yu, X., S. McCorkle, M. Wang, Y. Lee, J. Li, A.K. Saha et al. 2004. Leptinomimetic effects of the AMP kinase activator AICAR in leptin-resistant rats: Prevention of diabetes and ectopic lipid deposition. *Diabetologia* 47:2012-21.

Zhou, G., R. Myers, Y. Li, Y. Chen, X. Shen, J. Fenyk-Melody et al. 2001. Role of AMP-activated protein kinase in mechanism of metformin action. *J Clin Invest* 108:1167-74.

Chapter 14

Protein Kinase C and Insulin Resistance

Carsten Schmitz-Peiffer, PhD

Isoforms of the protein kinase C (PKC) family of signaling enzymes play wide-ranging roles in the regulation of cell function. Although certain PKCs have been implicated in the insulin-induced or exercise-induced stimulation of glucose uptake in muscle, others exert negative effects on insulin action and may be important contributors to the development of insulin resistance. This chapter examines the mechanisms by which particular PKCs may modulate muscle glucose metabolism, especially the negative regulation of insulin action that occurs in times of fat oversupply. Because of this inhibition, certain PKCs may be suitable sites for intervention to improve insulin sensitivity.

PKC Family of Serine and Threonine Kinases

At least 10 PKC isoforms exist. They have been grouped into 3 subfamilies based on their structural and regulatory properties (Mellor and Parker 1998; Parker and Murray-Rust 2004). They are of interest in obesity and lipid-induced insulin resistance because of their dependence on lipid second messengers, especially DAG, for activation.

The classical PKC isoforms (cPKCs) alpha, beta1, beta2, and gamma were the earliest characterized. They depend on both DAG and Ca^{++} due to the presence of fully functional C1 and C2 domains (figure 14.1a). These domains bind the activators after the kinase associates with cell membranes, inducing a conformational change that relieves the autoinhibitory action of the pseudosubstrate

region on the catalytic site (figure 14.1b). PKCγ appears to be expressed mostly in the central nervous system, whereas other cPKCs are far more widely expressed and have been detected in skeletal muscle (Avignon et al. 1996; Schmitz-Peiffer, Browne et al. 1997). A typical setting for the activation of cPKCs involves the stimulation of phospholipase C (PLC), for example through G protein–coupled receptors or receptor tyrosine kinases, which elevates intracellular levels of both DAG and also inositol-1,4,5-trisphosphate, a Ca^{++}-releasing agent (Rhee 2001).

The novel PKC isoforms (nPKCs) delta, epsilon, eta, and theta were subsequently discovered. They are DAG sensitive but independent of Ca^{++} for activation because their C2-related domains are unable to coordinate Ca^{++} ions (figure 14.1). Thus, although nPKCs can be activated by the acute release of DAG by PLC action, other circumstances in which intracellular DAG levels become elevated, such as obesity, also give rise to their activation. PKCδ and PKCϵ are ubiquitous, and PKCη is highly expressed in lung and skin (Hug and Sarre 1993). PKCθ is of particular interest in the study of skeletal muscle because it is selectively expressed in muscle and hematopoietic cells (Baier et al. 1994; Osada et al. 1992).

In addition to Ca^{++} and DAG, the phosphorylation of specific residues of cPKCs and nPKCs is also necessary for full activation. The details are still unclear, but it seems that maturation of cPKCs involves phosphorylation of a conserved activation loop in the catalytic domain, carried out by PDK1, and subsequent autophosphorylation at C-terminal

The author would like to acknowledge the support of the National Health and Medical Research Council of Australia and the Diabetes Australia Research Trust.

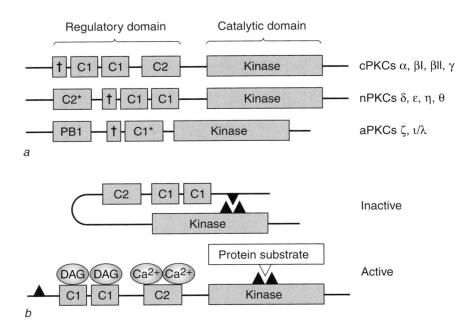

Figure 14.1 Structural features of PKC isoforms. *(a)* The PKC family consists of three groups: conventional, novel, and atypical. The pseudosubstrate regions (†), C1, C2, modified C2 (C2*), and PB1 domains of each group are indicated. *(b)* Under basal conditions, the pseudosubstrate region interacts with the catalytic site of the kinase domain in an autoinhibitory fashion. When activators such as DAG and Ca⁺⁺ are present, the enzyme undergoes a conformational change that allows phosphorylation of protein substrates.

sites, which holds the kinases in their final Ca⁺⁺- and DAG-responsive conformation (Newton 2003). Such phosphorylation events are also necessary in the case of nPKCs, but in certain systems they may occur during the acute activation process, rendering these enzymes sensitive to agents that activate the PI3K–PDK1 axis as well as to those promoting PLC activity (Newton 2003).

Atypical PKCs (aPKCs) PKCζ and PKCι/λ comprise the third subfamily of PKC. They depend on neither Ca⁺⁺ or DAG. They completely lack a C2 domain and have only a modified C1 domain, which does not bind DAG (figure 14.1). They are acutely activated upon PI3K stimulation by PDK1, which also phosphorylates aPKCs at the activation loop (Newton 2003). Because these kinases exhibit acidic amino acids that mimic the C-terminal phosphorylation of cPKCs and nPKCs, no subsequent autophosphorylation is necessary. In addition, activation by the association of protein binding partners is mediated through the N-terminal PB1 domain (figure 14.1). There is some confusion in the literature concerning the misnamed PKCλ, which is not a distinct isoform but merely the mouse homologue of PKCι (Selbie et al. 1993). The presence of aPKCs in muscle is well documented, but it

has been difficult to distinguish between PKCζ and PKCι/λ activity due to the lack of specificity of the available antibodies.

Both cPKCs and nPKCs, and to a lesser extent aPKCs, are recovered in different subcellular fractions upon stimulation. Thus, a commonly used indicator of PKC activation is the demonstration of such translocation, most often from a cytosolic fraction to a membrane fraction (Kraft and Anderson 1983). Redistribution has also been demonstrated by immunofluorescence microscopy (Wang et al. 1999), suggesting that PKCs phosphorylate their protein substrates at specific cellular locations. Although this may simply relate to the increased presence of DAG in membranes, it may also often reflect the action of various PKC-binding proteins, such as the receptors for activated C kinase (RACKs), which target the kinases to particular sites upon activation (Schechtman and Mochly-Rosen 2001).

The elevation of DAG levels is normally transient; however, more prolonged translocation of cPKC and nPKC isoforms can occur following physiological stimulation (e.g., through phospholipase D), which gives rise to further DAG generation (Kiley et al. 1991; Olivier and Parker 1994). In addition, potent activation of PKC by pharmacological means using phorbol esters (nonhydrolysable DAG analogues) also causes chronic translocation. In either case, PKC is subjected to proteolysis as a consequence of prolonged activation (Kiley et al. 1992; Olivier and Parker 1994; Young et al. 1987). PKC isoforms differ in their sensitivity to such downregulation, with PKCε appearing less susceptible (Borner et al. 1992; Huwiler, Fabbro, and Pfeilschifter 1991).

Roles for PKC in Normal Glucose Homeostasis

Given the importance of the PKC family in signal transduction and cellular regulation, it is not surprising that a number of isoforms have been implicated in the signaling pathways that promote glucose disposal. First, there has been much focus on defining the role of PKC in insulin-stimulated glucose uptake, and second, several studies have addressed the action of these kinases in mediating enhanced glucose uptake in response to exercise.

Isoform Involvement in Insulin Action

The dependence of PKCs on PI3K and PDK1, as well as the key role that PI3K plays in insulin signal transduction, may help to explain why many studies have reported the activation of one or more PKC isoforms upon insulin stimulation. The best-characterized activation is that of the aPKCs, which may then act in parallel with Akt to promote GLUT4 translocation and other aspects of glucose metabolism (Dugani and Klip 2005; Farese, Sajan, and Standaert 2005; Oriente et al. 2001). Although several substrates for Akt that may be involved in downstream insulin action have been described (Watson and Pessin 2006), few reports have identified corresponding substrates for aPKCs. It seems that PKCζ can phosphorylate vesicle-associated membrane protein 2 (VAMP2; Braiman et al. 2001a) and also form a complex with Munc18c (Hodgkinson, Mander, and Sale 2005), proteins involved in regulating GLUT4 vesicle trafficking. PKCζ may also promote GLUT4 translocation to the plasma membrane by mediating a remodeling of the submembrane actin mesh (Liu et al. 2006).

Determination of aPKC activity by immunoprecipitation and in vitro kinase assay or by immunoblotting with antibodies specific to the phosphorylation site of the activation loop has revealed defects in aPKC activation in response to insulin in muscle from subjects with type 2 diabetes and animal models of diabetes (Kanoh et al. 2003; Sajan et al. 2004). This most likely reflects a reduced activation of PI3K by the hormone; however, several studies failed to show an inhibition of Akt (Schmitz-Peiffer 2000), also PI3K dependent. This has been interpreted as a tighter coupling of aPKC to PI3K activation, perhaps due to the dependence of this kinase on higher levels of PIP_3, the product of PI3K (Taniguchi et al. 2006).

Although less widely reproduced, insulin has also been reported to stimulate the activity of various cPKC and nPKC isoforms. From one series of studies, it appears that in the presence of insulin, PKCδ associates with the insulin receptor and promotes proximal insulin signaling and hence glucose uptake in cultured muscle cells (Braiman, Alt et al. 1999; Braiman et al. 2001b; Braiman, Sheffi-Friedman et al. 1999). This may involve tyrosine phosphorylation of the kinase, perhaps by the insulin receptor itself or by Src tyrosine kinase. In other work, a role for PKCβ2 but not PKCβ1 has been described in the positive modulation of insulin-stimulated glucose transport (Chalfant et al. 1995, 1996). Further investigation, however, is needed to clarify the involvement of these kinases in insulin action and to determine the mechanisms of activation by the hormone.

PKC Activation in Contracting Muscle

In addition to mediating insulin-stimulated GLUT4 translocation, PKC activity has also been implicated in the enhanced glucose uptake observed in exercising muscle. Early reports involving isolated rat skeletal muscle indicated that Ca^{++}- and DAG-dependent PKC activity was stimulated or induced to translocate in exercising muscle (Cleland et al. 1989, 1990; Henriksen, Rodnick, and Holloszy 1989; Richter et al. 1987). Such activation was associated with the contraction-mediated increase in glucose uptake in these studies, whereas it was not associated with insulin-mediated glucose transport. On the other hand, activation of PKCδ has been linked to the exercise-induced improvements in upstream insulin signaling in insulin-resistant sand rats (Heled et al. 2002, 2003). Interestingly, PKCδ has now been suggested to mediate some of the positive effects of IL-6, a cytokine acutely elevated by exercise in muscle, on insulin signaling and glucose disposal (Weigert et al. 2006).

More recent work, including studies carried out with human subjects, has focused on the activation of aPKC by exercise (Beeson et al. 2003; Chen et al. 2002; Nielsen et al. 2003; Perrini et al. 2004; Richter et al. 2004; Rose et al. 2004). Again, it appears that the activation mechanism is independent of proximal insulin signaling, being insensitive to inhibition of PI3K, although the precise pathway has yet to be determined. From these findings it has been proposed that aPKC activity promotes glucose metabolism in exercising muscle, perhaps through the same enhancement of GLUT4 translocation that it elicits in insulin-stimulated cells. Alternatively, the effects of aPKC induced by contraction could be mediated through the AMP-activated kinase pathway, which also plays a role in exercise-stimulated glucose transport (Rose and Richter 2005) and which has been identified as a downstream target of PKCζ (Xie et al. 2006).

It is therefore conceivable that physical activity may exert at least part of its beneficial effects on glucose homeostasis through the activation of PKC isoforms, especially PKCζ. Though the details are still unclear, contraction-mediated enhancement of PKC activity may circumvent the defects in proximal insulin signaling that are responsible for insulin resistance, thereby enhancing GLUT4 translocation to the plasma membrane to promote glucose uptake into muscle cells.

PKC and Defective Glucose Disposal

Although the role of PKC isoforms in insulin- and contraction-stimulated glucose uptake is still emerging, there is a significant body of evidence linking chronic PKC activation to negative modulation of insulin-stimulated glucose disposal. This evidence comprises work carried out in cultured cells as well as in animal and human models of insulin resistance.

Isoforms Linked to the Generation of Insulin Resistance

The contribution of PKC to lipid-induced insulin resistance has been extensively investigated. Early work was mostly correlative, linking increased

activation of PKC to lipid availability and insulin resistance. Thus intracellular redistribution of PKCε and PKCθ was observed in the skeletal muscle of fat-fed rats, which also exhibited increased muscle triglyceride and DAG content as well as diminished glucose disposal during a euglycemic-hyperinsulinemic clamp (Schmitz-Peiffer, Browne et al. 1997). A reduction in the fat oversupply to skeletal muscle by treatment with rosiglitazone, a TZD that lowers lipid availability (Oakes et al. 1994), was able to prevent the changes in PKC distribution and improve insulin sensitivity (Schmitz-Peiffer, Oakes et al. 1997). Similarly, acute lipid infusion, which also leads to skeletal muscle insulin resistance, promoted translocation of PKCθ in rats (Griffin et al. 1999), whereas chronic glucose infusion elevated muscle lipid content and induced PKCε translocation, even after plasma glucose levels were normalized (Laybutt et al. 1999). Activation of these nPKC isoforms was also observed in genetic models of obesity and diabetes (Ikeda et al. 2001; Qu, Seale, and Donnelly 1999a; Shafrir and Ziv 1998).

Other isoforms have also been implicated. Acute insulin resistance caused by lipid infusion has been associated with translocation of PKCδ in liver (Lam et al. 2002) and PKCβ and PKCδ in muscle (Itani et al. 2002). Increased levels of PKCβ have been measured in human muscle from patients who were obese or diabetic (Itani et al. 2000). Finally, the use of phorbol esters and PKC inhibitors has provided preliminary evidence for a negative effect of PKC activation on glucose disposal in muscle and fat cells from insulin-resistant subjects (Cortright et al. 2000) and in rat muscle (Lin et al. 2001).

These studies are consistent with the hypothesis that inappropriate activation of one or more PKC isoforms, especially PKCε and PKCθ, can interfere with glucose disposal. Studies in our laboratory have shown that these PKC isoforms are preferentially activated in cultured muscle cells by unsaturated fatty acids, such as oleate and linoleate, rather than by saturated fatty acids, such as palmitate (Cazzolli et al. 2002). This has been confirmed in mouse skeletal muscle by comparing the effects of high-fat diets rich in either saturated or unsaturated fatty acids on PKC translocation (Schmitz-Peiffer, unpublished data). This selectivity most likely reflects the greater ability of DAG species that contain unsaturated acyl side chains to activate PKCs (Wakelam 1998), because DAG accumulates upon increased fatty acid availability (Schmitz-Peiffer, Browne et al. 1997; Yu et al. 2002). On the other hand, some studies have implicated PKC activation in the inhibition of insulin action by palmitate (Jove et al. 2005, 2006; Montell et al. 2001; Reynoso, Salgado, and Calderon 2003), suggesting that DAG molecules with saturated acyl side chains are not inert in this respect.

The downregulation of PKCθ that accompanies its translocation in models of lipid oversupply (Schmitz-Peiffer, Browne et al. 1997; Schmitz-Peiffer, Oakes et al. 1997; Yu et al. 2002) appears as a decrease in cytosolic protein, while membrane-associated levels remain unchanged. This has still been interpreted as chronic activation in response to elevated lipids, because the ratio of membrane-bound to cytosolic PKCθ increases. However, the downregulation could also indicate reduced potential for subsequent PKC activation, which would argue against an inhibitory role for PKCθ in the longer term. As discussed later in this chapter, supporting evidence for either interpretation has been reported (Cazzolli et al. 2002; Gao et al. 2007; Kim et al. 2004; Serra et al. 2003).

PKC activation has been implicated in several studies of insulin resistance and diabetes, even in the absence of lipid oversupply (Avignon et al. 1996; Considine et al. 1995; Donnelly et al. 1994; Qu, Seale, and Donnelly 1999b; Tang et al. 1993). In some cases this may indicate hyperinsulinemia or hyperglycemia. Insulin itself can elevate DAG levels (Chen et al. 1994; Heydrick et al. 1991; Standaert et al. 1996) and lead to alterations in activation and expression of PKC isoforms (Avignon et al. 1995; Yamada et al. 1995). Hyperglycemia has been linked to increased DAG levels by de novo synthesis and hence PKC activation (Koya and King 1998). However, even when PKC activity is increased as a secondary event, it may still contribute to further insulin resistance. It is therefore of great interest to determine whether the correlation between PKC activation and impaired insulin action is causal and to understand the mechanisms involved.

Cross Talk Among PKC Isoforms and Proximal Insulin Signaling

Some studies have suggested that PKC might modulate insulin action at distal sites, such as at the level of GLUT4 translocation (Condorelli et

al. 1998), or of metabolic enzymes, including GS (Ahmad et al. 1984; Eto, Karginov, and Brautigan 1999). However, most investigations have focused on proximal insulin signal transduction. PKC activity increases phosphorylation of the insulin receptor at various sites in vitro or in intact cells (Ahn, Donner, and Rosen 1993; Caruso et al. 1999; Chin et al. 1993; Koshio, Akanuma, and Kasuga 1989; Lewis et al. 1990; Liu and Roth 1994). However, the role of such phosphorylation is still unclear since receptor tyrosine kinase activity was not always affected (Ahn, Rosen, and Donner 1993; Bollag et al. 1986; Caruso et al. 1999; Coghlan and Siddle 1993; Takayama, White, and Kahn 1988). It is possible that PKC modulates other aspects of the interaction with insulin, such as receptor internalization and downregulation (Ikeda et al. 2001; Schmitz-Peiffer et al. 2007; Seedorf, Shearman, and Ullrich 1995).

Indirect Action of PKC on IRS1

More recent work has examined the role of PKC in the promotion of IRS1 serine phosphorylation, which aids in the negative regulation of insulin signaling by inhibiting IRS1 tyrosine phosphorylation and subsequent recruitment of downstream signaling components such as PI3K (Zick 2003). Pharmacological activation of PKC or overexpression of specific isoforms has been associated with IRS1 serine phosphorylation and diminished downstream signaling (Chin, Liu, and Roth 1994; Kellerer et al. 1998; Paz et al. 1997). However, several serine residues have been identified as inhibitory phosphorylation sites, and a number of kinases are also involved. It is therefore possible that PKCs can mediate direct and indirect phosphorylation of IRS1 to inhibit insulin signaling.

The best-characterized serine phosphorylation site of IRS1 is Ser-307 (mouse sequence, equivalent to Ser-312 in human IRS1), which is phosphorylated under conditions that promote insulin resistance, including obesity. This key site is close to the phosphotyrosine binding (PTB) domain of IRS1, which is important in its association with the insulin receptor. Ser-307 phosphorylation appears to be mediated by several alternative kinases, including mTOR, IKKβ, and JNK (Zick 2003). Interestingly, both PKCε and PKCθ act upstream of IKKβ (Khoshnan et al. 2000; Tojima et al. 2000) and JNK (Brandlin et al. 2002; Ghaffari-Tabrizi et al. 1999;

Werlen et al. 1998) in other systems. Thus, PKCθ has been proposed to inhibit insulin signaling through these kinases in adipocytes treated with fatty acid (Gao et al. 2004).

In addition, PKC may participate through the activation of MAPK by acting at the level of the upstream kinases Raf-1 or MEK (Schonwasser et al. 1998). Serines 612, 632, 662, and 731 of IRS1, four potential MAPK phosphorylation sites, are found close to tyrosine phosphorylation sites for the insulin receptor, which reside in binding sites for PI3K. Mutation of these serine residues to alanine increased IRS1-associated PI3K activity (Mothe and Van Obberghen 1996), suggesting that they play an inhibitory role. Activation of PKC in HEK 293 cells stimulated a kinase that can carry out phosphorylation of Ser-612, and livers from ob/ob mice also exhibited an increase in such an activity (De Fea and Roth 1997b). Furthermore, PKC activated MAPK in HEK 293 cells, and inhibition of MAPK blocked the PKC-induced inhibition of PI3K activation by insulin (De Fea and Roth 1997a). Taken together, these findings suggest that MAPK activated by PKC can inhibit IRS1 tyrosine phosphorylation and PI3K activation.

Direct Phosphorylation of IRS1

On the other hand, it has been shown that PKC may directly catalyze the phosphorylation of particular serine molecules on IRS1. Initial reports indicated that PKCζ could phosphorylate this protein in vitro and mediate its phosphorylation in intact cells to inhibit downstream signaling. This action most likely represents negative feedback after insulin stimulation of the kinase (Liu et al. 2001; Ravichandran et al. 2001), but the process may become inappropriately activated under conditions leading to insulin resistance.

Descriptions of direct IRS1 phosphorylation by different PKC isoforms now exist. PKCθ promotes Ser-1101 phosphorylation in vitro and in intact cells, which inhibits subsequent IRS1 tyrosine phosphorylation and downstream signaling (Li et al. 2004). Likewise, PKCβ can also phosphorylate IRS1 in vitro, and overexpressing PKCβ in adipocytes leads to inhibition of insulin-stimulated glucose uptake (Ishizuka et al. 2004). PKCδ phosphorylates a number of sites on IRS1 in vitro, of which Ser-307, Ser-323, and Ser-574 appear to mediate the

inhibitory effects of the kinase on insulin action in intact cells (Greene et al. 2004). Finally, the recently described phosphorylation of Ser-318 may be directly mediated by more than one PKC isoform, including PKCζ activated in response to insulin (Moeschel et al. 2004) and PKCδ recruited to IRS1 by elevated IL-6 levels (Weigert et al. 2006). This phosphorylation appears to acutely enhance insulin signaling, but in the longer term it promotes negative feedback, such that the insulin resistance caused by hyperinsulinemia may be partially mediated through this site (Moeschel et al. 2004; Weigert et al. 2005). Although the role of IL-6 in the modulation of muscle insulin sensitivity is controversial and requires clarification, the effect of Ser-318 phosphorylation may be time dependent. It has been suggested that both the short-term benefits of exercise, which elevates IL-6 levels, and the long-term downregulation of insulin action by leptin are mediated through the phosphorylation of this IRS1 site in muscle (Hennige et al. 2006; Weigert et al. 2006).

IRS1 Ser-24 has recently been described as a novel site for PKCα-mediated serine phosphorylation (Nawaratne et al. 2006). This site lies in the pleckstrin homology domain of the docking protein, which plays an important role in its intracellular localization by mediating the binding to phosphoinositide membrane lipids. Thus, phosphorylation at this site impaired lipid binding (Nawaratne et al. 2006). A separate study of the interaction between PKCα and IRS1 indicated that the kinase both inhibited insulin signaling and promoted insulin degradation. These effects were countered by the presence of the scaffolding protein 14-3-3ε, which inhibited PKCα activity (Oriente et al. 2005).

The in vitro phosphorylation of IRS1 by PKCs, and also the effects of overexpressing a particular isoform in intact cells, should be interpreted with caution, partly because the specificity of PKCs for protein substrates such as IRS1 is determined to a large degree by binding proteins. Binding proteins limit their cellular localization (Jaken and Parker 2000) and may be circumvented under these experimental conditions. Ideally, findings made in this way should be supported by the reciprocal effects of inhibition or deletion of particular PKCs. Unfortunately, pharmacological inhibitors of PKC isoforms have limited specificity, due in large part to the conservation of the catalytic site within the PKC family as well as within the larger superfamily of protein kinases. However, various studies have indicated that peptides derived from individual PKC regulatory domains can block the interaction of these kinases with their binding partners, preventing their normal function (Hool 2005; Mackay and Mochly-Rosen 2001). This approach has yet to be applied to the phosphorylation of IRS1, but it may provide new insights into the disruption of insulin signaling by specific PKC isoforms.

It is clear that PKC activity can inhibit insulin signaling through IRS1 by mediating phosphorylation at certain serine residues. The relative importance of these residues is unclear, and it is likely that certain sites such as Ser-307 act as gatekeepers, integrating the signals from different inhibitory pathways (Schmitz-Peiffer and Whitehead 2003), while other sites are involved in fine-tuning IRS1 signaling potential.

PKC-Mediated Inhibition of Akt

The phosphorylation and activation of Akt by insulin is an essential step in the translocation of GLUT4 to the plasma membrane, and a defect in this step has been found in models of lipid-induced insulin resistance (Chavez et al. 2003; Schmitz-Peiffer, Craig, and Biden 1999; Storz et al. 1999; Tremblay et al. 2001). Cells pretreated with agents that inhibit insulin-stimulated glucose disposal, such as saturated fatty acids (Powell et al. 2004; Schmitz-Peiffer, Craig, and Biden 1999) or TNFα (Teruel, Hernandez, and Lorenzo 2001), can exhibit diminished Akt phosphorylation in the absence of changes in IRS1 tyrosine phosphorylation and PI3K activation, indicating a direct effect at the level of Akt.

A similar finding has been made with cells treated with ceramide (Summers et al. 1998), a sphingolipid that can be generated by de novo synthesis from fatty acids and also released acutely by the action of sphingomyelinase in response to TNFα. Ceramide therefore mediates the actions of different inhibitory factors; however, more than one mechanism has been proposed for the effect of the lipid intermediate on Akt. Ceramide can activate the serine and threonine phosphatase PP2A, leading to increased Akt dephosphorylation (Cazzolli et al. 2001; Stratford et al. 2004), but it can also interfere with Akt membrane recruitment and

phosphorylation (Hajduch et al. 2001; Stratford et al. 2004). In addition, ceramide can activate PKCζ (Lozano et al. 1994; Powell et al. 2004), which also acts as a negative regulator of Akt (Bourbon, Sandirasegarane, and Kester 2002; Doornbos et al. 1999; Konishi, Kuroda, and Kikkawa 1994; Mao et al. 2000), preventing its association with specific membrane phospholipids by phosphorylation of its PH domain (Powell et al. 2003). This is consistent with the suggestion that aPKCζ acts as a molecular switch (Muller et al. 1995) that either promotes or inhibits glucose disposal depending on the presence of insulin and ceramide (Powell et al. 2003).

There are conflicting reports as to whether other PKC isoforms such as PKCε positively or negatively regulate Akt (Li et al. 2006; Matsumoto et al. 2001). Although some of the cPKC and nPKC isoforms are also activated by ceramide (Huwiler, Fabbro, and Pfeilschifter 1998), they have not been implicated in the effects of the sphingolipid on insulin action. As described earlier, these isoforms tend to be activated in the presence of unsaturated rather than saturated fatty acids, whereas de novo ceramide generation is a feature of saturated fatty acid oversupply (Powell et al. 2004; Schmitz-Peiffer, Craig, and Biden 1999). The importance of ceramide in intact tissue and whole-body insulin resistance, as well as its beneficial modulation by exercise, is now becoming apparent (Dobrzyn, Knapp, and Gorski 2004; Holland et al. 2007; Straczkowski et al. 2004; Summers and Nelson 2005).

Cross Talk With Other Signaling Pathways

In addition to the direct inhibition of insulin signaling components, PKCs may also mediate the activation of other pathways, which in turn interfere with insulin action. As mentioned, PKC isoforms have been placed upstream of IKKβ. This kinase could mediate increased serine phosphorylation of IRS1, but it also plays a key role in activating transcription dependent on NF-κβ. IKKβ phosphorylates cytosolic IκB to promote its degradation, enabling nuclear translocation of NF-κβ and the subsequent increase in expression of inflammatory genes that depend on NF-κβ.

Evidence for cross talk between PKC and NF-κβ has been obtained in studies of human skeletal muscle (Itani et al. 2002), cultured muscle cells (Jove et al. 2005, 2006), and rat liver (Boden et al. 2005). NF-κβ activation leads to the increased expression of inflammatory cytokines, such as TNFα, IL-1β, and IL-6 (Boden et al. 2005; Jove et al. 2005, 2006), and iNOS (Aktan 2004), each of which can negatively affect insulin action. NF-κβ-independent induction of IL-6 expression by PKCε has also been reported (Ohashi et al. 2005). The inhibitory signaling and proinflammatory events mediated by PKCs, which in turn may lead to insulin resistance, are summarized in figure 14.2.

In contrast to NF-κβ activation, other events mediated by PKC are less well characterized. PKCδ may contribute to oxidative stress-induced insulin resistance through the activation of NADPH oxidase under conditions of lipid excess (Talior et al. 2003, 2005), although a direct phosphorylation remains to be shown. The substrates for PKC isoforms are poorly defined in general; however, PKCθ but not PKCα or PKCε has recently been shown to phosphorylate N-myc downstream-regulated protein 2 (NDRG2) in vitro and in intact muscle cells, thus preventing its phosphorylation by Akt in response to insulin (Burchfield et al. 2004). NDRG2 may help regulate cell differentiation and tumor suppression; however, its role in insulin action remains to be elucidated.

Effects of PKC Isoform Deletion on Whole-Body Insulin Action

Given the evidence supporting a role for PKC in the negative regulation of insulin action, as well as the wide range of mechanisms proposed, there have been relatively few studies rigorously addressing causation as opposed to association in whole-animal models of diabetes. This is in part because specific pharmacological inhibitors have yet to be characterized for most PKC isoforms, and even the commonly used general PKC inhibitors are not without effect on other classes of kinases (Davies et al. 2000). The most compelling studies concerning a causative role for a particular isoform in the regulation of insulin action have therefore compared glucose disposal in wild-type and PKC KO or transgenic mice. The first reports of insulin action in such animals concerned the cPKCs. PKCα KO mice exhibited enhanced insulin signaling and glucose uptake in muscle and fat cells (Leitges et al. 2002). However, whole-body glucose tolerance and a possible protection against insulin resistance were not addressed. Similarly, though PKCβ KO mice also showed minor improvements in glucose tolerance, it is unknown whether these mice are less susceptible to insulin resistance (Standaert et al. 1999).

On the other hand, the effects of functional ablation of PKCθ on the susceptibility to lipid-induced

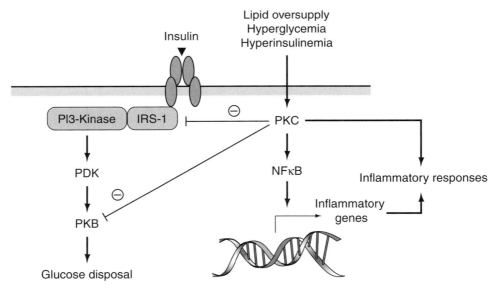

Figure 14.2 Inhibitory actions of PKC. Agents that chronically activate PKCs can lead to inhibition of insulin signaling at the level of IRS1 and Akt. Alternatively, PKC activation can affect insulin action more indirectly, such as by promoting inflammatory responses that induce insulin resistance, in some cases by activating NFκβ.

insulin resistance have been examined in more detail. First, the effects of transgenic overexpression of a kinase dead mutant of PKCθ in muscle were investigated (Serra et al. 2003). This mutant was expected to compete for PKCθ substrates and thus inhibit the endogenous PKCθ in a dominant negative manner, protecting against the inhibitory effect of the kinase on insulin signaling (Baier-Bitterlich et al. 1996). Surprisingly, given the association of PKCθ translocation with insulin resistance in skeletal muscle, the transgene had adverse effects on body weight and insulin sensitivity. The transgenic mice exhibited hyperinsulinemia, indicative of insulin resistance, at 4 mo; became obese due to accumulation of visceral adipose tissue at 6 to 7 mo; and demonstrated impaired glucose homeostasis as assessed by glucose tolerance tests.

These results suggest that the association of diminished cytosolic PKCθ with insulin resistance represents a reduced capacity for PKCθ function rather than an activation of the kinase with inhibitory consequences. In support of this, we have failed to reverse the insulin resistance of glycogen synthesis in lipid-pretreated myotubes by the overexpression of dominant negative mutants of PKC (Cazzolli et al. 2002).

Also in agreement are the results of work involving fat-fed PKCθ KO mice, which exhibited increased susceptibility to obesity and dietary insulin resistance, most likely due to reduced energy expenditure (Gao et al. 2007). In a separate study involving the same whole-body PKCθ KO mice, however, deletion of the kinase prevented defects in insulin signaling through IRS1 and glucose disposal in muscle caused by acute lipid infusion, leading to protection from whole-body insulin resistance (Kim et al. 2004). This latter finding is in agreement with the more common interpretation of the correlative data, which argues that an increased proportion of membrane-associated PKCθ, irrespective of any change in total PKCθ, indicates a chronic PKC activation that interferes with insulin action, most likely through IRS1 serine and threonine phosphorylation.

Since these studies differ in their ablation of PKCθ activity and in the time frame of study, it is difficult to draw firm conclusions about the role of PKCθ in glucose homeostasis. However, the conflicting reports could be partly reconciled by the interpretation that acute activation of PKCθ by increased lipid availability leads to inhibitory effects on insulin signaling, whereas over the longer term the kinase plays a positive role in muscle development, as recently described (D'Andrea et al. 2006), and hence in glucose metabolism.

A causative role for PKCε in insulin resistance has not been as widely addressed, even though at least three PKCε KO mouse lines have been generated (Castrillo et al. 2001; Gruber et al. 2005; Hodge et al. 1999). Unlike PKCθ, this enzyme is not significantly downregulated upon chronic activation (Schmitz-Peiffer, Browne et al. 1997), and it should be easier to interpret the correlative evidence concerning its role in insulin resistance. Recent work, however, has indicated that although fat-fed PKCε KO mice do indeed show improved glucose tolerance compared with wild-type littermates, this effect is not explained by protection against skeletal muscle or hepatic insulin resistance but by enhanced insulin availability (Schmitz-Peiffer et al. 2007). Our findings contrast with those made using a short term model of insulin resistance, which suggested that PKCε can mediate the acute inhibitory effects of lipids on insulin action (Samuel et al. 2007). Therefore, either this isoform does not play a role in the generation of insulin resistance by chronic fat oversupply, or other kinases can compensate functionally for the lack of this isoform in skeletal muscle. Again, support for the former has been reported in studies of lipid-treated myotubes (Cazzolli et al. 2002).

There have been no reports of altered glucose metabolism or protection from insulin resistance in mice deficient in either PKCγ or PKCδ, most likely due to the limited evidence for the involvement of these kinases. Finally, although whole-body knockout of aPKC isoforms is lethal, embryonic stem cells deficient in PKCι/λ that are subsequently induced to differentiate into adipocytes demonstrate a requirement for this enzyme in insulin-stimulated glucose transport (Bandyopadhyay et al. 2004). A similar finding has now been made using muscle-specific PKCι/λ knockout mice (Farese et al. 2007), again in agreement with earlier work indicating the role of this isoform in normal insulin action.

Concluding Remarks

Understanding of the modulation of glucose disposal by PKC has progressed considerably from early observations that specific isoforms are chronically activated in association with defective insulin action. Although aPKCs are emerging as positive regulators, correlative studies have established PKCθ and PKCε as the strongest candidates for mediating insulin resistance. Mechanistic details are becoming apparent, with a focus on increased serine phosphorylation of IRS1 and the upregulation of inflammatory genes. From the results of genetic manipulation, it is clear that PKCθ has important regulatory effects on glucose homeostasis that may not be limited to the inhibition of insulin action in skeletal muscle, as first hypothesized.

A potential for both positive and negative roles may not be limited to this isoform. This is indicated by the actions of PKCζ, which appears to promote glucose uptake acutely at the level of GLUT4 translocation but also downregulates prolonged insulin action by increasing IRS1 serine phosphorylation and also independently attenuating Akt activation. These inhibitory mechanisms most likely become activated inappropriately under conditions that induce insulin resistance. If PKCζ plays a role in the beneficial effects of exercise on glucose homeostasis, then presumably the acute actions of the enzyme are switched from IRS1 phosphorylation and Akt inhibition to GLUT4 translocation.

The relationship between the activation of PKC and the inhibitory effects of other kinases such as JNK and IKKβ, for which causative roles in insulin resistance have been more clearly demonstrated (Arkan et al. 2005; Cai et al. 2005; Hirosumi et al. 2002), is intriguing. These effectors may lie downstream of PKC signaling in specific systems, which could simplify the targeting of inhibitory pathways for therapeutic intervention. Major challenges to the development of small molecule inhibitors of PKC isoforms include the specificity and potency of such compounds. With the exception of LY333531, which shows specificity toward PKCβ (Jirousek et al. 1996) and may be useful in the treatment of diabetic complications due to hyperglycemia (Kelly et al. 2003), there are currently no specific inhibitors available for PKC isoforms. Future inhibitors could include molecules designed to block the interaction of the kinases with their binding proteins (Begley et al. 2004; Stallings-Mann et al. 2006), which may allow a level of specificity unattained by catalytic site antagonists. Indeed, such an approach could intervene in particular PKC-mediated pathways while leaving other pathways intact, which may be advantageous given the ubiquitous nature of these kinases.

In summary, the targeting of particular PKC isoforms is likely to modulate glucose homeostasis in a positive manner. However, further work is required to delineate the downstream signaling from these enzymes in skeletal muscle and other tissues involved in glucose disposal and to develop suitably specific small molecule inhibitors.

References

Ahmad, Z., F.T. Lee, R.A. DePaoli, and P.J. Roach. 1984. Phosphorylation of glycogen synthase by the Ca^{2+}- and phospholipid-activated protein kinase (protein kinase C). *J Biol Chem* 259:8743-7.

Ahn, J., O.M. Rosen, and D.B. Donner. 1993. Human insulin receptor mutated at threonine 1336 functions normally in Chinese hamster ovary cells. *J Biol Chem* 268:16839-44.

Ahn, J.C., D.B. Donner, and O.M. Rosen. 1993. Interaction of the human insulin receptor tyrosine kinase from the baculovirus expression system with protein kinase-C in a cell-free system. *J Biol Chem* 268:7571-6.

Aktan, F. 2004. iNOS-mediated nitric oxide production and its regulation. *Life Sci* 75:639-53.

Arkan, M.C., A.L. Hevener, F.R. Greten, S. Maeda, Z.W. Li, J.M. Long, A. Wynshaw-Boris, G. Poli, J. Olefsky, and M. Karin. 2005. IKK-beta links inflammation to obesity-induced insulin resistance. *Nat Med* 11:191-8.

Avignon, A., M.L. Standaert, K. Yamada, H. Mischak, B. Spencer, and R.V. Farese. 1995. Insulin increases mRNA levels of protein kinase C-alpha and -beta in rat adipocytes and protein kinase C-alpha, -beta and -theta in rat skeletal muscle. *Biochem J* 308:181-7.

Avignon, A., K. Yamada, X.P. Zhou, B. Spencer, O. Cardona, S. Sabasiddique, L. Galloway, M.L. Standaert, and R.V. Farese. 1996. Chronic activation of protein kinase C in soleus muscles and other tissues of insulin-resistant type II diabetic Goto-Kakizaki (GK), obese/aged, and obese/Zucker rats—a mechanism for inhibiting glycogen synthesis. *Diabetes* 45:1396-404.

Baier, G., G. Baier-Bitterlich, N. Meller, K.M. Coggeshall, L. Giampa, D. Telford, N. Isakov, and A. Altman. 1994. Expression and biochemical characterization of human protein kinase C-theta. *Eur J Biochem* 225:195-203.

Baier-Bitterlich, G., F. Uberall, B. Bauer, F. Fresser, H. Wachter, H. Grunicke, G. Utermann, A. Altman, and G. Baier. 1996. Protein kinase C-theta isoenzyme selective stimulation of the transcription factor complex AP-1 in T lymphocytes. *Mol Cell Biol* 16:1842-50.

Bandyopadhyay, G., M.L. Standaert, M.P. Sajan, Y. Kanoh, A. Miura, U. Braun, F. Kruse, M. Leitges, and R.V. Farese. 2004. Protein kinase C-lambda knockout in embryonic stem cells and adipocytes impairs insulin-stimulated glucose transport. *Mol Endocrinol* 18:373-83.

Beeson, M., M.P. Sajan, M. Dizon, D. Grebenev, J. Gomez-Daspet, A. Miura, Y. Kanoh, J. Powe, G. Bandyopadhyay, M.L. Standaert, and R.V. Farese. 2003. Activation of protein kinase C-zeta by insulin and phosphatidylinositol-3,4,5-(PO4)(3) is defective in muscle in type 2 diabetes and impaired glucose tolerance—amelioration by rosiglitazone and exercise. *Diabetes* 52:1926-34.

Begley, R., T. Liron, J. Baryza, and D. Mochly-Rosen. 2004. Biodistribution of intracellularly acting peptides conjugated reversibly to Tat. *Biochem Biophys Res Comm* 318:949-54.

Boden, G., P. She, M. Mozzoli, P. Cheung, K. Gumireddy, P. Reddy, X. Xiang, Z. Luo, and N. Ruderman. 2005. Free fatty acids produce insulin resistance and activate the proinflammatory nuclear factor-κB pathway in rat liver. *Diabetes* 54:3458-65.

Bollag, G.E., R.A. Roth, J. Beaudoin, R.D. Mochly, and D.J. Koshland. 1986. Protein kinase C directly phosphorylates the insulin receptor in vitro and reduces its protein-tyrosine kinase activity. *Proc Natl Acad Sci USA* 83:5822-4.

Borner, C., S.N. Guadagno, D. Fabbro, and I.B. Weinstein. 1992. Expression of four protein kinase C isoforms in rat fibroblasts. Distinct subcellular distribution and regulation by calcium and phorbol esters. *J Biol Chem* 267:12892-9.

Bourbon, N.A., L. Sandirasegarane, and M. Kester. 2002. Ceramide-induced inhibition of Akt is mediated through protein kinase C zeta—implications for growth arrest. *J Biol Chem* 277:3286-92.

Braiman, L., A. Alt, T. Kuroki, M. Ohba, A. Bak, T. Tennenbaum, and S.R. Sampson. 1999. Protein kinase C delta mediates insulin-induced glucose transport in primary cultures of rat skeletal muscle. *Mol Endocrinol* 13:2002-12.

Braiman, L., A. Alt, T. Kuroki, M. Ohba, A. Bak, T. Tennenbaum, and S.R. Sampson. 2001a. Activation of protein kinase C zeta induces serine phosphorylation of VAMP2 in the GLUT4 compartment and increases glucose transport in skeletal muscle. *Mol Cell Biol* 21:7852-61.

Braiman, L., A. Alt, T. Kuroki, M. Ohba, A. Bak, T. Tennenbaum, and S.R. Sampson. 2001b. Insulin induces specific interaction between insulin receptor and protein kinase C delta in primary cultured skeletal muscle. *Mol Endocrinol* 15:565-74.

Braiman, L., L. Sheffi-Friedman, A. Bak, T. Tennenbaum, and S.R. Sampson. 1999. Tyrosine phosphorylation of specific protein kinase C isoenzymes participates in insulin stimulation of glucose transport in primary cultures of rat skeletal muscle. *Diabetes* 48:1922-9.

Brandlin, I., T. Eiseler, R. Salowsky, and F.J. Johannes. 2002. Protein kinase C mu regulation of the JNK pathway is triggered via phosphoinositide-dependent kinase 1 and protein kinase C epsilon. *J Biol Chem* 277:45451-7.

Burchfield, J.G., A.J. Lennard, S. Narasimhan, W.E. Hughes, V.C. Wasinger, G.L. Corthals, T. Okuda, H. Kondoh, T.J. Biden, and C. Schmitz-Peiffer. 2004. Akt mediates insulin-stimulated phosphorylation of Ndrg2—evidence for cross-talk with protein kinase C theta. *J Biol Chem* 279:18623-32.

Cai, D.S., M.S. Yuan, D.F. Frantz, P.A. Melendez, L. Hansen, J. Lee, and S.E. Shoelson. 2005. Local and systemic insulin resistance resulting from hepatic activation of IKK-beta and NF-kappa B. *Nat Med* 11:183-90.

Caruso, R., C. Miele, F. Oriente, A. Maitan, G. Bifulco, F. Andreozzi, G. Condorelli, P. Formisano, and F. Beguinot. 1999. In L6 skeletal muscle cells, glucose induces cytosolic translocation of protein kinase C-alpha and trans-activates the insulin receptor kinase. *J Biol Chem* 274:28637-44.

Castrillo, A., D.J. Pennington, F. Otto, P.J. Parker, M.J. Owen, and L. Bosca. 2001. Protein kinase C epsilon is required for macrophage activation and defense against bacterial infection. *J Exp Med* 194:1231-42.

Cazzolli, R., L. Carpenter, T.J. Biden, and C. Schmitz-Peiffer. 2001. A role for protein phosphatase 2A-like activity, but not atypical protein kinase C zeta, in the inhibition of protein kinase B/Akt and glycogen synthesis by palmitate. *Diabetes* 50:2210-8.

Cazzolli, R., D.L. Craig, T.J. Biden, and C. Schmitz-Peiffer. 2002. Inhibition of glycogen synthesis by fatty acid in C2C12 muscle cells is independent of PKC-alpha , -epsilon , and -theta. *Am J Physiol* 282:E1204-13.

Chalfant, C.E., H. Mischak, J.E. Watson, B.C. Winkler, J. Goodnight, R.V. Farese, and D.R. Cooper. 1995. Regulation of alternative splicing of protein kinase C beta by insulin. *J Biol Chem* 270:13326-32.

Chalfant, C.E., S. Ohno, Y. Konno, A.A. Fisher, L.D. Bisnauth, J.E. Watson, and D.R. Cooper. 1996. A carboxy-terminal deletion mutant of protein kinase C beta II inhibits insulin-stimulated 2-deoxyglucose uptake in L6 rat skeletal muscle cells. *Mol Endocrinol* 10:1273-81.

Chavez, J.A., T.A. Knotts, L.P. Wang, G. Li, R.T. Dobrowsky, G.L. Florant, and S.A. Summers. 2003. A role for ceramide, but not diacylglycerol, in the antagonism of insulin signal transduction by saturated fatty acids. *J Biol Chem* 278:10297-303.

Chen, H.C., G. Bandyopadhyay, M.P. Sajan, Y. Kanoh, M. Standaert, and R.V. Farese. 2002. Activation of the ERK pathway and atypical protein kinase C isoforms in exercise- and aminoimidazole-4-carboxamide-1-beta-D-riboside (AICAR)-stimulated glucose transport. *J Biol Chem* 277:23554-62.

Chen, K.S., S.J. Heydrick, M.L. Brown, J.C. Friel, and N.B. Ruderman. 1994. Insulin increases a biochemically distinct pool of diacylglycerol in the rat soleus muscle. *Am J Physiol* 266:E479-85.

Chin, J.E., M. Dickens, J.T. Tavare, and R.A. Roth. 1993. Overexpression of protein kinase-C isoenzymes-alpha, beta-I, gamma and epsilon in cells overexpressing the insulin receptor— effects on receptor phosphorylation and signalling. *J Biol Chem* 268:6338-47.

Chin, J.E., F. Liu, and R.A. Roth. 1994. Activation of protein kinase C alpha inhibits insulin-stimulated tyrosine phosphorylation of insulin receptor substrate-1. *Mol Endocrinol* 8:51-8.

Cleland, P.J., K.C. Abel, S. Rattigan, and M.G. Clark. 1990. Long-term treatment of isolated rat soleus muscle with phorbol ester leads to loss of contraction-induced glucose transport. *Biochem J* 267:659-63.

Cleland, P.J.F., G.J. Appleby, S. Rattigan, and M.G. Clark. 1989. Exercise-induced translocation of protein kinase C and production of diacylglycerol and phosphatidic acid in rat skeletal muscle in vivo. *J Biol Chem* 264:17704-11.

Coghlan, M.P., and K. Siddle. 1993. Phorbol esters induce insulin receptor phosphorylation in transfected fibroblasts without affecting tyrosine kinase activity. *Biochem Biophys Res Commun* 193:371-7.

Condorelli, G., G. Vigliotta, C. Iavarone, M. Caruso, C.G. Tocchetti, F. Andreozzi, A. Cafieri, M.F. Tecce, P. Formisano, L. Beguinot, and F. Beguinot. 1998. Ped/pea-15 gene controls glucose transport and is overexpressed in type 2 diabetes mellitus. *EMBO J* 17:3858-66.

Considine, R.V., M.R. Nyce, L.E. Allen, L.M. Morales, S. Triester, J. Serrano, J. Colberg, S. Lanzajacoby, and J.F. Caro. 1995. Protein kinase C is increased in the liver of humans and rats with noninsulin-dependent diabetes mellitus: An alteration not due to hyperglycemia. *J Clin Invest* 95:2938-44.

Cortright, R.N., J.L. Azevedo, Q. Zhou, M. Sinha, W.J. Pories, S.I. Itani, and G.L. Dohm. 2000. Protein kinase C modulates insulin action in human skeletal muscle. *Am J Physiol* 278:E553-62.

D'Andrea, M., A. Pisaniello, C. Serra, M.I. Senni, L. Castaldi, M. Molinaro, and M. Bouche. 2006. Protein kinase C theta co-operates with calcineurin in the activation of slow muscle genes in cultured myogenic cells. *J Cell Physiol* 207:379-88.

Davies, S.P., H. Reddy, M. Caivano, and P. Cohen. 2000. Specificity and mechanism of action of some commonly used protein kinase inhibitors. *Biochem J* 351:95-105.

De Fea, K., and R.A. Roth. 1997a. Modulation of insulin receptor substrate-1 tyrosine phosphorylation and function by mitogen-activated protein kinase. *J Biol Chem* 272:31400-6.

De Fea, K., and R.A. Roth. 1997b. Protein kinase C modulation of insulin receptor substrate-1 tyrosine phosphorylation requires serine 612. *Biochemistry* 36:12939-947.

Dobrzyn, A., M. Knapp, and J. Gorski. 2004. Effect of acute exercise and training on metabolism of ceramide in the heart muscle of the rat. *Acta Physiol Scand* 181:313-9.

Donnelly, R., M.J. Reed, S. Azhar, and G.M. Reaven. 1994. Expression of the major isoenzyme of protein kinase-C in skeletal muscle, nPKC theta, varies with muscle type and in response to fructose-induced insulin resistance. *Endocrinology* 135:2369-74.

Doornbos, R.P., M. Theelen, D.H.P. van der Hoeven, W.J. van Blitterswijk, A.J. Verkleij, and P. Henegouwen. 1999. Protein kinase C zeta is a negative regulator of protein kinase B activity. *J Biol Chem* 274:8589-96.

Dugani, C.B., and A. Klip. 2005. Glucose transporter 4: Cycling, compartments and controversies. *EMBO Rep* 6:1137-42.

Eto, M., A. Karginov, and D.L. Brautigan. 1999. A novel phosphoprotein inhibitor of protein type-1 phosphatase holoenzymes. *Biochemistry* 38:16952-7.

Farese, R.V., M.P. Sajan, and M.L. Standaert. 2005. Atypical protein kinase C in insulin action and insulin resistance. *Biochem Soc Trans* 33:350-3.

Farese, R.V., M.P. Sajan, H. Yang, P. Li, S. Mastorides, W.R. Gower, Jr., S. Nimal, C.S. Choi, S. Kim, G.I. Shulman, C.R. Kahn, U. Braun, and M. Leitges. 2007. Muscle-specific knockout of PKC-lambda impairs glucose transport and induces metabolic and diabetic syndromes. *J Clin Invest* 117:2289-301.

Gao, Z., Z. Wang, X. Zhang, A.A. Butler, A. Zuberi, B. Gawronska-Kozak, M. Lefevre, D. York, E. Ravussin, H.R. Berthoud, O. McGuinness, W.T. Cefalu, and J. Ye. 2007. Inactivation of PKCtheta leads to increased susceptibility to obesity and dietary insulin resistance in mice. *Am J Physiol* 292(1): E84-91.

Gao, Z.G., X.Y. Zhang, A. Zuberi, D. Hwang, M.J. Quon, M. Lefevre, and J.P. Ye. 2004. Inhibition of insulin sensitivity by free fatty acids requires activation of multiple serine kinases in 3T3-L1 adipocytes. *Mol Endocrinol* 18:2024-34.

Ghaffari-Tabrizi, N., B. Bauer, A. Villunger, G. Baier-Bitterlich, A. Altman, G. Utermann, F. Uberall, and G. Baier. 1999. Protein kinase C theta, a selective upstream regulator of JNK/SAPK and IL-2 promoter activation in Jurkat T cells. *Eur J Immunol* 29:132-42.

Greene, M.W., N. Morrice, R.S. Garofalo, and R.A. Roth. 2004. Modulation of human insulin receptor substrate-1 tyrosine phosphorylation by protein kinase C delta. *Biochem J* 378:105-16.

Griffin, M.E., M.J. Marcucci, G.W. Cline, K. Bell, N. Barucci, D. Lee, L.J. Goodyear, E.W. Kraegen, M.F. White, and G.I. Shulman. 1999. Free fatty acid-induced insulin resistance is associated with activation of protein kinase C q and alterations in the insulin signaling cascade. *Diabetes* 48:1270-4.

Gruber, T., N. Thuille, N. Hermann-Kleiter, M. Leitges, and G. Baier. 2005. Protein kinase Cepsilon is dispensable for TCR/CD3-signaling. *Mol Immunol* 42:305-10.

Hajduch, E., A. Balendran, I.H. Batty, G.J. Litherland, A.S. Blair, C.P. Downes, and H.S. Hundal. 2001. Ceramide impairs the insulin-dependent membrane recruitment of protein kinase B leading to a loss in downstream signalling in L6 skeletal muscle cells. *Diabetologia* 44:173-83.

Heled, Y., Y. Shapiro, Y. Shani, D.S. Moran, L. Langzam, L. Braiman, S.R. Sampson, and J. Meyerovitch. 2002. Physical exercise prevents the development of type 2 diabetes mellitus in Psammomys obesus. *Am J Physiol* 282:E370-5.

Heled, Y., Y. Shapiro, Y. Shani, D.S. Moran, L. Langzam, L. Braiman, S.R. Sampson, and J. Meyerovitch. 2003. Physical exercise enhances protein kinase C delta activity and insulin receptor tyrosine phosphorylation in diabetes-prone Psammomys obesus. *Metabolism* 52:1028-33.

Hennige, A.M., N. Stefan, K. Kapp, R. Lehmann, C. Weigert, A. Beck, K. Moeschel, J. Mushack, E. Schleicher, and H.U. Haring. 2006. Leptin down-regulates insulin action through phosphorylation of serine-318 in insulin receptor substrate 1. *FASEB J* 20:1206-8.

Henriksen, E.J., K.J. Rodnick, and J.O. Holloszy. 1989. Activation of glucose transport in skeletal muscle by phospholipase C and phorbol ester. Evaluation of the regulatory roles of protein kinase C and calcium. *J Biol Chem* 264:21536-43.

Heydrick, S.J., N.B. Ruderman, T.G. Kurowski, H.B. Adams, and K.S. Chen. 1991. Enhanced stimulation of diacylglycerol and lipid synthesis by insulin in denervated muscle—altered protein kinase-C activity and possible link to insulin resistance. *Diabetes* 40:1707-11.

Hirosumi, J., G. Tuncman, L.F. Chang, C.Z. Gorgun, K.T. Uysal, K. Maeda, M. Karin, and G.S. Hotamisligil. 2002. A central role for JNK in obesity and insulin resistance. *Nature* 420:333-6.

Hodge, C.W., K.K. Mehmert, S.P. Kelley, T. McMahon, A. Haywood, M.F. Olive, D. Wang, A.M. Sanchez-Perez, and R.O. Messing. 1999. Supersensitivity to allosteric GABA(A) receptor modulators and alcohol in mice lacking PKCepsilon. *Nat Neurosci* 2:997-1002.

Hodgkinson, C.P., A. Mander, and G.J. Sale. 2005. Protein kinase-zeta interacts with munc18c: Role in GLUT4 trafficking. *Diabetologia* 48:1627-36.

Holland, W.L., J.T. Brozinick, L.P. Wang, E.D. Hawkins, K.M. Sargent, Y. Liu, K. Narra, K.L. Hoehn, T.A. Knotts, A. Siesky, D.H. Nelson, S.K. Karathanasis, G.K. Fontenot, M.J. Birnbaum, and S.A. Summers. 2007. Inhibition of ceramide synthesis ameliorates glucocorticoid-, saturated-fat-, and obesity-induced insulin resistance. *Cell Metab* 5:167-79.

Hool, L.C. 2005. Protein kinase C isozyme selective peptides—a current view of what they tell us about location and function of isozymes in the heart. *Curr Pharm Des* 11:549-59.

Hug, H., and T.F. Sarre. 1993. Protein kinase-C isoenzymes—divergence in signal transduction. Biochem J 291:329-43.

Huwiler, A., D. Fabbro, and J. Pfeilschifter. 1991. Differential recovery of protein kinase-C-alpha and kinase-C-epsilon isozymes after long-term phorbol ester treatment in rat renal mesangial cells. *Biochem Biophys Res Commun* 180:1422-8.

Huwiler, A., D. Fabbro, and J. Pfeilschifter. 1998. Selective ceramide binding to protein kinase C-alpha and -delta isoenzymes in renal mesangial cells. *Biochemistry* 37:14556-62.

Ikeda, Y., G.S. Olsen, E. Ziv, L.L. Hansen, A.K. Busch, B.F. Hansen, E. Shafrir, and L. Mosthaf-Seedorf. 2001. Cellular mechanism of nutritionally induced insulin resistance in Psammomys obesus—overexpression of protein kinase Ce in skeletal muscle precedes the onset of hyperinsulinemia and hyperglycemia. *Diabetes* 50:584-92.

Ishizuka, T., K. Kajita, Y. Natsume, Y. Kawai, Y. Kanoh, A. Miura, M. Ishizawa, Y. Uno, H. Morita, and K. Yasuda. 2004. Protein kinase C (PKC) beta modulates serine phosphorylation of insulin receptor substrate-1 (IRS-1)—effect of overexpression of PKC beta on insulin signal transduction. *Endocr Res* 30:287-99.

Itani, S.I., N.B. Ruderman, F. Schmieder, and G. Boden. 2002. Lipid-induced insulin resistance in human muscle is associated with changes in diacylglycerol, protein kinase C, and IκB-a. *Diabetes* 51:2005-11.

Itani, S.I., Q. Zhou, W.J. Pories, K.G. MacDonald, and G.L. Dohm. 2000. Involvement of protein kinase C in human skeletal muscle insulin resistance and obesity. *Diabetes* 49:1353-8.

Jaken, S., and P.J. Parker. 2000. Protein kinase C binding partners [Review]. *Bioessays* 22:245-54.

Jirousek, M.R., J.R. Gillig, C.M. Gonzalez, W.F. Heath, J.H. Mcdonald, D.A. Neel, C.J. Rito, U. Singh, L.E. Stramm, A. Melikianbadalian, M. Baevsky, L.M. Ballas, S.E. Hall, L.L. Winneroski, and M.M. Faul. 1996. (S)-13-[(dimethylamino)methyl]-10,11,14,15-tetrahydro-4,9:16,21-dimetheno-1H,13H-dibenzo[e,k]pyrrolo [3,4-H][1,4,13]oxadiazacyclohexadecene-1,3(2H)-dione (LY333531) and related analogues: Isozyme selective inhibitors of protein kinase C beta. *J Med Chem* 39:2664-71.

Jove, M., A. Planavila, J.C. Laguna, and M. Vazquez-Carrera. 2005. Palmitate-induced interleukin 6 production is mediated by protein kinase C and nuclear-factor kappa B activation and leads to glucose transporter 4 down-regulation in skeletal muscle cells. *Endocrinology* 146:3087-95.

Jove, M., A. Planavila, R.M. Sanchez, M. Merlos, J.C. Laguna, and M. Vazquez-Carrera. 2006. Palmitate induces tumor necrosis factor-alpha expression in C2C12 skeletal muscle cells by a mechanism involving protein kinase C and nuclear factor-kappa B activation. *Endocrinology* 147:552-61.

Kanoh, Y., M.P. Sajan, G. Bandyopadhyay, A. Miura, M.L. Standaert, and R.V. Farese. 2003. Defective activation of atypical protein kinase C zeta and lambda by insulin and phosphatidylinositol-3,4,5-(PO₄)(3) in skeletal muscle of rats following high-fat feeding and streptozotocin-induced diabetes. *Endocrinology* 144:947-54.

Kellerer, M., J. Mushack, E. Seffer, H. Mischak, A. Ullrich, and H.U. Haring. 1998. Protein kinase C isoforms alpha, delta and theta require insulin receptor substrate-1 to inhibit the tyrosine kinase activity of the insulin receptor in human kidney embryonic cells (hek 293 cells). *Diabetologia* 41:833-8.

Kelly, D.J., Y. Zhang, C. Hepper, R.M. Gow, K. Jaworski, B.E. Kemp, J.L. Wilkinson-Berka, and R.E. Gilbert. 2003. Protein kinase C beta inhibition attenuates the progression of experimental diabetic nephropathy in the presence of continued hypertension. *Diabetes* 52:512-8.

Khoshnan, A., D. Bae, C.A. Tindell, and A.E. Nel. 2000. The physical association of protein kinase C theta with a lipid raft-associated inhibitor of kappa B factor kinase (IKK) complex plays a role in the activation of the NF-kappa B cascade by TCR and CD28. *J Immunol* 165:6933-40.

Kiley, S.C., P.J. Parker, D. Fabbro, and S. Jaken. 1991. Differential regulation of protein kinase C isozymes by thyrotropin-releasing hormone in GH4C1 cells. *J Biol Chem* 266:23761-8.

Kiley, S.C., P.J. Parker, D. Fabbro, and S. Jaken. 1992. Hormone- and phorbol ester-activated protein kinase C isozymes mediate a reorganization of the actin cytoskeleton associated with prolactin secretion in GH4C1 cells. *Mol Endocrinol* 6:120-31.

Kim, J.K., J.J. Fillmore, M.J. Sunshine, B. Albrecht, T. Higashimori, D.-W. Kim, Z.-X. Liu, T.J. Soos, G.W. Cline, W.R. O'Brien, D.R. Littman, and G.I. Shulman. 2004. PKC-q knockout mice are protected from fat-induced insulin resistance. *J Clin Invest* 114:823-7.

Konishi, H., S. Kuroda, and U. Kikkawa. 1994. The pleckstrin homology domain of RAC protein kinase associates with the regulatory domain of protein kinase C zeta. *Biochem Biophys Res Commun* 205:1770-5.

Koshio, O., Y. Akanuma, and M. Kasuga. 1989. Identification of a phosphorylation site of the rat insulin receptor catalyzed by protein kinase C in an intact cell. *FEBS Lett* 254:22-4.

Koya, D., and G.L. King. 1998. Protein kinase C activation and the development of diabetic complications [Review]. *Diabetes* 47:859-66.

Kraft, A.S., and W.B. Anderson. 1983. Phorbol esters increase the amount of Ca²⁺, phospholipid-dependent protein kinase associated with plasma membrane. *Nature* 301:621-3.

Lam, T.K.T., H. Yoshii, C.A. Haber, E. Bogdanovic, L. Lam, I.G. Fantus, and A. Giacca. 2002. Free fatty acid-induced hepatic insulin resistance: A potential role for protein kinase C-delta. *Am J Physiol* 283:E682-91.

Laybutt, D.R., C. Schmitz-Peiffer, A.K. Saha, N.B. Ruderman, T.J. Biden, and E.W. Kraegen. 1999. Muscle lipid accumulation and protein kinase C activation in the insulin-resistant chronically glucose-infused rat. *Am J Physiol* 277:E1070-6.

Leitges, M., M. Plomann, M.L. Standaert, G. Bandyopadhyay, M.P. Sajan, Y. Kanoh, and R.V. Farese. 2002. Knockout of PKC alpha enhances insulin signaling through PI3K. *Mol Endocrinol* 16:847-58.

Lewis, R.E., L. Cao, D. Perregaux, and M.P. Czech. 1990. Threonine 1336 of the human insulin receptor is a major target for phosphorylation by protein kinase C. *Biochemistry* 29:1807-13.

Li, L.W., K. Sampat, N. Hu, J. Zakari, and S.H. Yuspa. 2006. Protein kinase C negatively regulates Akt activity and modifies UVC-induced apoptosis in mouse keratinocytes. *J Biol Chem* 281:3237-43.

Li, Y., T.J. Soos, X.H. Li, J. Wu, M. DeGennaro, X.J. Sun, D.R. Littman, M.J. Birnbaum, and R.D. Polakiewicz. 2004. Protein kinase C theta inhibits insulin signaling by phosphorylating IRS1 at Ser(1101). *J Biol Chem* 279:45304-7.

Lin, Y.S., S.I. Itani, T.G. Kurowski, D.J. Dean, Z.J. Luo, G.C. Yaney, and N.B. Ruderman. 2001. Inhibition of insulin signaling and glycogen synthesis by phorbol dibutyrate in rat skeletal muscle. *Am J Physiol* 281:E8-15.

Liu, F., and R.A. Roth. 1994. Identification of serines-1035/1037 in the kinase domain of the insulin receptor as protein kinase C alpha mediated phosphorylation sites. *FEBS Lett* 352:389-92.

Liu, L.Z., H.L. Zhao, J. Zuo, S.K.S. Ho, J.C.N. Chan, Y. Meng, F.D. Fang, and P.C.Y. Tong. 2006. Protein kinase C xi mediates insulin-induced glucose transport through actin remodeling in L6 muscle cells. *Mol Biol Cell* 17:2322-30.

Liu, Y.F., K. Paz, A. Herschkovitz, A. Alt, T. Tennenbaum, S.R. Sampson, M. Ohba, T. Kuroki, D. LeRoith, and Y. Zick. 2001. Insulin stimulates PKC zeta-mediated phosphorylation of insulin receptor substrate-1 (IRS-1)—a self-attenuated mechanism to negatively regulate the function of IRS proteins. *J Biol Chem* 276:14459-65.

Lozano, J., E. Berra, M.M. Municio, M.T. Diazmeco, I. Dominguez, L. Sanz, and J. Moscat. 1994. Protein kinase C zeta isoform is critical for kappa B-dependent promoter activation by sphingomyelinase. *J Biol Chem* 269:19200-2.

Mackay, K., and D. Mochly-Rosen. 2001. Localization, anchoring, and functions of protein kinase C isozymes in the heart [Review]. *J Mol Cell Cardiol* 33:1301-7.

Mao, M.L., X.J. Fang, Y.L. Lu, R. LaPushin, R.C. Bast, and G.B. Mills. 2000. Inhibition of growth-factor-induced phosphorylation and activation of protein kinase B/Akt by atypical protein kinase C in breast cancer cells. *Biochem J* 352:475-82.

Matsumoto, M., M. Ogawa, Y. Hino, K. Furukawa, Y. Ono, M. Takahashi, M. Ohba, T. Kuroki, and M. Kasuga. 2001. Inhibition of insulin-induced activation of Akt by a kinase-deficient mutant of the epsilon isozyme of protein kinase C. *J Biol Chem* 276:14400-6.

Mellor, H., and P.J. Parker. 1998. The extended protein kinase C superfamily [Review]. *Biochem J* 332:281-92.

Moeschel, K., A. Beck, C. Weigert, R. Lammers, H. Kalbacher, W. Voelter, E.D. Schleicher, H.U. Haring, and R. Lehmann. 2004. Protein kinase C-zeta-induced phosphorylation of Ser(318) in insulin receptor substrate-1 (IRS-1) attenuates the interaction with the insulin receptor and the tyrosine phosphorylation of IRS-1. *J Biol Chem* 279:25157-63.

Montell, E., M. Turini, M. Marotta, M. Roberts, V. Noe, C.J. Ciudad, K. Mace, and A.M. Gomez-Foix. 2001. DAG accumulation from saturated fatty acids desensitizes insulin stimulation of glucose uptake in muscle cells. *Am J Physiol* 280:E229-37.

Mothe, I., and E. Van Obberghen. 1996. Phosphorylation of insulin receptor substrate-1 on multiple serine residues, 612, 632, 662, and 731, modulates insulin action. *J Biol Chem* 271:11222-7.

Muller, G., M. Ayoub, P. Storz, J. Rennecke, D. Fabbro, and K. Pfizenmaier. 1995. PKC zeta is a molecular switch in signal transduction of TNF-alpha, bifunctionally regulated by ceramide and arachidonic acid. *EMBO J* 14:1961-9.

Nawaratne, R., A. Gray, C.H. Jorgensen, C.P. Downes, K. Siddle, and J.K. Sethi. 2006. Regulation of insulin receptor substrate 1 pleckstrin homology domain by protein kinase C: Role of serine 24 phosphorylation. *Mol Endocrinol* 20:1838-52.

Newton, A.C. 2003. Regulation of the ABC kinases by phosphorylation: Protein kinase C as a paradigm. *Biochem J* 370:361-71.

Nielsen, J.N., C. Frosig, M.P. Sajan, A. Miura, M.L. Standaert, D.A. Graham, J.F.P. Wojtaszewski, R.V. Farese, and E.A. Richter. 2003. Increased atypical PKC activity in endurance-trained human skeletal muscle. *Biochem Biophys Res Commun* 312:1147-53.

Oakes, N.D., C.J. Kennedy, A.B. Jenkins, D.R. Laybutt, D.J. Chisholm, and E.W. Kraegen. 1994. A new antidiabetic agent, BRL 49653, reduces lipid availability and improves insulin action and glucoregulation in the rat. *Diabetes* 43:1203-10.

Ohashi, K., A. Kanazawa, S. Tsukada, and S. Maeda. 2005. PKC epsilon, induces interleukin-6 expression through the MAPK pathway in 3T3-L1 adipocytes. *Biochem Biophys Res Commun* 327:707-12.

Olivier, A.R., and P.J. Parker. 1994. Bombesin, platelet-derived growth factor, and diacylglycerol induce selective membrane association and down-regulation of protein kinase-C isotypes in Swiss 3T3-cells. *J Biol Chem* 269:2758-63.

Oriente, F., F. Andreozzi, C. Romano, G. Perruolo, A. Perfetti, F. Fiory, C. Miele, F. Beguinot, and P. Formisano. 2005. Protein kinase C-alpha regulates insulin action and degradation by interacting with insulin receptor substrate-1 and 14-3-3 epsilon. *J Biol Chem* 280:40642-9.

Oriente, F., P. Formisano, C. Miele, F. Fiory, M.A. Maitan, G. Vigliotta, A. Trencia, S. Santopietro, M. Caruso, E. Van Obberghen, and F. Beguinot. 2001. Insulin receptor substrate-2 phosphorylation is necessary for protein kinase C zeta activation by insulin in L6hIR cells. *J Biol Chem* 276:37109-19.

Osada, S., K. Mizuno, T.C. Saido, K. Suzuki, T. Kuroki, and S. Ohno. 1992. A new member of the protein kinase-C family, nPKCtheta, predominantly expressed in skeletal muscle. *Mol Cell Biol* 12:3930-8.

Parker, P.J., and J. Murray-Rust. 2004. PKC at a glance. *J Cell Sci* 117:131-2.

Paz, K., R. Hemi, D. Leroith, A. Karasik, E. Elhanany, H. Kanety, and Y. Zick. 1997. A molecular basis for insulin resistance. Elevated serine and threonine phosphorylation of IRS-1 and IRS-2 inhibits their binding to the juxtamembrane region of the insulin receptor and impairs their ability to undergo insulin-induced tyrosine phosphorylation. *J Biol Chem* 272:29911-8.

Perrini, S., J. Henriksson, J.R. Zierath, and U. Widegren. 2004. Exercise-induced protein kinase C isoform-specific activation in human skeletal muscle. *Diabetes* 53:21-4.

Powell, D.J., E. Hajduch, G. Kular, and F.S. Hundal. 2003. Ceramide disables 3-phosphoinositide binding to the pleckstrin homology domain of protein kinase B (PKB)/Akt by a PKC zeta-dependent mechanism. *Mol Cell Biol* 23:7794-808.

Powell, D.J., S. Turban, A. Gray, E. Hajduch, and H.S. Hundal. 2004. Intracellular ceramide synthesis and protein kinase C zeta activation play an essential role in palmitate-induced insulin resistance in rat L6 skeletal muscle cells. *Biochem J* 382:619-29.

Qu, X., J.P. Seale, and R. Donnelly. 1999a. Tissue and isoform-selective activation of protein kinase C in insulin-resistant obese Zucker rats—effects of feeding. *J Endocrinol* 162:207-14.

Qu, X.Q., J.P. Seale, and R. Donnelly. 1999b. Tissue- and isoform-specific effects of aging in rats on protein kinase C in insulin-sensitive tissues. *Clin Sci* 97:355-61.

Ravichandran, L.V., D.L. Esposito, J. Chen, and M.J. Quon. 2001. Protein kinase C-zeta phosphorylates insulin receptor substrate-1 and impairs its ability to activate phosphatidylinositol 3-kinase in response to insulin. *J Biol Chem* 276:3543-9.

Reynoso, R., L.M. Salgado, and V. Calderon. 2003. High levels of palmitic acid lead to insulin resistance due to changes in the level of phosphorylation of the insulin receptor and insulin receptor substrate-1. *Mol Cell Biochem* 246:155-62.

Rhee, S.G. 2001. Regulation of phosphoinositide-specific phospholipase C. *Annu Rev Biochem* 70:281-312.

Richter, E.A., P.J. Cleland, S. Rattigan, and M.G. Clark. 1987. Contraction-associated translocation of protein kinase C in rat skeletal muscle. *FEBS Lett* 217:232-6.

Richter, E.A., B. Vistisen, S.J. Maarbjerg, M. Sajan, R.V. Farese, and B. Kiens. 2004. Differential effect of bicycling exercise intensity on activity and phosphorylation of atypical protein kinase C and extracellular signal-regulated protein kinase in skeletal muscle. *J Physiol* 560:909-18.

Rose, A.J., B.J. Michell, B.E. Kemp, and M. Hargreaves. 2004. Effect of exercise on protein kinase C activity and localization in human skeletal muscle. *J Physiol* 561:861-70.

Rose, A.J., and E.A. Richter. 2005. Skeletal muscle glucose uptake during exercise: How is it regulated? *Physiology (Bethesda)* 20:260-70.

Sajan, M.P., M.L. Standaert, A. Miura, G. Bandyopadhyay, P. Vollenweider, D.M. Franklin, R. Lea-Currie, and R.V. Farese. 2004. Impaired activation of protein kinase C-zeta by insulin and phosphatidylinositol-3,4,5-(PO$_4$)(3) in cultured preadipocyte-derived adipocytes and myotubes of obese subjects. *J Clin Endocrinol Metab* 89:3994-8.

Samuel, V.T., Z.X. Liu, A. Wang, S.A. Beddow, J.G. Geisler, M. Kahn, X.M. Zhang, B.P. Monia, S. Bhanot, and G.I. Shulman. 2007. Inhibition of protein kinase Cε prevents hepatic insulin resistance in nonalcoholic fatty liver disease. *J Clin Invest* 117:739-45.

Schechtman, D., and D. Mochly-Rosen. 2001. Adaptor proteins in protein kinase C-mediated signal transduction. *Oncogene* 20:6339-47.

Schmitz-Peiffer, C. 2000. Signalling aspects of insulin resistance in skeletal muscle: Mechanisms induced by lipid oversupply. *Cell Signal* 12:583-94.

Schmitz-Peiffer, C., C.L. Browne, N.D. Oakes, A. Watkinson, D.J. Chisholm, E.W. Kraegen, and T.J. Biden. 1997. Alterations in the expression and cellular localization of protein kinase C isozymes epsilon and theta are associated with insulin resistance in skeletal muscle of the high-fat-fed rat. *Diabetes* 46:169-78.

Schmitz-Peiffer, C., D.L. Craig, and T.J. Biden. 1999. Ceramide generation is sufficient to account for the inhibition of the insulin-stimulated PKB pathway in C2C12 skeletal muscle cells pretreated with palmitate. *J Biol Chem* 274:24202-10.

Schmitz-Peiffer, C., Laybutt, D.R., Burchfield, J.G., Gurisik, E., Narasimhan, S., Mitchell, C.J., Pedersen, D.J., Braun, U., Cooney, G.J., Leitges, M., Biden, T.J. 2007. Inhibition of PKC Improves Glucose-Stimulated Insulin Secretion and Reduces Insulin Clearance. *Cell Metab* (In Press).

Schmitz-Peiffer, C., N.D. Oakes, C.L. Browne, E.W. Kraegen, and T.J. Biden. 1997. Reversal of chronic alterations of skeletal muscle protein kinase C from fat-fed rats by BRL-49653. *Am J Physiol* 273:E915-21.

Schmitz-Peiffer, C., and J.P. Whitehead. 2003. IRS-1 regulation in health and disease. *IUBMB Life* 55:367-74.

Schonwasser, D.C., R.M. Marais, C.J. Marshall, and P.J. Parker. 1998. Activation of the mitogen-activated protein kinase/extracellular signal-regulated kinase pathway by conventional, novel, and atypical protein kinase C isotypes. *Mol Cell Biol* 18:790-8.

Seedorf, K., M. Shearman, and A. Ullrich. 1995. Rapid and long term effects of protein kinase C on receptor tyrosine kinase phosphorylation and degradation. *J Biol Chem* 270:18953-60.

Selbie, L.A., C. Schmitz-Peiffer, Y.H. Sheng, and T.J. Biden. 1993. Molecular cloning and characterization of PKC(iota), an atypical isoform of protein kinase-C derived from insulin-secreting cells. *J Biol Chem* 268:24296-302.

Serra, C., M. Federici, A. Buongiorno, M.I. Senni, S. Morelli, E. Segratella, M. Pascuccio, C. Tiveron, E. Mattei, L. Tatangelo, R. Lauro, M. Molinaro, A. Giaccari, and M. Bouche. 2003. Transgenic mice with dominant negative PKC-theta in skeletal muscle: A new model of insulin resistance and obesity. *J Cell Physiol* 196:89-97.

Shafrir, E., and E. Ziv. 1998. Cellular mechanism of nutritionally induced insulin resistance: The desert rodent Psammomys obesus and other animals in which insulin resistance leads to detrimental outcome. *J Basic Clin Physiol Pharmacol* 9:347-85.

Stallings-Mann, M., L. Jamieson, R.P. Regala, C. Weems, N.R. Murray, and A.P. Fields. 2006. A novel small-molecule inhibitor of protein kinase C iota blocks transformed growth of non-small-cell lung cancer cells. *Cancer Res* 66:1767-74.

Standaert, M.L., G. Bandyopadhyay, L. Galloway, J. Soto, Y. Ono, U. Kikkawa, R.V. Farese, and M. Leitges. 1999. Effects of knockout of the protein kinase C beta gene on glucose transport and glucose homeostasis. *Endocrinology* 140:4470-7.

Standaert, M.L., G. Bandyopadhyay, X.P. Zhou, L. Galloway, and R.V. Farese. 1996. Insulin stimulates phospholipase D-dependent phosphatidylcholine hydrolysis, Rho translocation, de novo phospholipid synthesis, and diacylglycerol/protein kinase C signaling in L6 myotubes. *Endocrinology* 137:3014-20.

Storz, P., H. Doppler, A. Wernig, K. Pfizenmaier, and G. Muller. 1999. Cross-talk mechanisms in the development of insulin resistance of skeletal muscle cells—palmitate rather than tumour necrosis factor inhibits insulin-dependent protein kinase B (PKB)/Akt stimulation and glucose uptake. *Eur J Biochem* 266:17-25.

Straczkowski, M., I. Kowalska, A. Nikolajuk, S. Dzienis-Straczkowska, I. Kinalska, M. Baranowski, M. Zendzian-Piotrowska, Z. Brzezinska, and J. Gorski. 2004. Relationship between insulin sensitivity and sphingomyelin signaling pathway in human skeletal muscle. *Diabetes* 53:1215-21.

Stratford, S., K.L. Hoehn, F. Liu, and S.A. Summers. 2004. Regulation of insulin action by ceramide—dual mechanisms linking ceramide accumulation to the inhibition of Akt/protein kinase B. *J Biol Chem* 279:36608-15.

Summers, S.A., L.A. Garza, H.L. Zhou, and M.J. Birnbaum. 1998. Regulation of insulin-stimulated glucose transporter GLUT4 translocation and Akt kinase activity by ceramide. *Mol Cell Biol* 18:5457-64.

Summers, S.A., and D.H. Nelson. 2005. A role for sphingolipids in producing the common features of type 2 diabetes, metabolic syndrome X, and Cushing's syndrome. *Diabetes* 54:591-602.

Takayama, S., M.F. White, and C.R. Kahn. 1988. Phorbol ester-induced serine phosphorylation of the insulin receptor decreases its tyrosine kinase activity. *J Biol Chem* 263:3440-7.

Talior, I., T. Tennenbaum, T. Kuroki, and H. Eldar-Finkelman. 2005. PKC-delta-dependent activation of oxidative stress in adipocytes of obese and insulin-resistant mice: Role for NADPH oxidase. *Am J Physiol* 288:E405-11.

Talior, I., M. Yarkoni, N. Bashan, and H. Eldar-Finkelman. 2003. Increased glucose uptake promotes oxidative stress and PKC-delta activation in adipocytes of obese, insulin-resistant mice. *Am J Physiol* 285:E295-302.

Tang, E.Y., P.J. Parker, J. Beattie, and M.D. Houslay. 1993. Diabetes induces selective alterations in the expression of protein kinase-C isoforms in hepatocytes. *FEBS Lett* 326:117-23.

Taniguchi, C.M., T. Kondo, M. Sajan, J. Luo, R. Bronson, T. Asano, R. Farese, L.C. Cantley, and C.R. Kahn. 2006. Divergent regulation of hepatic glucose and lipid metabolism by phosphoinositide 3-kinase via Akt and PKC lambda/zeta. *Cell Metab* 3:343-53.

Teruel, T., R. Hernandez, and M. Lorenzo. 2001. Ceramide mediates insulin resistance by tumor necrosis factor-alpha in brown adipocytes by maintaining Akt in an inactive dephosphorylated state. *Diabetes* 50:2563-71.

Tojima, Y., A. Fujimoto, M. Delhase, Y. Chen, S. Hatakeyama, K. Nakayama, Y. Kaneko, Y. Nimura, N. Motoyama, K. Ikeda, M. Karin, and M. Nakanishi. 2000. NAK is an IkappaB kinase-activating kinase. *Nature* 404:778-82.

Tremblay, F., C. Lavigne, H. Jacques, and A. Marette. 2001. Defective insulin-induced GLUT4 translocation in skeletal muscle of high fat-fed rats is associated with alterations in both Akt/protein kinase B and atypical protein kinase C (zeta/lambda) activities. *Diabetes* 50:1901-10.

Wakelam, M. 1998. Diacylglycerol—when is it an intracellular messenger? [Review]. *Biochim Biophys Acta* 1436:117-26.

Wang, Q.J., D. Bhattacharyya, S. Garfield, K. Nacro, V.E. Marquez, and P.M. Blumberg. 1999. Differential localization of protein kinase C delta by phorbol esters and related compounds using a fusion protein with green fluorescent protein. *J Biol Chem* 274:37233-9.

Watson, R.T., and J.E. Pessin. 2006. Bridging the GAP between insulin signaling and GLUT4 translocation. *Trends Biochem Sci* 31:215-22.

Weigert, C., A.M. Hennige, T. Brischmann, A. Beck, K. Moeschel, M. Schauble, K. Brodbeck, H.U. Haring, E.D. Schleicher, and R. Lehmann. 2005. The phosphorylation of Ser(318) of insulin receptor substrate 1 is not per se inhibitory in skeletal muscle cells but is necessary to trigger the attenuation of the insulin-stimulated signal. *J Biol Chem* 280:37393-9.

Weigert, C., A.M. Hennige, R. Lehmann, K. Brodbeck, F. Baumgartner, M. Schauble, H.U. Haring, and E.D. Schleicher. 2006. Direct cross-talk of interleukin-6 and insulin signal transduction via insulin receptor substrate-1 in skeletal muscle cells. *J Biol Chem* 281:7060-7.

Werlen, G., E. Jacinto, Y. Xia, and M. Karin. 1998. Calcineurin preferentially synergizes with PKC-theta to activate JNK and IL-2 promoter in T lymphocytes. *EMBO J* 17:3101-11.

Xie, Z., Y. Dong, M. Zhang, M.Z. Cui, R.A. Cohen, U. Riek, D. Neumann, U. Schlattner, and M.H. Zou. 2006. Activation of protein kinase Czeta by peroxynitrite regulates LKB1-dependent AMP-activated protein kinase in cultured endothelial cells. *J Biol Chem* 281:6366-75.

Yamada, K., A. Avignon, M.L. Standaert, D.R. Cooper, B. Spencer, and R.V. Farese. 1995. Effects of insulin on the translocation of protein kinase C-theta and other protein kinase C isoforms in rat skeletal muscles. *Biochem J* 308:177-80.

Young, S., P.J. Parker, A. Ullrich, and S. Stabel. 1987. Down-regulation of protein kinase C is due to an increased rate of degradation. *Biochem J* 244:775-9.

Yu, C., Y. Chen, H. Zong, Y. Wang, R. Bergeron, J.K. Kim, G.W. Cline, S.W. Cushman, G.J. Cooney, B. Atcheson, M.F. White, E.W. Kraegen, and G.I. Shulman. 2002. Mechanism by which fatty acids inhibit insulin activation of IRS-1 associated phosphatidylinositol 3-kinase activity in muscle. *J Biol Chem* 277:50230-6.

Zick, Y. 2003. Role of Ser/Thr kinases in the uncoupling of insulin signaling. *Int J Obes* 27:S56-60.

Chapter 15

Evidence for Prescribing Exercise as a Therapy for Treating Patients With Type 2 Diabetes

Sarah J. Lessard, PhD; and John A. Hawley, PhD, FACSM

The alarming increase in insulin resistance and type 2 diabetes and its staggering effects on worldwide health and health care costs have been discussed in chapters 1 and 2. Despite the global burden of these metabolic disorders, there have been relatively few advances in their prevention and treatment over the last decade. Current treatments fall short in their ability to prevent type 2 diabetes and its secondary complications, as evidenced by the relentlessly increasing rates of diabetes.

As outlined by Booth, Chakravarthy, and Laye in chapter 2, there is convincing evidence that inactivity contributes significantly to the onset of insulin resistance and type 2 diabetes. Thus, it follows that lifestyle modification in the form of exercise training may be highly effective in preventing and treating these disease states. In this chapter we examine the evidence for exercise training as a therapeutic intervention for insulin resistance and diabetes and suggest how it may fit into the current treatment paradigm for type 2 diabetes. This chapter also summarizes the molecular and epidemiological evidence for exercise training as a therapy for type 2 diabetes with respect to its ability to reverse or prevent the potential molecular defects leading to insulin resistance.

Options for Treating Insulin Resistance and Type 2 Diabetes

Insulin resistance precedes type 2 diabetes by several decades, and the complications of diabetes are often irreversible (Saltiel 2000). Accordingly, it seems prudent to attempt to improve insulin resistance before the onset of diabetes and its secondary complications.

Pharmacological Therapies for Type 2 Diabetes

In response to the rising prevalence of type 2 diabetes, the use of oral antidiabetic drugs is rapidly increasing (Wysowski, Armstrong, and Governale 2003). Currently, there are several treatment options for patients with type 2 diabetes, depending on the stage and progression of their symptoms (Inzucchi 2002; Wysowski, Armstrong, and Governale 2003). Sulfonylurea drugs (glyburide) and other secretagogues (repaglinide) improve glucose homeostasis by stimulating insulin secretion by the pancreatic beta cells. Other treatment modalities include biguanides (metformin), which decrease hepatic

Work from our laboratory discussed in this chapter on the effects of training and drug interventions on insulin sensitivity was supported by the Australian Research Council (DP0663863); Masterfoods Australia-New Zealand, a Mars, Incorporated company; and the RMIT VRII scheme.

glucose production, and alpha-glucosidase inhibitors (acarbose), which decrease gut carbohydrate absorption. A recent therapeutic option for people with insulin resistance or type 2 diabetes includes the thiazolidinediones (TZDs), which increase peripheral insulin sensitivity and glucose disposal (Malinowski and Bolesta 2000).

All of these therapies are effective at lowering blood glucose concentrations via distinct mechanisms of action and molecular targets, at least initially (Inzucchi 2002; Wysowski, Armstrong, and Governale 2003). However, their capacity to maintain whole-body glucose homeostasis as insulin resistance and diabetes progress is limited, prompting the search for more effective molecular targets to combat increasing blood glucose and its associated complications. Furthermore, none is without adverse side effects, and the consequences of their long-term use are largely unknown.

Exercise Training as a Therapy for Type 2 Diabetes

The limited effectiveness and adverse side effects associated with current oral antihyperglycemic agents has prompted a search for new drug targets for treating type 2 diabetes (Moller 2001). Since lifestyle factors leading to obesity appear to be major contributors to the development of type 2 diabetes, it follows that lifestyle modification may be prescribed as an effective therapy for this disease (Moller 2001).

Indeed, lifestyle modification through increased physical activity and dietary control has been successful in the prevention of type 2 diabetes in patients with impaired glucose tolerance (Eriksson et al. 1999; Molitch et al. 2003; Pan et al. 1997; Tuomilehto et al. 2001). In a randomized, placebo-controlled trial of more than 3,000 people with impaired glucose tolerance, lifestyle intervention (reduced caloric intake combined with 150 min of moderate physical exercise each week) was two-fold more effective than metformin treatment (850 mg twice daily) in reducing the incidence of type 2 diabetes over 2.8 y (Knowler et al. 2002).

Although these investigations (Eriksson et al. 1999; Knowler et al. 2002; Molitch et al. 2003; Pan et al. 1997; Tuomilehto et al. 2001) studied a combination of lifestyle adjustments that may reduce the incidence of type 2 diabetes, there is convincing evidence to support the prescription of exercise alone as a therapy for insulin resistance and type 2 diabetes (Booth et al. 2000, 2002; Hawley 2004; Pedersen and Saltin 2006). In this regard, Houmard and coworkers (2004) demonstrated a dose–response relationship between volume of exercise training (min/wk) and insulin sensitivity in individuals who were sedentary, obese, or overweight and were randomly assigned to exercise programs of varying volumes and intensities for 6 mo (Houmard et al. 2004). Houmard and colleagues concluded that an exercise prescription of 170 min of walking or jogging each week was more effective than 115 min of walking or jogging for improving whole-body insulin sensitivity, regardless of exercise intensity (although both volumes of exercise improved insulin sensitivity compared with sedentary controls).

Molecular Evidence for Prescribing Exercise Training

Since skeletal muscle is the major source for insulin-stimulated glucose uptake (DeFronzo et al. 1985), any treatment to improve glucose uptake in this tissue will improve whole-body glucose homeostasis. The observation that exercise training increases both insulin-dependent and -independent glucose transport in skeletal muscle is well established (see chapter 3; Garetto et al. 1984; Holloszy 2005; Holloszy and Narahara 1965; Ivy et al. 1983; Richter et al. 1982, 1984; Wallberg-Henriksson et al. 1988). Similar observations regarding exercise-induced improvements in skeletal muscle glucose transport have been made in insulin-resistant rats (Cortez et al. 1991), diabetic rats (Wallberg-Henriksson and Holloszy 1984), and humans with insulin resistance and type 2 diabetes (Christ-Roberts et al. 2003).

Increased glucose transport in response to a bout of exercise or muscle contraction may be mediated by a variety of intramyocellular signaling events, including AMPK activation, Akt phosphorylation, NO production, and calcium-mediated mechanisms involving CaMK and PKC (Hawley 2004; Jessen and Goodyear 2005; Richter et al. 2004; Sakamoto and Goodyear 2002). However, the insulin-sensitizing effects of an acute exercise bout are short-lived and may persist for only 48 h if another bout of exercise is not undertaken (Etgen Jr. et al. 1993; Ivy et al. 1983). In contrast, chronic exercise training may produce metabolic adaptations that result in sustained improvements in whole-body and muscle insulin sensitivity (Booth et al. 2000, 2002; Hawley 2004; Pedersen and Saltin 2006). The remainder of this review focuses on the chronic adaptations in skeletal muscle that may result in improved insulin sensitivity following chronic exercise training. A summary of these improvements is shown in figure 15.1.

Upregulation of Insulin Signaling Proteins

Exercise-induced increases in muscle insulin sensitivity may be attributed to increased expression or activity of signaling proteins involved in the regulation of skeletal muscle glucose uptake (Zierath 2002). Perhaps the most consistently observed effect of exercise in healthy and insulin-resistant skeletal muscle is increased expression of GLUT4 protein

(Henriksen 2002; Holloszy 2005; Zierath 2002). In addition, insulin sensitivity following 7 d of exercise training in previously sedentary men was associated with increased insulin-stimulated PI3K activity (Houmard et al. 1999). However, the mechanism for this increase is unknown; it was not accompanied by increased protein levels of PI3K or upstream components of the insulin-signaling cascade.

We have recently observed that 4 wk of exercise training normalizes impairments in the protein expression of Akt, GLUT4, and AS160 caused by high-fat feeding in rat skeletal muscle (Lessard et al. 2007). The exercise-induced changes in protein expression were associated with improved insulin action in the hind-limb muscles of exercise-trained rats. The collective results from several studies suggest that upregulation of the activity or content of insulin signaling proteins is one mechanism by which exercise training improves insulin action and glucose uptake in skeletal muscle.

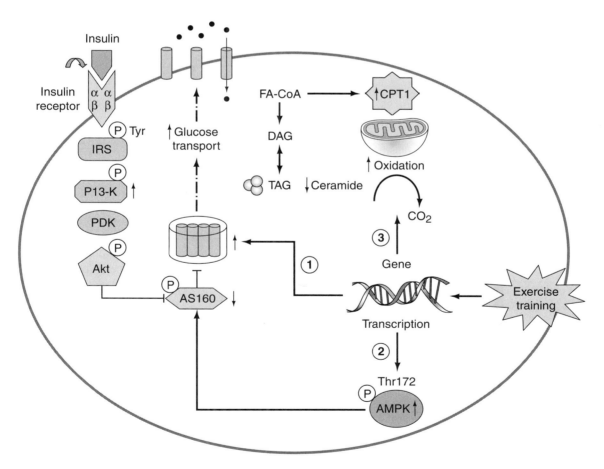

Figure 15.1 Exercise training results in the transcription of genes involved in (1) insulin signaling and GLUT4 translocation to the sarcolemma, (2) upregulation of AMPK, and (3) increased skeletal muscle oxidative capacity.

Chronic AMPK Activation

In addition to acute activation of AMPK due to muscle contraction (see chapters 11 and 13), chronic exercise training may result in upregulation of AMPK protein levels. In individuals who are healthy, 3 wk of endurance training increased the skeletal muscle protein content of the AMPK α1, β2, and γ1 subunits (Frosig et al. 2004). Similarly, 7 wk of exercise training (treadmill running) in obese Zucker rats resulted in a 1.5-fold increase in AMPKα1 protein expression and restored impaired AMPK activation to the level of lean controls (Sriwijitkamol et al. 2006). Pold and colleagues (2005) observed that 8 wk of treadmill running in ZDF rats produced improvements in insulin sensitivity similar to those seen with daily AICAR administration. However, unlike leptin-deficient (ob/ob mouse) and leptin-receptor-deficient (fa/fa Zucker rat) rodent models of diabetes (Yu et al. 2004), humans with type 2 diabetes do not exhibit decreased AMPK subunit expression or activation compared with healthy controls (Wojtaszewski et al. 2005). Even so, Wojtaszewski and coworkers (2005) observed a comparable increase in the expression of the alpha1, beta2, and gamma3 subunits of AMPK in response to 6 wk of resistance training in individuals with or without type 2 diabetes.

Exercise training induces a chronic upregulation of AMPK (Frosig et al. 2004; Sriwijitkamol et al. 2006; Wojtaszewski et al. 2005), and chronic activation of AMPK through AICAR results in improved skeletal muscle insulin sensitivity (see chapter 13). Thus, it is plausible that the insulin-sensitizing effects of exercise training are at least partially due to increased skeletal muscle AMPK activation. Chronic AMPK activation, as seen with exercise training, may improve insulin sensitivity by regulating the expression of specific genes involved in glucose and lipid homeostasis (Jorgensen, Richter, and Wojtaszewski 2006). AMPK activation through daily AICAR injections (1 mg/g) for 4 wk was associated with increased expression of GLUT4, HK2, and mitochondrial proteins (i.e., CS, cytochrome C) in rodent skeletal muscle (Winder et al. 2000; Zheng et al. 2001).

It is also possible that exercise-induced upregulation of AMPK mediates its effects through distal components of the insulin signaling cascade. Recently, AICAR was shown to induce AS160 phos-phorlyation in mouse skeletal muscle by a mechanism that was independent of insulin but at least partly attributable to the AMPKα2 isoform (Bruss et al. 2005; Kramer et al. 2006; Treebak et al. 2006). Interestingly, muscle contraction also stimulated AS160 phosphorylation by an unknown mechanism that did not involve AMPK or Akt activation (Bruss et al. 2005; Kramer et al. 2006). In this regard, we have recently demonstrated that improved skeletal muscle insulin sensitivity following 4 wk of treadmill running in high-fat-fed rats was associated with increased AMPKα1 activity and protein content as well as decreased AS160 protein content (Lessard et al. 2007).

In summary, exercise-induced increases in skeletal muscle AMPK activity may improve glucose homeostasis by several distinct mechanisms: stimulating insulin-independent glucose transport, enhancing insulin signal transduction at the level of AS160, and increasing the expression of proteins that facilitate glucose metabolism, such as HK2, GLUT4, and oxidative enzymes.

Increased Oxidative Capacity and Decreased Lipid Accumulation in Muscle

As described previously, the dysregulation of fatty acid uptake (chapter 3) and storage (chapter 4) in skeletal muscle is associated with impaired whole-body insulin sensitivity. Evidence linking the accumulation of lipid metabolites with reduced mitochondrial oxidative capacity and insulin resistance has also been presented (chapter 10). Thus, the regulation of lipid turnover and utilization is another potential mechanism by which exercise training may improve insulin sensitivity (Bruce and Hawley 2004; Hawley and Lessard 2007).

Exercise training increases the oxidative capacity of skeletal muscle by upregulating the expression of proteins involved in mitochondrial biogenesis, such as PGC-1, PPARα, and NRF1 (Gollnick and Saltin 1982; Irrcher et al. 2003). Oxidative enzyme capacity is lower in people with insulin resistance, which is thought to contribute to a metabolic inflexibility that does not allow for easy transition between fasting and postprandial states (Kelley 2002). In turn, this inflexibility is thought to contribute to the aberrant skeletal muscle glucose and lipid metabolism that is associated with insulin resistance and type 2 diabetes. Furthermore, the maximal activities of several skeletal muscle oxidative enzymes (i.e., CS) are good predictors of whole-body insulin sensitivity, suggesting that treatments that increase oxidative capacity may also improve insulin sensitivity (Bruce et al. 2003).

In support of this contention, Goodpaster, Katsiaras, and Kelley (2003) demonstrated that the strongest predictor of insulin sensitivity following endurance training in individuals who are obese was enhanced whole-body lipid oxidation. In addition, we found that 4 wk of exercise training in high-fat-fed rats increased PGC-1 protein expression and the rate of palmitate oxidation in perfused hind-limb muscles, and it was associated with improved insulin-sensitive glucose uptake (Lessard et al., forthcoming). Furthermore, increased oxidative capacity following exercise training was recently associated with increased CPT1 activity and decreased ceramide and DAG content in the muscle of obese subjects (Bruce et al. 2006). The findings by Bruce and coworkers (2006) and our lab (forthcoming) suggest that exercise training may improve muscle insulin sensitivity by increasing the proportion of lipids targeted for oxidation, thereby reducing the accumulation of lipid species that inhibit insulin signal transduction.

In summary, there are several means by which exercise training may improve skeletal muscle glucose uptake (see figure 15.2). These include upregulation of proteins that facilitate glucose

transport and metabolism in muscle, chronic activation of AMPK, and increased oxidative capacity, which may benefit lipid utilization and turnover. Furthermore, exercise training has the potential to ameliorate several other conditions associated with metabolic syndrome, including obesity, hypertension, and cardiovascular disease, and is not associated with adverse metabolic side effects (Pedersen and Saltin 2006; Roberts and Barnard 2005). Therefore, exercise training should be considered an effective therapy for the treatment of insulin resistance in skeletal muscle.

Exercise and Drug Combination Therapy

As described in preceding chapters, the pathophysiology of type 2 diabetes is complex, and its progression involves several organs and tissues, including the liver, adipose tissue, pancreas, and skeletal muscle. However, drugs for the treatment of type 2 diabetes are often specific to one target molecule or tissue. Thus, it is not surprising that the currently available antihyperglycemic agents often fail to help patients reach their targets for glycemic control (Lebovitz 2004; Ménard et al. 2005).

In order to attain better glycemic control in people with type 2 diabetes, combination therapies are becoming more common (Lebovitz 2004). Most

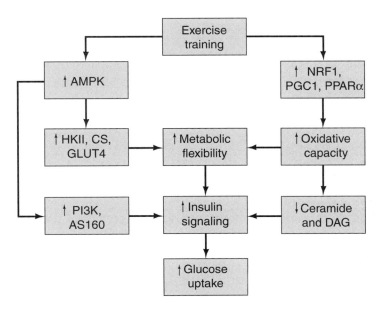

Figure 15.2   Chronic adaptations that may contribute to exercise-induced improvements in skeletal muscle insulin sensitivity.

new combination therapies involve the simultaneous administration of two or more oral antihyperglycemic agents with different molecular targets. For example, the single-tablet combination of a TZD (rosiglitazone) and a biguanide (metformin), which have different modes of action, has a greater glucose-lowering effect than either treatment alone has (Bailey and Day 2004). Alternatively, the prescription of an oral antihyperglycemic agent in conjunction with insulin can also be effective (Yki-Jarvinen 2002).

Given the potential for pharmacological therapies and exercise training to improve glucose homeostasis by different mechanisms, it is plausible that their combined treatment would produce additive beneficial effects. In support of this contention, Hevener, Reichart, and Olefsky (2000) observed that treating obese Zucker (fa/fa) rats with the TZD troglitazone in combination with exercise training (treadmill running) for 3 wk was more effective in improving insulin sensitivity than either treatment alone was. Troglitazone and exercise training produced an additive increase in skeletal muscle GLUT4 content, suggesting that the whole-body insulin sensitization was at least partially attributable to increased muscle glucose uptake (Hevener, Reichart, and Olefsky 2000).

In contrast, Lessard and coworkers (2007) found that exercise training (4 wk treadmill running), but not rosiglitazone treatment (2 mg · kg^{-1} · d^{-1}) improved insulin-stimulated glucose uptake in skeletal muscle during hind-limb perfusion in high-fat-fed rats. Rosiglitazone treatment further impaired lipid-induced insulin resistance in skeletal muscle in this model, which was associated with increased muscle lipid storage and transport. Importantly, exercise training prevented the rosiglitazone-induced impairments in muscle lipid storage and insulin sensitivity when both treatments were combined. It was concluded that the liver and adipose tissue, rather than the skeletal muscle, may be more important targets for the insulin-sensitizing actions of rosiglitazone (Lessard et al., 2007). However, the results from both studies (Hevener, Reichart, and Olefsky 2000; Lessard et al., 2007) provide evidence that the combined prescription of TZDs and exercise training is likely to be more effective than either treatment alone.

In the short term, moderate-intensity exercise may decrease plasma glucose and insulin concentrations (Hubinger, Franzen, and Gries 1987; Larsen et al. 1997). Therefore, exercise training may interact with sulfonylurea derivatives such as glibenclamide, which also decrease plasma glucose concentration by stimulating insulin secretion. Accordingly, Larsen and coworkers (1999) found that in postabsorptive individuals with type 2 diabetes, the hypoglycemic actions of glibenclamide and exercise are enhanced when the treatments are combined. Reduced plasma glucose levels observed with this combination treatment were attributed to reduced exercise-stimulated hepatic glucose production as a result of glibenclamide-induced increases in insulin secretion (Larsen et al. 1999).

There is also evidence that combining exercise with unconventional treatments for insulin resistance may produce additive benefits. The relationship between oxidative stress and insulin resistance, described by White and Marette in chapter 7, raises the possibility that antioxidants may play a role in the treatment of this disease. In obese Zucker rats, the combination of the antioxidant compound, α-lipoic acid, and exercise training increased glucose transport in skeletal muscle in an additive fashion (Saengsirisuwan et al. 2001; Saengsirisuwan et al. 2004). These improvements were associated with enhanced IRS1-dependent insulin signaling and may have resulted from a reduction in oxidative stress (Henriksen 2006; Henriksen and Saengsirisuwan 2003). Interestingly, an additional benefit of α-lipoic acid was not observed in exercise-trained lean Zucker rats that were insulin sensitive (Saengsirisuwan et al. 2002).

In people with metabolic syndrome, cardiovascular risk factors such as hypertension are strongly linked with insulin resistance. Thus, interventions that can improve both hypertension and insulin resistance, such as exercise training and angiotensin-converting enzyme (ACE) inhibitors, would be valuable. Indeed, the treatment of obese Zucker rats with both exercise training (6 wk treadmill running) and the ACE inhibitor trandolapril (1 mg · kg^{-1} · d^{-1}) resulted in greater increases in glucose tolerance and skeletal muscle insulin sensitivity than either treatment alone achieved (Steen et al. 1999). The enhancements in muscle insulin action following combination treatment observed by Steen and colleagues (1999) were associated with increased muscle GLUT4 and HK content.

The precise mechanisms by which both conventional (TZDs) and unconventional (antioxidants and antihypertensives) skeletal muscle insulin sensitizers are able to work synergistically with exercise to improve insulin sensitivity are unclear. Considering the potential for a combination of pharmacological therapy and exercise training to produce greater improvements to glucose homeostasis than monotherapy produces, further investigation is warranted.

Exercise-Like Effects of Current Antihyperglycemic Drugs

The argument that exercise training should be a prescribed therapy for type 2 diabetes is further strengthened by the fact that many effective antihyperglycemic agents have exercise-like effects. In addition, the search for new drug targets is aimed at molecules that have the ability to activate signaling systems that are activated by muscle contraction or exercise training (Moller 2001). For example, there is significant interest in the discovery of new drug targets that activate the AMPK signaling pathway because of the important role this energy-sensing molecule plays in whole-body glucose and lipid metabolism (Musi and Goodyear 2002; Winder 2000). As described by Winder in chapter 13, several currently available antihyperglycemic agents such as TZDs and metformin are thought to work at least in part by activation of AMPK in skeletal muscle and liver. AICAR is also effective in improving glucose homeostasis in animal models; however, the doses required are likely too high to be practical and safe for use in humans, prompting the search for more potent AMPK activators. As outlined earlier, exercise training is an effective option for the chronic activation of AMPK and improvement of glucose homeostasis and generally has no dangerous side effects.

A typical response to exercise and AMPK activation in skeletal muscle is an increase in the expression of mitochondrial UCP3 (Zhou et al. 2000). In this regard, Brunmair and coworkers (2004) reported that TZDs (rosiglitazone, pioglitazone) acutely increase UCP3 mRNA levels in skeletal muscle in healthy rats within 6 h of exposure. These authors (Brunmair et al. 2004) described the ability of TZDs to activate both AMPK and UCP3 as an exercise-like effect. We have also demonstrated that chronic rosiglitazone treatment in insulin-resistant obese Zucker rats results in increased skeletal muscle AMPKα2 and UCP3 protein expression (Lessard et al. 2006).

Molecular adaptations to exercise training include increased mitochondrial oxidative capacity and a transition to a more oxidative (Type I) muscle fiber profile (Wang et al. 2004). Such adaptations are achieved through the exercise-induced upregulation of transcription factors such as PPARα and PPARδ, which promote the transcription of genes involved in mitochondrial oxidative and Type I fiber gene expression, respectively. Given the positive association between increased oxidative capacity and insulin sensitivity in muscle (Luquet et al. 2003; Ryder et al. 2003), it is not surprising that agonists of PPARα and PPARδ are now pharmacological targets for the treatment of insulin resistance (Buse et al. 2005; Fagerberg et al. 2005; Takahashi et al. 2006). The experimental PPARδ agonist, GW501516, can prevent the development of obesity in db/db mice, promote fatty acid oxidation in muscle, and improve lipid-induced insulin resistance in animal models (Takahashi et al. 2006). Pharmacological PPARα and PPARγ agonists such as tesaglitazar and muraglitazar can also improve glucose and lipid metabolism as well as insulin sensitivity in obese Zucker rats (Oakes et al. 2005) and humans who are insulin resistant (Buse et al. 2005; Fagerberg et al. 2005).

In summary, it appears that several existing and experimental oral antihyperglycemic agents function through pathways that are typically activated via exercise training. It is not surprising that these molecular targets are being chosen, given the efficacy of exercise training in improving insulin action and glucose homeostasis. However, it is ironic that significant effort is being put into the search for drugs that mimic exercise training when exercise training itself is an available, practical, and economical therapeutic option with several beneficial effects and few, if any, adverse effects.

Prescribing Exercise Training: Practical Considerations

Significant molecular and epidemiological evidence supports exercise training as a treatment for insulin resistance and type 2 diabetes. Less clear, however, are the dose and mode of exercise required to elicit optimal therapeutic effects. Findings by Houmard and colleagues (2004) indicate that an exercise prescription of 170 min/wk improves insulin sensitivity to a greater degree than a prescription of 115 min/wk. Although both high-intensity (running at 65%-85% $\dot{V}O_2$max) and moderate-intensity (walking at 40%-55% $\dot{V}O_2$max) exercise were examined by Houmard and colleagues (2004), the insulin-sensitizing benefits at a given training volume occurred independently of exercise intensity. In contrast, a meta-analysis of 14 controlled clinical trials published by Boulé and coworkers (2001) examining the effects of at least 8 wk of exercise training (endurance or resistance) on glycemic control concluded that exercise intensity, but not exercise volume, correlated significantly with improvements in the clinical marker of glycemic control, Hb_{A1c}. Thus, it remains unclear whether the volume or intensity of exercise training is more important when prescribing exercise training for insulin sensitization (Pedersen and Saltin 2006).

Another important consideration when formulating an exercise prescription is the type of exercise (i.e., resistance or endurance training) to be undertaken. Although there is more information regarding the effect of endurance training on insulin sensitization, there is accumulating evidence that resistance training (see chapter 12) is also an effective treatment for insulin resistance and type 2 diabetes (Pedersen and Saltin 2006). Cauza and coworkers (2005) concluded that resistance training (6 sets per muscle group each week) is more effective than endurance training (90 min/wk) in improving glycemic control in people with type 2 diabetes. Furthermore, Holten and colleagues (2004) demonstrated that improved insulin sensitivity following resistance training (30 min, 3 times/wk) was associated with increased skeletal muscle GLUT4 content and enhanced insulin signaling in people with or without type 2 diabetes. It has also been demonstrated that supervised circuit training combining both resistance and aerobic intervals can significantly improve glycemic control (Maiorana et al. 2002). Thus, it appears that endurance training, resistance training, or a combination of both may be effective treatment for type 2 diabetes.

Current guidelines from the American Diabetes Association recommend that people with impaired glucose tolerance perform at least 150 min/wk of moderate to vigorous physical activity (Sigal et al. 2006). It is also recommended that this weekly exercise prescription include 3 sessions of resistance training. These guidelines are in line with those recommended by Diabetes Australia and the Danish Diabetes Association (Pedersen and Saltin 2006).

However, in order for exercise guidelines to be effective they must first be recommended to patients by health care professionals. Results from the 2002 Medical Expenditure Survey, which included >25,000 U.S. adults, indicated that 73% of adults with diabetes were advised by a health care professional to exercise more (Morrato et al. 2006). Despite this, the proportion of adults with diabetes who perform regular physical activity is believed to be low (Morrato et al. 2006; Sullivan et al. 2005). Perhaps compliance with exercise recommendations would improve if patients were provided with more specific guidelines, resources, and support to achieve glycemic control with exercise training, rather than simply being told to exercise more. As with any prescription, specific guidelines for use and patient compliance are necessary for therapeutic success.

Concluding Remarks

Although the primary defects leading to the development of insulin resistance and type 2 diabetes remain unclear, significant progress has been made during the past decade in furthering our understanding of the molecular basis underlying the beneficial effects of exercise training in stimulating the entry of glucose into insulin-sensitive tissues. Exercise-induced increases in muscle insulin sensitivity can be attributed to several factors, including (a) increased expression or activity of signaling proteins involved in the regulation of glucose uptake (i.e., GLUT4, PI3K), (b) chronic activation of AMPK, (c) increased oxidative capacity of skeletal muscle via upregulation of the expression of proteins involved in mitochondrial biogenesis (i.e., PGC-1), and (d) increases in muscle and whole-body lipid utilization and turnover.

Even though the case for a causal link between the rise in physical inactivity and the increase in insulin resistance is compelling, the use of antidiabetic drugs continues to increase. These drugs are effective at lowering blood glucose concentration, at least in the short to medium term; however, they function through molecular and biochemical pathways that are typically activated as a result of exercise. Furthermore, none of these agents is without adverse side effects and the consequences of their long-term use are largely unknown. Hence, it seems ironic that significant effort is expended in the search for drugs that mimic exercise training when exercise itself is a readily available, practical, and economical therapeutic option with many beneficial effects, and few, if any, adverse side effects.

Given the strong evidence that physical activity helps prevent insulin resistance, and the fact that exercise training increases mitochondrial biogenesis and improves glucose tolerance and insulin action in people with insulin resistance and type 2 diabetes, the question of why such a potent modulator of these conditions is not more commonly prescribed is perplexing and should be of utmost concern to health care professionals worldwide. Investing almost exclusively in strategies that target secondary and tertiary treatment of chronic disease states (i.e., pharmaceutical interventions) is extremely shortsighted; primary defense mechanisms (i.e., exercise, diet, and lifestyle interventions) will decrease disease prevalence by preventing these conditions in the first place. Determination of the underlying biological mechanisms that result from exercise training is essential in order to define the precise variations in physical activity that lead to the most desired effects on targeted risk factors, as well as to aid in the development of such interventions.

References

Bailey, C.J., and C. Day. 2004. Avandamet: Combined metformin-rosiglitazone treatment for insulin resistance in type 2 diabetes. *Int J Clin Pract* 58:867-76.

Booth, F.W., M.V. Chakravarthy, S.E. Gordon, and E.E. Spangenburg. 2002. Waging war on physical inactivity: Using modern molecular ammunition against an ancient enemy. *J Appl Physiol* 93:3-30.

Booth, F.W., S.E. Gordon, C.J. Carlson, and M.T. Hamilton. 2000. Waging war on modern chronic diseases: Primary prevention through exercise biology. *J Appl Physiol* 88:774-87.

Boulé, N.G., E. Haddad, G.P. Kenny, G.A. Wells, and R.J. Sigal. 2001. Effects of exercise on glycemic control and body mass in type 2 diabetes mellitus: a meta-analysis of controlled clinical trials. *JAMA* 286: 1218-27.

Bruce, C.R., M.J. Anderson, A.L. Carey, D.G. Newman, A. Bonen, A.D. Kriketos, G.J. Cooney, and J.A. Hawley. 2003. Muscle oxidative capacity is a better predictor of insulin sensitivity than lipid status. *J Clin Endocrinol Metab* 88:5444-51.

Bruce, C.R., and J.A. Hawley. 2004. Improvements in insulin resistance with aerobic exercise training: A lipocentric approach. *Med Sci Sports Exerc* 36:1196-201.

Bruce, C.R., A.B. Thrush, V.A. Mertz, V. Bezaire, A. Chabowski, G.J. Heigenhauser, and D.J. Dyck. 2006. Endurance training in obese humans improves glucose tolerance and mitochondrial fatty acid oxidation and alters muscle lipid content. *Am J Physiol Endocrinol Metab* 291:E99-107.

Brunmair, B., F. Gras, L. Wagner, M. Artwohl, B. Zierhut, W. Waldhausl, and C. Furnsinn. 2004. Expression of uncoupling protein-3 mRNA in rat skeletal muscle is acutely stimulated by thiazolidinediones: An exercise-like effect? *Diabetologia* 47:1611-4.

Bruss, M.D., E.B. Arias, G.E. Lienhard, and G.D. Cartee. 2005. Increased phosphorylation of Akt substrate of 160 kDa (AS160) in rat skeletal muscle in response to insulin or contractile activity. *Diabetes* 54:41-50.

Buse, J.B., C.J. Rubin, R. Frederich, K. Viraswami-Appanna, K.C. Lin, R. Montoro, G. Shockey, and J.A. Davidson. 2005. Muraglitazar, a dual (alpha/gamma) PPAR activator: A randomized, double-blind, placebo-controlled, 24-week monotherapy trial in adult patients with type 2 diabetes. *Clin Ther* 27:1181-95.

Cauza, E., U. Hanusch-Enserer, B. Strasser, K. Kostner, A. Dunky, and P. Haber. 2005. Strength and endurance training lead to different post exercise glucose profiles in diabetic participants using a continuous subcutaneous glucose monitoring system. *Eur J Clin Invest* 35:745-51.

Christ-Roberts, C.Y., T. Pratipanawatr, W. Pratipanawatr, R. Berria, R. Belfort, and L.J. Mandarino. 2003. Increased insulin receptor signaling and glycogen synthase activity contribute to the synergistic effect of exercise on insulin action. *J Appl Physiol* 95:2519-29.

Cortez, M.Y., C.E. Torgan, J.T. Brozinick Jr., and J.L. Ivy. 1991. Insulin resistance of obese Zucker rats exercise trained at two different intensities. *Am J Physiol* 261:E613-9.

DeFronzo, R.A., R. Gunnarsson, O. Bjorkman, M. Olsson, and J. Wahren. 1985. Effects of insulin on peripheral and splanchnic glucose metabolism in non-insulin-dependent (type II) diabetes mellitus. *J Clin Invest* 76:149-55.

Eriksson, J., J. Lindstrom, T. Valle, S. Aunola, H. Hamalainen, P. Ilanne-Parikka, S. Keinanen-Kiukaanniemi, M. Laakso, M. Lauhkonen, P. Lehto, A. Lehtonen, A. Louheranta, M. Mannelin, V. Martikkala, M. Rastas, J. Sundvall, A. Turpeinen, T. Viljanen, M. Uusitupa, and J. Tuomilehto. 1999. Prevention of type II diabetes in subjects with impaired glucose tolerance: The Diabetes Prevention Study (DPS) in Finland. Study design and 1-year interim report on the feasibility of the lifestyle intervention programme. *Diabetologia* 42:793-801.

Etgen Jr., G.J., J.T. Brozinick Jr., H.Y. Kang, and J.L. Ivy. 1993. Effects of exercise training on skeletal muscle glucose uptake and transport. *Am J Physiol* 264:C727-33.

Fagerberg, B., S. Edwards, T. Halmos, J. Lopatynski, H. Schuster, S. Stender, G. Stoa-Birketvedt, S. Tonstad, S. Halldorsdottir, and I. Gause-Nilsson. 2005. Tesaglitazar, a novel dual peroxisome proliferator-activated receptor alpha/gamma agonist, dose-dependently improves the metabolic abnormalities associated with insulin resistance in a non-diabetic population. *Diabetologia* 48:1716-25.

Frosig, C., S.B. Jorgensen, D.G. Hardie, E.A. Richter, and J.F. Wojtaszewski. 2004. 5'-AMP-activated protein kinase activity and protein expression are regulated by endurance training in human skeletal muscle. *Am J Physiol Endocrinol Metab* 286:E411-7.

Garetto, L.P., E.A. Richter, M.N. Goodman, and N.B. Ruderman. 1984. Enhanced muscle glucose metabolism after exercise in the rat: The two phases. *Am J Physiol* 246: E471-5.

Gollnick, P.D., and B. Saltin. 1982. Significance of skeletal muscle oxidative enzyme enhancement with endurance training. *Clin Physiol* 2:1-12.

Goodpaster, B.H., A. Katsiaras, and D.E. Kelley. 2003. Enhanced fat oxidation through physical activity is associated with improvements in insulin sensitivity in obesity. *Diabetes* 52:2191-7.

Hawley, J.A. 2004. Exercise as a therapeutic intervention for the prevention and treatment of insulin resistance. *Diabetes Metab Res Rev* 20:383-93.

Hawley, J.A., and S.J. Lessard. 2007. Mitochondrial function: Use it or lose it. *Diabetologia* 50:699-702.

Henriksen, E.J. 2002. Invited review: Effects of acute exercise and exercise training on insulin resistance. *J Appl Physiol* 93:788-96.

Henriksen, E.J. 2006. Exercise training and the antioxidant alpha-lipoic acid in the treatment of insulin resistance and type 2 diabetes. *Free Radic Biol Med* 40:3-12.

Henriksen, E.J., and V. Saengsirisuwan. 2003. Exercise training and antioxidants: Relief from oxidative stress and insulin resistance. *Exerc Sport Sci Rev* 31:79-84.

Hevener, A.L., D. Reichart, and J. Olefsky. 2000. Exercise and thiazolidinedione therapy normalize insulin action in the obese Zucker fatty rat. *Diabetes* 49:2154-9.

Holloszy, J.O. 2005. Exercise-induced increase in muscle insulin sensitivity. *J Appl Physiol* 99:338-43.

Holloszy, J.O., and H.T. Narahara. 1965. Studies of tissue permeability. X. Changes in permeability to 3-methylglucose associated with contraction of isolated frog muscle. *J Biol Chem* 240:3493-500.

Holten, M.K., M. Zacho, M. Gaster, C. Juel, J.F. Wojtaszewski, and F. Dela. 2004. Strength training increases insulin-mediated glucose uptake, GLUT4 content, and insulin signaling in skeletal muscle in patients with type 2 diabetes. *Diabetes* 53:294-305.

Houmard, J. A., C.D. Shaw, M.S. Hickey, and C.J. Tanner. 1999. Effect of short-term exercise training on insulin-stimulated PI 3-kinase activity in human skeletal muscle. *Am J Physiol* 277:E1055-60.

Houmard, J.A., C.J. Tanner, C.A. Slentz, B.D. Duscha, J.S. McCartney, and W.E. Kraus. 2004. Effect of the volume and intensity of exercise training on insulin sensitivity. *J Appl Physiol* 96:101-6.

Hubinger, A., A. Franzen, and F.A. Gries. 1987. Hormonal and metabolic response to physical exercise in hyperinsulinemic and non-hyperinsulinemic type 2 diabetics. *Diabetes Res* 4:57-61.

Inzucchi, S.E. 2002. Oral antihyperglycemic therapy for type 2 diabetes: Scientific review. *JAMA* 287:360-72.

Irrcher, I., P.J. Adhihetty, A.M. Joseph, V. Ljubicic, and D.A. Hood. 2003. Regulation of mitochondrial biogenesis in muscle by endurance exercise. *Sports Med* 33:783-93.

Ivy, J.L., J. C. Young, J.A. McLane, R.D. Fell, and J.O. Holloszy. 1983. Exercise training and glucose uptake by skeletal muscle in rats. *J Appl Physiol* 55:1393-6.

Jessen, N., and L.J. Goodyear. 2005. Contraction signaling to glucose transport in skeletal muscle. *J Appl Physiol* 99:330-7.

Jorgensen, S.B., E.A. Richter, and J.F. Wojtaszewski. 2006. Role of AMPK in skeletal muscle metabolic regulation and adaptation in relation to exercise. *J Physiol* 574:17-31.

Kelley, D.E. 2002. Skeletal muscle triglycerides: An aspect of regional adiposity and insulin resistance. *Ann NY Acad Sci* 967:135-45.

Knowler, W.C., E. Barrett-Connor, S.E. Fowler, R.F. Hamman, J.M. Lachin, E.A. Walker, and D.M. Nathan. 2002. Reduction in the incidence of type 2 diabetes with lifestyle intervention or metformin. *New Engl J Med* 346:393-403.

Kramer, H.F., C.A. Witczak, N. Fujii, N. Jessen, E.B. Taylor, D.E. Arnolds, K. Sakamoto, M.F. Hirshman, and L.J. Goodyear. 2006. Distinct signals regulate AS160 phosphorylation in response to insulin, AICAR, and contraction in mouse skeletal muscle. *Diabetes* 55:2067-76.

Larsen, J.J., F. Dela, M. Kjaer, and H. Galbo. 1997. The effect of moderate exercise on postprandial glucose homeostasis in NIDDM patients. *Diabetologia* 40:447-53.

Larsen, J.J., F. Dela, S. Madsbad, J. Vibe-Petersen, and H. Galbo. 1999. Interaction of sulfonylureas and exercise on glucose homeostasis in type 2 diabetic patients. *Diabetes Care* 22:1647-54.

Lebovitz, H.E. 2004. Oral antidiabetic agents: 2004. *Med Clin North Am* 88:847-63, ix-x.

Lessard, S.J., D.A. Rivas, Z.P. Chen, A. Bonen, M.A. Febbraio, D.W. Reeder, B.E. Kemp, B.B. Yaspelkis III, and J. A. Hawley. 2007. Tissue-specific effects of Rosiglitazone and exercise in the treatment of lipid-induced insulin resistance. *Diabetes* 56:1856-64.

Lessard, S.J., Z.P. Chen, M.J. Watt, M. Hashem, J.J. Reid, M.A. Febbraio, B.E. Kemp, and J.A. Hawley. 2006. Chronic rosiglitazone treatment restores AMPKalpha2 activity in insulin-resistant rat skeletal muscle. *Am J Physiol Endocrinol Metab* 290:E251-7.

Luquet, S., J. Lopez-Soriano, D. Holst, A. Fredenrich, J. Melki, M. Rassoulzadegan, and P.A. Grimaldi. 2003. Peroxisome proliferator-activated receptor delta controls muscle development and oxidative capability. *FASEB J* 17:2299-301.

Maiorana, A., G. O'Driscoll, C. Goodman, R. Taylor, and D. Green. 2002. Combined aerobic and resistance exercise improves glycemic control and fitness in type 2 diabetes. *Diabetes Res Clin Pract* 56:115-23.

Malinowski, J.M., and S. Bolesta. 2000. Rosiglitazone in the treatment of type 2 diabetes mellitus: A critical review. *Clin Ther* 22:1149-68.

Ménard, J., H. Payette, J.P. Baillargeon, P. Maheux, S. Lepage, D. Tessier, and J.L. Ardilouze. 2005. Efficacy of intensive multitherapy for patients with type 2 diabetes mellitus: a randomized controlled trial. *CMAJ* 173: 1457-66.

Molitch, M.E., W. Fujimoto, R.F. Hamman, and W.C. Knowler. 2003. The diabetes prevention program and its global implications. *J Am Soc Nephrol* 14:S103-7.

Moller, D.E. 2001. New drug targets for type 2 diabetes and the metabolic syndrome. *Nature* 414:821-7.

Morrato, E.H., J.O. Hill, H.R. Wyatt, V. Ghushchyan, and P.W. Sullivan. 2006. Are health care professionals advising patients with diabetes or at risk for developing diabetes to exercise more? *Diabetes Care* 29:543-8.

Musi, N., and L.J. Goodyear. 2002. Targeting the AMP-activated protein kinase for the treatment of type 2 diabetes. *Curr Drug Targets Immune Endocr Metabol Disord* 2:119-27.

Oakes, N.D., P. Thalen, T. Hultstrand, S. Jacinto, G. Camejo, B. Wallin, and B.L. Jung. 2005. Tesaglitazar, a dual PPARα/γ agonist, ameliorates glucose and lipid intolerance in obese

Zucker rats. *Am J Physiol Regul Integr Comp Physiol* 289: R938-46.

Pan, X.R., G.W. Li, Y.H. Hu, J.X. Wang, W.Y. Yang, Z.X. An, Z.X. Hu, J. Lin, J.Z. Xiao, H.B. Cao, P.A. Liu, X.G. Jiang, Y.Y. Jiang, J.P. Wang, H. Zheng, H. Zhang, P.H. Bennett, and B.V. Howard. 1997. Effects of diet and exercise in preventing NIDDM in people with impaired glucose tolerance. The Da Qing IGT and Diabetes Study. *Diabetes Care* 20:537-44.

Pedersen, B.K., and B. Saltin. 2006. Evidence for prescribing exercise as therapy in chronic disease. *Scand J Med Sci Sports* 16(Suppl. no. 1): 3-63.

Pold, R., L.S. Jensen, N. Jessen, E.S. Buhl, O. Schmitz, A. Flyvbjerg, N. Fujii, L.J. Goodyear, C.F. Gotfredsen, C.L. Brand, and S. Lund. 2005. Long-term AICAR administration and exercise prevents diabetes in ZDF rats. *Diabetes* 54:928-34.

Richter, E.A., L.P. Garetto, M.N. Goodman, and N.B. Ruderman. 1982. Muscle glucose metabolism following exercise in the rat: Increased sensitivity to insulin. *J Clin Invest* 69:785-93.

Richter, E.A., L.P. Garetto, M.N. Goodman, and N.B. Ruderman. 1984. Enhanced muscle glucose metabolism after exercise: Modulation by local factors. *Am J Physiol* 246: E476-82.

Richter, E.A., J.N. Nielsen, S.B. Jorgensen, C. Frosig, J.B. Birk, and J.F. Wojtaszewski. 2004. Exercise signaling to glucose transport in skeletal muscle. *Proc Nutr Soc* 63:211-6.

Roberts, C.K., and R.J. Barnard. 2005. Effects of exercise and diet on chronic disease. *J Appl Physiol* 98:3-30.

Ryder, J.W., R. Bassel-Duby, E.N. Olson, and J.R. Zierath. 2003. Skeletal muscle reprogramming by activation of calcineurin improves insulin action on metabolic pathways. *J Biol Chem* 278:44298-304.

Saengsirisuwan, V., T.R. Kinnick, M.B. Schmit, and E.J. Henriksen. 2001. Interactions of exercise training and lipoic acid on skeletal muscle glucose transport in obese Zucker rats. *J Appl Physiol* 91:145-53.

Saengsirisuwan, V., F.R. Perez, T.R. Kinnick, and E.J. Henriksen. 2002. Effects of exercise training and antioxidant R-ALA on glucose transport in insulin-sensitive rat skeletal muscle. *J Appl Physiol* 92:50-8.

Saengsirisuwan, V., F.R. Perez, J.A. Sloniger, T. Maier, and E.J. Henriksen. 2004. Interactions of exercise training and alpha-lipoic acid on insulin signaling in skeletal muscle of obese Zucker rats. *Am J Physiol Endocrinol Metab* 287:E529-36.

Sakamoto, K., and L.J. Goodyear. 2002. Invited review: Intracellular signaling in contracting skeletal muscle. *J Appl Physiol* 93:369-83.

Saltiel, A.R. 2000. Series introduction: The molecular and physiological basis of insulin resistance: Emerging implications for metabolic and cardiovascular diseases. *J Clin Invest* 106:163-4.

Sigal, R.J., G.P. Kenny, D.H. Wasserman, C. Castaneda-Sceppa, and R.D. White. 2006. Physical activity/exercise and type 2 diabetes: A consensus statement from the American Diabetes Association. *Diabetes Care* 29:1433-8.

Sriwijitkamol, A., J.L. Ivy, C. Christ-Roberts, R.A. DeFronzo, L.J. Mandarino, and N. Musi. 2006. LKB1-AMPK signaling in muscle from obese insulin-resistant Zucker rats and effects of training. *Am J Physiol Endocrinol Metab* 290:E925-32.

Steen, M.S., K.R. Foianini, E.B. Youngblood, T.R. Kinnick, S. Jacob, and E.J. Henriksen. 1999. Interactions of exercise training and ACE inhibition on insulin action in obese Zucker rats. *J Appl Physiol* 86:2044-51.

Sullivan, P.W., E.H. Morrato, V. Ghushchyan, H.R. Wyatt, and J.O. Hill. 2005. Obesity, inactivity, and the prevalence of diabetes and diabetes-related cardiovascular comorbidities in the U.S., 2000-2002. *Diabetes Care* 28:1599-603.

Takahashi, S., T. Tanaka, T. Kodama, and J. Sakai. 2006. Peroxisome proliferator-activated receptor delta (PPARdelta), a novel target site for drug discovery in metabolic syndrome. *Pharmacol Res* 53:501-7.

Treebak, J.T., S. Glund, A. Deshmukh, D.K. Klein, Y.C. Long, T.E. Jensen, S.B. Jorgensen, B. Viollet, L. Andersson, D. Neumann, T. Wallimann, E.A. Richter, A.V. Chibalin, J.R. Zierath, and J.F. Wojtaszewski. 2006. AMPK-mediated AS160 phosphorylation in skeletal muscle is dependent on AMPK catalytic and regulatory subunits. *Diabetes* 55:2051-8.

Tuomilehto, J., J. Lindstrom, J.G. Eriksson, T.T. Valle, H. Hamalainen, P. Ilanne-Parikka, S. Keinanen-Kiukaanniemi, M. Laakso, A. Louheranta, M. Rastas, V. Salminen, and M. Uusitupa. 2001. Prevention of type 2 diabetes mellitus by changes in lifestyle among subjects with impaired glucose tolerance. *New Engl J Med* 344:1343-50.

Wallberg-Henriksson, H., S.H. Constable, D.A. Young, and J.O. Holloszy. 1988. Glucose transport into rat skeletal muscle: Interaction between exercise and insulin. *J Appl Physiol* 65:909-13.

Wallberg-Henriksson, H., and J.O. Holloszy. 1984. Contractile activity increases glucose uptake by muscle in severely diabetic rats. *J Appl Physiol* 57:1045-9.

Wang, Y.X., C.L. Zhang, R.T. Yu, H.K. Cho, M.C. Nelson, C.R. Bayuga-Ocampo, J. Ham, H. Kang, and R.M. Evans. 2004. Regulation of muscle fiber type and running endurance by PPARdelta. *PLoS Biol* 2:e294.

Winder, W.W. 2000. AMP-activated protein kinase: Possible target for treatment of type 2 diabetes. *Diabetes Technol Ther* 2:441-8.

Winder, W.W., B.F. Holmes, D.S. Rubink, E.B. Jensen, M. Chen, and J.O. Holloszy. 2000. Activation of AMP-activated protein kinase increases mitochondrial enzymes in skeletal muscle. *J Appl Physiol* 88:2219-26.

Wojtaszewski, J.F., J.B. Birk, C. Frosig, M. Holten, H. Pilegaard, and F. Dela. 2005. 5'AMP activated protein kinase expression in human skeletal muscle: Effects of strength training and type 2 diabetes. *J Physiol* 564:563-73.

Wysowski, D.K., G. Armstrong, and L. Governale. 2003. Rapid increase in the use of oral antidiabetic drugs in the United States, 1990-2001. *Diabetes Care* 26:1852-5.

Yki-Jarvinen, H. 2002. Combination therapy with insulin and oral agents: Optimizing glycemic control in patients with type 2 diabetes mellitus. *Diabetes Metab Res Rev* 18(Suppl. no. 3): S77-81.

Yu, X., S. McCorkle, M. Wang, Y. Lee, J. Li, A.K. Saha, R.H. Unger, and N.B. Ruderman. 2004. Leptinomimetic effects of the AMP kinase activator AICAR in leptin-resistant rats: Prevention of diabetes and ectopic lipid deposition. *Diabetologia* 47:2012-21.

Zheng, D., P.S. MacLean, S.C. Pohnert, J.B. Knight, A.L. Olson, W.W. Winder, and G.L. Dohm. 2001. Regulation of muscle GLUT-4 transcription by AMP-activated protein kinase. *J Appl Physiol* 91:1073-83.

Zhou, M., B.Z. Lin, S. Coughlin, G. Vallega, and P.F. Pilch. 2000. UCP-3 expression in skeletal muscle: Effects of exercise, hypoxia, and AMP-activated protein kinase. *Am J Physiol Endocrinol Metab* 279:E622-9.

Zierath, J.R. 2002. Invited review: Exercise training-induced changes in insulin signaling in skeletal muscle. *J Appl Physiol* 93:773-81.

Index

About the Editors

John A. Hawley, PhD, is professor and head of the Exercise Metabolism and Diabetes Research Group in the School of Medical Sciences at the Royal Melbourne Institute of Technology in Melbourne, Australia, where he has a postgraduate research program comprising eight postdoctoral and doctoral students. His areas of research include the regulation of fat and carbohydrate metabolism, with a particular emphasis on insulin resistance and type 2 diabetes, and the role of exercise training in alleviating the metabolic syndrome.

A fellow of the American College of Sports Medicine and a member of the American Physiological Society, Hawley serves as an editorial board member for the *American Journal of Physiology: Endocrinology and Metabolism, Sports Medicine,* the *International Journal of Sport Nutrition and Exercise Metabolism,* the *International Journal of Sports Physiology and Performance,* and the *Malaysian Journal of Sport Science and Recreation.* Hawley is also a regular reviewer for many international journals.

In 1990, Hawley received the Medical Research Council (MRC) Scholarship for Outstanding Foreign Researcher from the South Africa MRC (1990-1992), which is awarded to assist doctoral studies in medical physiology. Hawley completed his PhD in physiology in 1993 while studying at the University of Cape Town Medical School, South Africa.

Hawley has published more than 150 papers in medical, biochemical, and sport science journals, three books, and 15 book chapters and has served as a visiting lecturer for the University of Otago, New Zealand; the African International Olympic Committee Sports Medicine Program; and the International Olympic Committee Sports Medicine Program. As an invited speaker at conferences and symposiums throughout Europe, the United States, Australia, New Zealand, and Malaysia, Hawley speaks on a range of subjects, including exercise as a therapy for the prevention of metabolic syndrome, mechanisms for improvements in insulin resistance after physical activity, the relationship of exercise to insulin resistance and diabetes, and nutritional strategies and exercise performance.

Juleen R. Zierath, PhD, is professor of physiology and head of the section of integrative physiology in the department of surgical science, Karolinska Institutet, Stockholm, Sweden, and an adjunct professor of biochemistry at Boston University School of Medicine.

Zierath leads an active research group consisting of members representing 10 countries. Through clinical and experimental research approaches, her group has unraveled the signaling mechanisms that mediate hormone action to promote glucose and lipid metabolism. In collaboration with a leading pharmaceutical company, she has contributed to the discovery of a nonprotein insulin receptor agonist that may offer a new type of oral treatment for people with diabetes. Her group collaborates with leading research groups from Scandinavia, Europe, Asia, and North America and is primarily funded by the Swedish Research Council, the Swedish Strategic Research Foundation, and the European Union.

She has published more than 150 peer-reviewed scientific papers, including 35 review articles in journals focused on endocrinology, metabolism, diabetes mellitus, and exercise physiology. She has also coauthored a textbook with Harriet Wallberg-Henriksson on the subject of skeletal muscle metabolism.

Zierath is the recipient of numerous awards, including the Minkowski Award from the European Association for the Study of Diabetes, the Fernström Award from Karolinska Institutet, and a Future Research Leader Award from the Foundation for Strategic Research, Sweden.